Titanium Alloys: Microstructure, Properties and Applications

Titanium Alloys: Microstructure, Properties and Applications

Editor: Keith Liverman

New York

Published by NY Research Press
118-35 Queens Blvd., Suite 400,
Forest Hills, NY 11375, USA
www.nyresearchpress.com

Titanium Alloys: Microstructure, Properties and Applications
Edited by Keith Liverman

International Standard Book Number: 978-1-64725-435-3 (Hardback)

Cataloging-in-Publication Data

Titanium alloys : microstructure, properties and applications / edited by Keith Liverman.
p. cm.
Includes bibliographical references and index.
ISBN 978-1-64725-435-3
1. Titanium alloys. 2. Alloys. 3. Titanium. I. Liverman, Keith.
TN693.T5 T58 2023
620.189 322--dc23

Contents

Preface

It is often said that books are a boon to mankind. They document every progress and pass on the knowledge from one generation to the other. They play a crucial role in our lives. Thus I was both excited and nervous while editing this book. I was pleased by the thought of being able to make a mark but I was also nervous to do it right because the future of students depends upon it. Hence, I took a few months to research further into the discipline, revise my knowledge and also explore some more aspects. Post this process, I begun with the editing of this book.

Titanium alloys are typically stronger than aluminum alloys and are lighter than steel. They have exceptional corrosion resistance and high tensile strength at extreme temperatures. They are used in numerous industries such as transportation industry, chemical industry, aerospace industry, and power generation industry. They can be easily welded and have exceptional formability. Titanium alloys have long been utilized for treating traumatic bone fractures. They are also widely utilized in the production of bone plates surgeries and metal orthopedic joint replacements. These alloys are a good substitute for stainless steel because of their easier machining capabilities, superior corrosion resistance, and lower stiffness. This book contains some path-breaking studies on titanium alloys. The readers would gain knowledge that would broaden their perspective on their microstructure, properties, and applications.

I thank my publisher with all my heart for considering me worthy of this unparalleled opportunity and for showing unwavering faith in my skills. I would also like to thank the editorial team who worked closely with me at every step and contributed immensely towards the successful completion of this book. Last but not the least, I wish to thank my friends and colleagues for their support.

Editor

1

The Effect of Initial Structure on Phase Transformation in Continuous Heating of a TA15 Titanium Alloy

Xiaoguang Fan *, Qi Li, Anming Zhao, Yuguo Shi and Wenjia Mei

State Key Laboratory of Solidification Processing, School of Materials Science and Engineering, Northwestern Polytechnical University, Xi'an 710072, China; rickey@mail.nwpu.edu.cn (Q.L.); zhaoanming117@126.com (A.Z.); shiygupc@163.com (Y.S.); mwj725@163.com (W.M.)

* Correspondence: fxg3200@nwpu.edu.cn

Academic Editor: Mark T. Whittaker

Abstract: The effect of initial structure on phase evolution in continuous heating of a near-α TA15 titanium alloy (Ti-6Al-2Zr-1Mo-1V) was experimentally investigated. To this end; three microstructures were obtained by multiple heat treatment: I-bimodal structure with 50% equaixed α, II-bimodal structure with 15% equiaxed α, III-trimodal structure with 18% equiaxed α and 25% lamellar α. Differential scanning calorimetry (DSC), dilatometry and quantitative metallography were carried out on specimens with the three initial structures at heating rates from 5 to 40 °C/min. The transformation kinetics was modeled with the Johnson–Mehl–Avrami (JMA) approach under non-isothermal condition. It was found that there exists a four-stage transformation for microstructures I and III. The secondary and third stages overlap for microstructure II. The four stages of phase transformation overlap with increasing heating rate. In the presence of α laths, the phase transformation kinetics is affected by the composition difference between lamellar α and primary equiaxed α. Phase transformation is controlled by the growth of existing large β phase.

Keywords: titanium alloy; phase transformation; microstructure; DSC; dilatometry

---

1. Introduction

Titanium alloys have been gaining more applications in many industry fields due to the high specific strength, good thermal stability and excellent corrosion resistance [1]. The near-α TA15 titanium alloy which has moderate strength up to 400 °C, excellent thermal stability, good weldability and low growth rate of fatigue crack has been widely used to manufacture structural components in airplanes. The mechanical properties of the titanium alloy are largely dependent on the microstructure [1–3]. The equiaxed and bimodal structures are commonly used for traditional α + β titanium alloys due to a balance in strength, ductility, creep and fatigue resistance (Table 1). The trimodal structure which consists of 10–20% equiaxed α, 30–50% lamellar α and transformed β matrix may also be required after secondary working due to its superior low-cycle fatigue resistance [4]. The diversity in microstructure results from the α-β phase transformation along with deformation induced morphology evolution. Thus, the microstructure can be modulated by optimizing hot working parameters (e.g., heating rate, heating path, heating temperature, strain, strain rate and cooling path).

The phase evolution in hot working involves the α-to-β transformation in heating, the β-to-α transformation in cooling as well as the stress induced transformation during deformation. Numerous researches have been carried out on the phase transformation in cooling. Tang et al. [5] and Sun et al. [6] found that the α lamellae can nucleate in a sympathetic way or by interface instability during slow cooling. He et al. [7] examine the orientation relationship between α and β phase after β working and found that the Burger's orientation relationship is strictly obeyed. Though deformation has little influence on the orientation relationship, the deformation induced texture can result in variation selection and causes a strong texture of β phase [8]. Kherrouba et al. [9] examined the transformation kinetics of Ti-6Al-4V alloy by Johnson–Mehl–Avrami (JMA) model and suggested that growth of β lamellae may be controlled by the combination of solute diffusion and interface migration. The applied stress in machining and deformation can also cause phase transformation. Liu et al. [10] found that can trigger the formation of ω phase can be triggered even before plastic deformation. Jonas et al. [11] reported that dynamic transformation from α to β occurs in hot deformation of several titanium alloys, resulting in significant flow softening during deformation.

Table 1. The mechanical properties of different microstructures for TA15 titanium alloy (compared to equiaxed structure).

Microstructure	Yield Strength	Elongation	Fracture Toughness	Creep Strength	HCF Strength	LCF Strength
Bimodal	++	−/o	+	+	+/o	+
Trimodal	+	−/o	+	+	+	++

From the point of microstructure control, the phase transformation in heating is as important as that in cooling and deformation. Wang et al. [12] used the dilatometry to investigate phase evolution in a TC21 alloy during continuous heating. They found that the phase transformation includes three stages, residual β → acicular α, acicular α → β and equiaxed α → β. The activation energy for α → β transformation was also estimated with the classical JMA equation. A similar transformation behavior was also reported for Ti-6Al-4V alloy by Sha and Guo [13]. Barriobero-Vila et al. [14] examined the element partitioning and related phase transformation kinetics in heating of a bimodal Ti-6Al-6V-2Sn alloy by in-situ high energy synchrotron X-ray diffraction. They found that partitioning of solutes leads to nonlinear variation of the lattice parameters of the β phase. Elmer et al. [15] examined the phase transition during welding of a Ti-6Al-4V alloy. They found that a large superheat is necessary for the completion of α-to-β transformation due to the high heating rate. The overall transformation mechanism may be diverse because the starting assumption determines the calculated parameters in JMA equation. Guo et al. [16] investigated the microstructural developments by deformation induced temperature rise in TA15 titanium alloy. A diffusion model was developed to depict the variation of particle size and volume fraction of equiaxed α under different heating rates. Chen et al. [17] investigated the phase transformation in continuous heating of a near-β titanium alloy. The phase transformation sequence and dominate mechanism were determined. The transformation kinetics and microstructural development in heating are not only dependent on the heating rate, but also greatly affected by the microstructure prior to heating, which needs further investigation.

In this work, the effect of initial structure on the phase evolution in continuous heating of TA15 titanium alloy was investigated. To this end, multiple heat treatments were employed to obtain three different microstructures from a wrought billet. The phase transformation process was determined by metallographic observation, differential scanning calorimetry (DSC) and dilatometry. The microstructure evolution under different heating rate and initial structures was studied. The effect of initial structure on phase transformation kinetics was also measured. The results can be used for quantative control of phase constitution in hot working.

2. Material and Procedures

2.1. Material

The as-received TA15 titanium alloy was a 1000 mm × 400 mm × 100 mm hot forged bar produced by western superconducting technologies Co., Xi'an, China. The chemical composition of the alloy was measured to be 6.69Al, 2.25Zr, 1.77Mo, 2.25V, 0.14Fe, 0.12O, 0.002H and balanced Ti (wt %) by wet chemical analysis. The β transus temperature was determined to be 985 °C by metallography. The as-received material was annealed at 820 °C to obtain a bimodal structure consisting of about 50% equiaxed α within transformed β matrix (microstructure I, Figure 1a). The measured fraction and particle size of the equiaxed α are about 0.5 and 12.5 μm, respectively.

Two other microstructures were obtained by heat treatments from microstructure I. Specimen with initially microstructure I was heated at the rate of 10 °C/min up to 970 °C, held for 30 min and then cooled in the air to get a bimodal structure composed of about 15% equiaxed α within transformed β matrix (microstructure II, Figure 1b). The grain size of equiaxed α phases decreased to about 9.5 μm.

Usually, the trimodal structure can be obtained by a near β hot working and a subsequent heat treatment in the α + β region. The hot working in the near β region aims to control the fraction of equiaxed α. The subsequent heat treatment produces the α lamellae. Thus, the material with bimodal structure was reheated to 940 °C and air cooled to obtain the trimodal structure (microstructure III, Figure 1c). The volume fraction of primary equiaxed and lamellar α were about 18% and 25%, respectively.

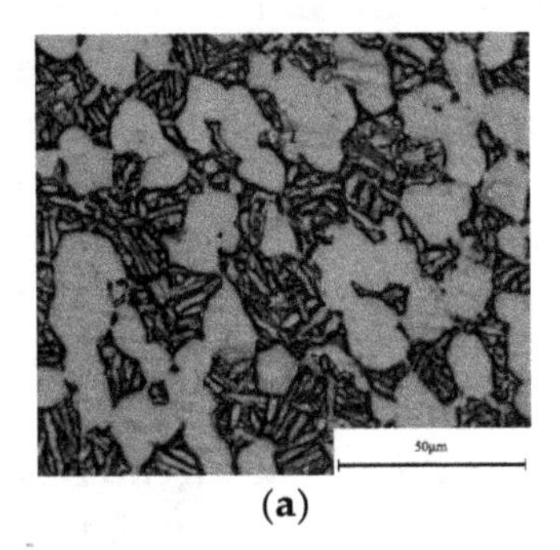
(**a**)

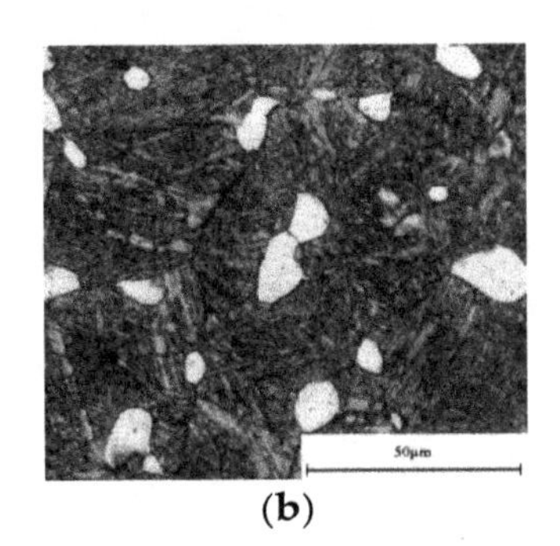
(**b**)

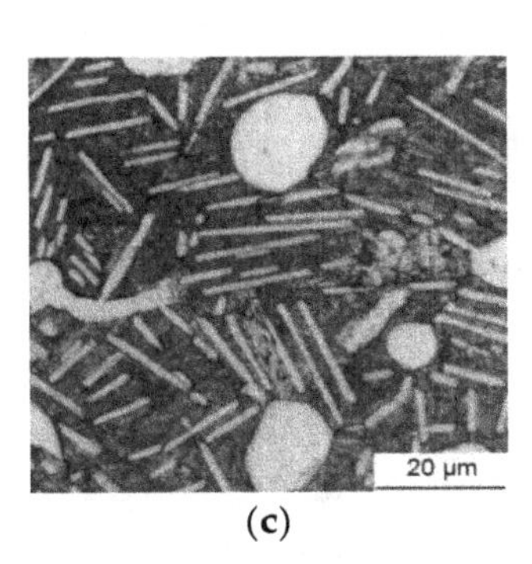

(**c**)

Figure 1. Optical microscopy images of the microstructures prior to heating: (**a**) microstructure I; (**b**) microstructure II; (**c**) microstructure III.

2.2. *Metallographic Examination*

The ϕ 10 mm × 10 mm cylinder specimens were heated at the rates of 5 and 10 °C/min to the preset temperatures with a resistance furnace, and then water quenched to freeze the high temperature microstructure. The specimen was coated with glass lubricant to prevent oxidation at high temperature. The selected quenching temperatures were 700, 750, 800, 860, 900, 940, 970 and 1100 °C. The specimens were electrical discharge cut, mechanical grinded and polished, and etched in a solution of 13% HNO_3, 7% HF and 80% H_2O. Micrographs were taken on a LECIA DMI3000 microscope (LECIA, Shanghai, China). Moreover, the fine secondary lamellar α phases were examined using a scanning electron microscopy TESCAN VEGA3 LMU (TESCAN, Shanghai, China). The volume fraction and size of the α phase were determined by image-pro plus software. The fraction of the α phase was measured by the ratio of the area of α phase to the overall area on a micrograph. The Optical microscopy micrographs at 1000 times magnification were used to measure the fraction of equiaxed α (f_1). Scanning electronic microscopy (SEM, TESCAN, Shanghai, China) images at 5000 times magnification were employed to measure the fraction of α lamellae in β matrix (f_2). The overall volume fraction of α is calculated by $f = f_1 + (1 - f_1) \times f_2$. The grain size of equiaxed α is measured to be the average diameter of the equiaxed α particles on the micrograph. Due to the limitation of radiation heating, phase transformation at higher heating rate was examined by differential scanning calorimetry (DSC) and dilatometry.

2.3. *Differential Scanning Calorimetry*

The characteristic temperatures for phase transformation were measured by the differential scanning calorimetry with a Netzsch DSC-404 calorimeter (Netzsch, Shanghai, China). The specimens were ϕ 4 mm × 0.5 mm disks with different initial structures. The specimens were heated at 10, 20 and 40 °C/min up to 1100 °C respectively. All tests were conducted under the shielding of high purity Ar flow.

2.4. Dilatometry

Dilatometry was carried out on a Netzsch DIL402C dilatometer (Netzsch, Shanghai, China). The specimens used in the experiment were ϕ 6 mm × 25 mm cylinders. The specimens were heated at 5, 10 and 20 °C/min up to 1100 °C, respectively. The whole process was shielded under high purity Ar flow with the flow rate of 50 mL/min. The change in the length of the specimen was recorded and used to determine the linear thermal expansion.

3. Results and Discussion

3.1. Microstructure Observations

In continuously heating, there often exists a three stage phase transformation, including β decompostion at low temperature, lamellar α to β at intermediate temperature and equiaxed α to β at high temperature. For initially microstructure I which was annealed at low temperature, the β-to-α transformation at low temperatures has little influence on the microstructure. The microstructure after heated up to 700 °C is close to the initial structure, as shown in Figure 2a. With increasing temperature, the primary equiaxed α remains unchanged. Meanwhile, the lamellar α is slightly thickened (Figure 2b). Significant β-to-α transformation occurs with further heating. Blocks of β phases firstly appear between the equiaxed and lamellar α phases (Figure 2c). The volume fraction of equiaxed α decreases slightly while the lamellar α are shortened significantly (Figure 2d). The transformation rate of lamellar α is so high that there exists a large fraction of equiaxed α when the lamellar α has already dissolute (Figure 2e). The size of equiaxed α particles becomes more inhomogeneously distributed when the equiaxed α began to dissolve. For diffusion controlled phase transformation process, the dissolution rate of a secondary particle is inverse proportional to its radius. Therefore, the difference in particle size is strengthened during α-to-β transformation (Figure 2f).

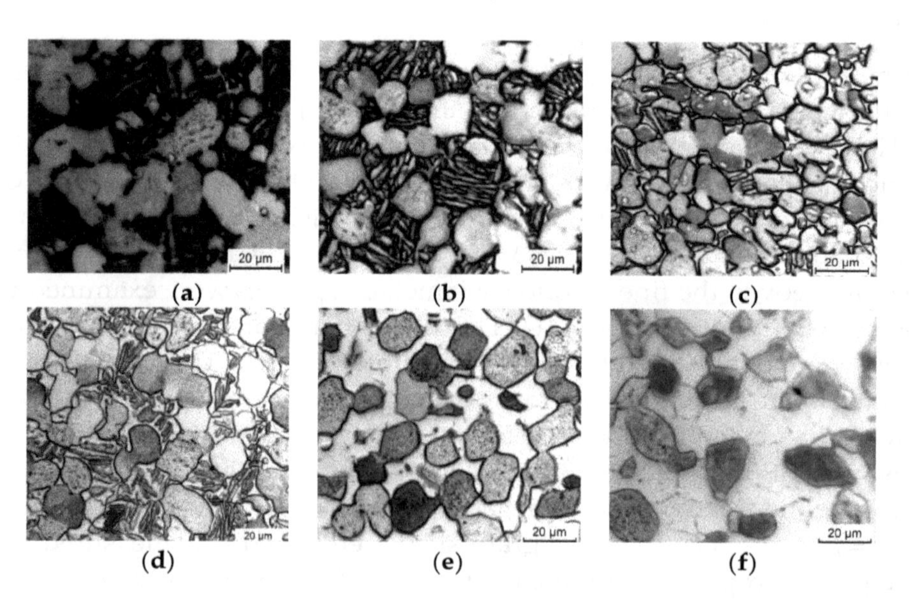

Figure 2. Optical microscopy micrographs of samples with initially microstructure I after heated up to (**a**) 700 °C; (**b**) 800 °C; (**c**) 860 °C; (**d**) 900 °C; (**e**) 940 °C and (**f**) 970 °C at 10 °C/min.

Microstructure II was obtained by high temperature annealing followed by a rapid cooling. The microstructure is more affected by the phase transformation in heating. Metallographic observation suggests that the volume fraction of equiaxed α varies little until the heating temperature is high enough to transform all α lamellae, as shown in Figure 3. On the other hand, the fraction, size and morphology of the lamellar α change significantly.

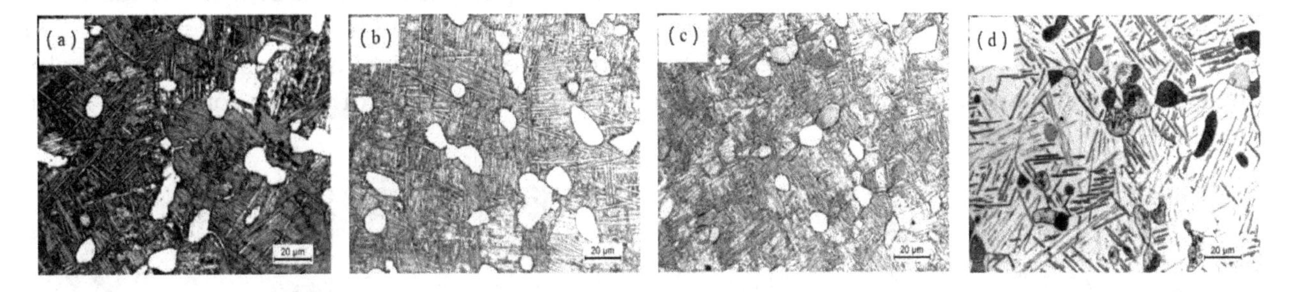

Figure 3. Optical microscopy micrographs of specimens with initially microstructure II after heated up to (**a**) 750 °C; (**b**) 800 °C; (**c**) 860 °C; (**d**) 940 °C at 10 °C/min.

Figure 4 shows the change of lamellar α phase during continuous heating with initially microstructure II. The lamellae varied little up to 700 °C (Figure 4a). For Ti-6Al-4V alloy, the β-to-α transformation occurs around 500 °C. The metastable β phase may have already taken place before 700 °C. However, it is hard to measure it by metallographic observation. Barriobero-Vila et al. [14] reported an increase of 3% in the volume fraction of α phase at a heating rate of 5 °C/min for the Ti-6Al-4V alloy. The increase in volume fraction decreases to 1% at the heating rate of 20 °C/min. It can be found from Figure 3a that the secondary grain boundary α becomes more continuous, which indicates the decomposition of residual β phase. However, the increase in α fraction is trivial comparing to the overall volume fraction of α lamellae (>60%). Meanwhile, the low heating temperature prohibits the coarsening of α lamellae. As a result, the microstructural change is not significant.

During temperature range of 700 to 900 °C, the volume fraction of lamellar α decreases slightly with temperature and the fine lamellar α disappeared (Figure 4b). They either transform to β phase or merge to become thicker lamellae, resulting in relatively thick α lamellae. The average thickness of the lamellar α increases significantly. Above 900 °C, the transformation of lamellar α is greatly accelerated. Thin and short lamellae transform faster than the long and thick ones. The lamellae become isolated by the β matrix. The thickness of the lamellae increases first and then decreases (Figure 4c,d)). The dissolution of secondary α laths is dominated by the shortening along the major axis, which can be taken as a reverse process of the growth of the Widmanstatten α.

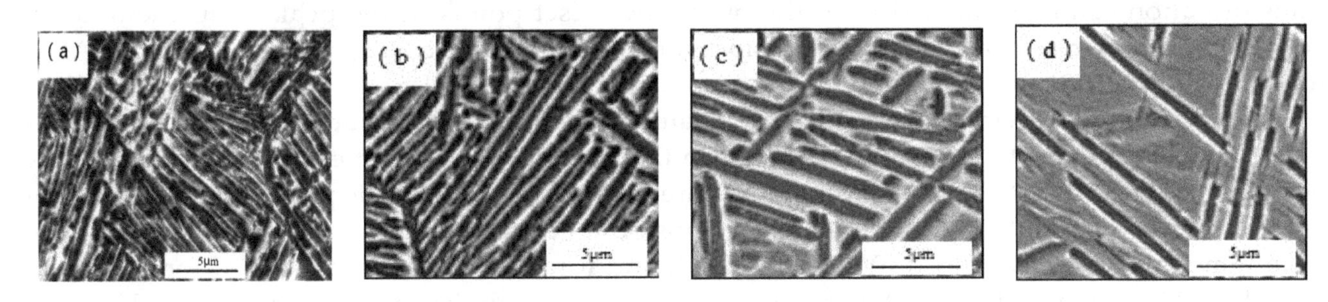

Figure 4. Secondary electron images of the lamellar α from initially microstructure II after heated up to (**a**) 700 °C; (**b**) 900 °C; (**c**) 940 °C; (**d**) 970 °C.

In continuous heating of the trimodal structure (microstructure III), the primary equiaxed α and lamellar α are unchanged when the heating temperature is below 700 °C (Figure 5a). With increasing temperature, the significant increase in the fraction of secondary lamellar α is observed (Figure 5b), which is often interpreted in terms of lamellae thickening. The residual β between the lamellar α consists of a large fraction of thin and disordered secondary α laths (Figure 6a). These α lathes transform preferentially to β phase with increasing temperature, as shown in Figure 6b. Though the secondary α laths are stabilized by annealing at 940 °C, the lower α stabilizer and the higher specific surface area, they transform more rapidly than the primary equiaxed α (Figure 6c,d)). The dissolution behavior of the α laths is similar to that of initially microstructure II. The α laths in microstructure III are formed by additional heat treatment. Their properties (chemical composition, morphology, interfacial coherency, etc.) are similar to the high temperature α lamellae formed in continuous heating of microstructure II.

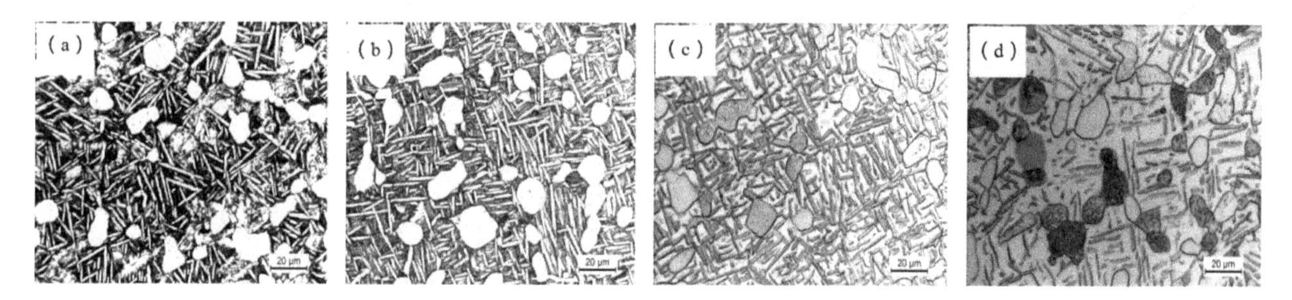

Figure 5. Optical microscopy images of the specimens after heated up from initially microstructure III: (**a**) 700 °C, Water quenched (WQ); (**b**) 800 °C, WQ; (**c**) 940 °C, WQ; (**d**) 970 °C, WQ.

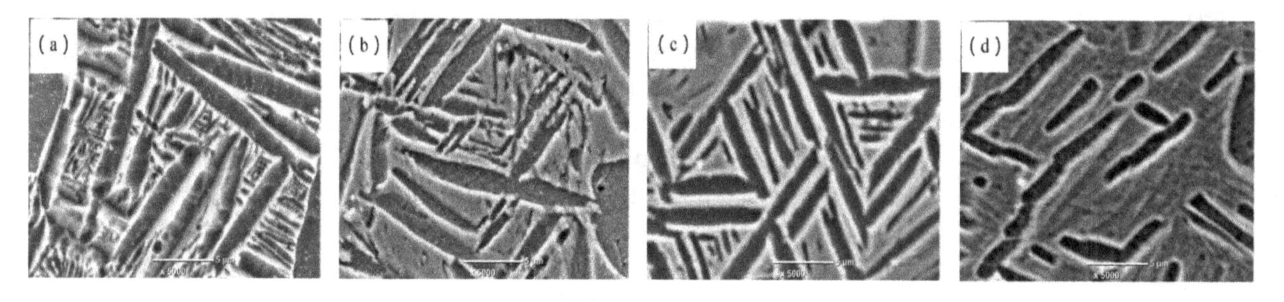

Figure 6. Secondary electron images of the lamellar α with initially microstructure III: (**a**) 800 °C, WQ; (**b**) 860 °C, WQ; (**c**) 900 °C, WQ; (**d**) 970 °C, WQ.

3.2. Analysis of Differential Scanning Calorimetry and Dilatometry

α-β phase transformations involve heat absorption or heat release, which correlate to the endothermic or exothermic peaks in DSC curve. The starting and ending temperatures of the transformation can be estimated from the onset and offset points of the peak. This method has been employed to investigate the phase transformation in continuous cooling and heating of Ti-6Al-4V titanium alloy [13,18].

DSC curves of different initial structure heated at the rate of 10, 20 and 40 °C/min are given in Figure 7. From the DSC curves, it can be seen that there exists a slight exothermic peak around 500 °C at the heating rate of 10 °C/min for all three initial structures. This peak corresponds to the decomposition of residual β phase. The β-to-α transformation temperature ranges from 430 to 530 °C for a bimodal two-phase Ti-6Al-6V-2Sn alloy [14] and it ranges from 590 to 735 °C for a bimodal two-phase TC21 alloy [12] at a heating rate of 5 °C/min, which are close to the current study. However, as the exothermic peak is not significant, it is impossible to determine the temperature range by DSC curves.

After the first exothermic peak, the heat flux increases continuously and reach the first endothermic peak at 760 to 780 °C. The starting temperatures of endothermic peak for the bimodal and trimodal structures are similar (<600 °C), which are much lower than that for the equiaxed structure (about 730 °C). Metallographic observation suggested that α-to-β phase transformation has begun in this temperature range. The tiny α phase inside residual β phase transforms prior to other α phases, as shown in Figure 7. Microstructure III has the highest exothermic peak because there are plenty of tiny α phase inside β matrix (Figure 7a). On the other hand, microstructure I was annealed at the lowest temperature, which stabilizes the constituent phase. As a result, it has the lowest peak and highest starting temperature.

A second endothermic peak appears at 850 to 870 °C. The tiny α phases inside residual β matrix has transformed completely (Figure 7b). As this peak is not significant for microstructure II, it may correspond to the rapid transformation of larger secondary α lamellae. The lamellar α in microstructure II is not as stable as that in microstructure I and III because it is formed during cooling from high temperature annealing. Actually, it is the same to the tiny α phase in microstructure III. The transformation process would be smooth and continuous for microstructure II in this temperature range.

A third endothermic peak is observed at 950 to 970 °C. Apparently, this corresponds to the transformation of equiaxed α to β phase. Microstructure I has the highest peak and the corresponding temperature for the third peak is the lowest, which may be attributed to the large volume fraction of primary equiaxed α phase. The finishing points of the peaks are all around 1000 °C, indicating the β transus temperature at a specific heating rate may not be affected by the initial structure.

No matter what initial structure is, there are three obvious endothermic peaks between the room temperature and the β-transus temperature, which correspond to the transformation of tiny secondary α lamellae, the coarse secondary α lamellae and the primary equiaxed α. The multiple forms of the α phase result in such a three stage α-to-β transformation behavior. The starting and ending points of each stage is affected by the initial structure. As the three stages of transformation overlap, it is impossible to obtain the specific starting and ending temperatures.

The three endothermic peaks can also be observed at the heating rate of 20 °C/min. The temperature for each peak increases as expected. However, the temperature interval between different peaks is deceased, indicating the overlapping of the three stages of phase transformation. When the heating rate increases to 40 °C/min, only the first peak can be clearly observed. The high heating rate increases the driving force for α-to-β transformation, which promotes the transformation of equiaxed α and the coarse lamellar α.

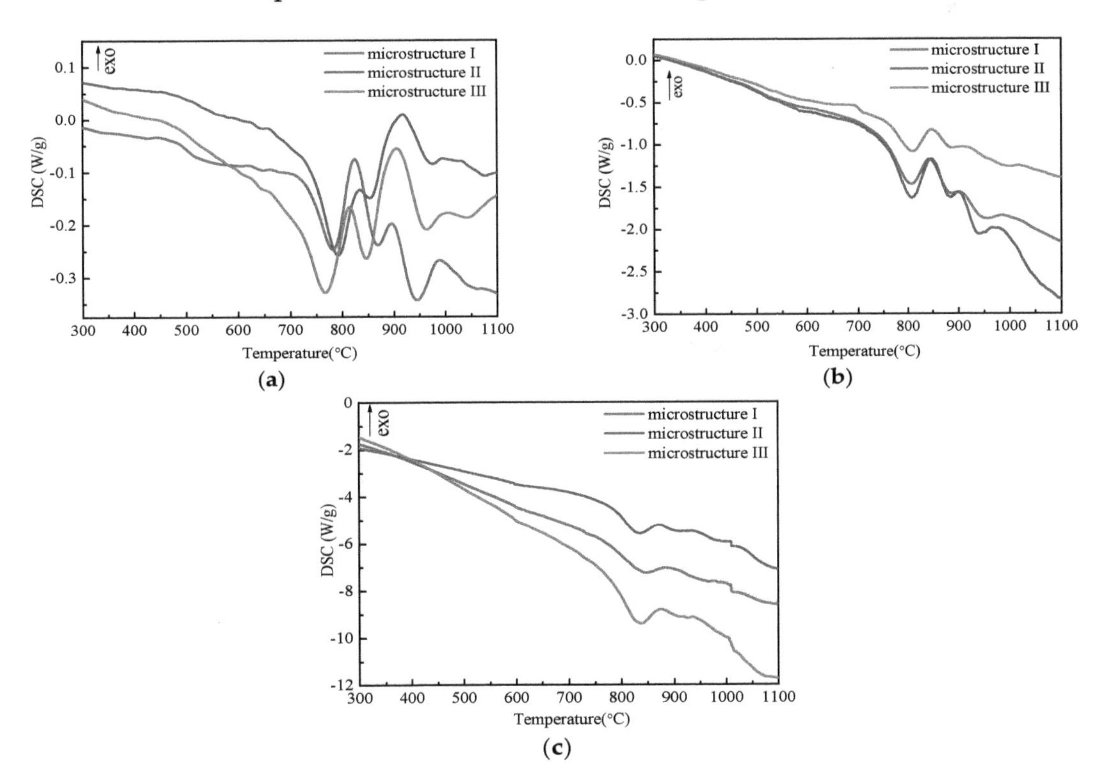

Figure 7. DSC curves of the different initially structures at the heating rate of (**a**) 10 °C/min; (**b**) 20 °C/min; (**c**) 40 °C/min.

The β-to-α phase transformation is accompanied by a volume contraction of 0.15–0.3% [19]. In two-phase titanium alloy, the overall expansion was related to the expansions of the constituent phases and the relative volume fractions. During continuous heating, the length change of the specimens was influenced in the following aspects: the dilatation by lattice change, the thermal expansion of the lattice and additional expansion of the β lattice due to the impoverishment of β stabilizers [20]. Because the sample length variation during phase transformation was not apparent in two-phase titanium alloy, the derivatives of length change with respect to temperature (dL/dT) were used to investigate β-α-β phase transformation, as shown in Figure 8. During continuous heating, the metastable β matrix firstly decomposed to acicular α phases and an increase of length change may be observed in dilatometric curves [12]. However, due to the element partitioning, the lattice of the β phase also contracts [20]. As a result, the thermal expansion rate varies little below 600 °C irrespective of initial structure and heating rate in this work. Due to the lattice expansion of β phase, the TA15 titanium alloy shows a slight increase in expansion rate when the heating temperature is above 600 °C. A similar behavior was also observed in the Ti-6Al-4V alloy. At the lowest heating rate of 5 °C/min, a substantial increase in expansion rate was observed for the initially microstructure I and III, while the variation in expansion rate was low for microstructure II. This is because the residual β phase and lamellar α are cooled from the high temperature β phase, which has a relatively low content of V. The lattice expansion would be low as the element partitioning is not significant. On the other hand, when the equiaxed α transforms to β phase, β stabilizers would diffuse over a long distance as the deviation of β stabilizer content in equiaxed α and β phase is very large. So the degree of enrichment of β stabilizers in β matrix would severely decrease, which increased the lattice parameter of β phase. Therefore, there was an apparent expansion in dilatometric curves in the third stage.

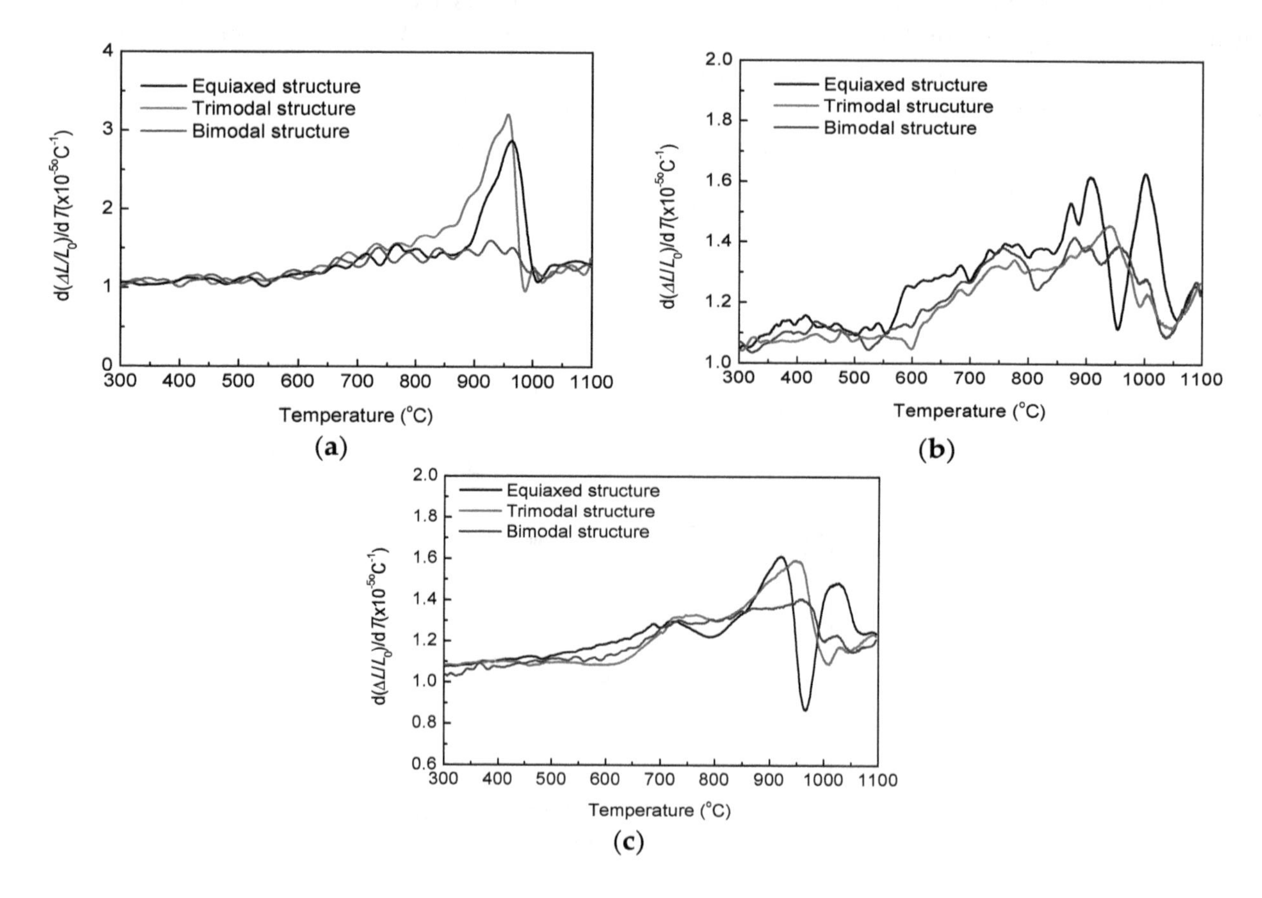

Figure 8. Derivatives of length change with respect to temperature (dL/dT) of the different initially structures at the heating rate of (**a**) 5 °C/min; (**b**) 10 °C/min; (**c**) 20 °C/min.

A sudden decrease in expansion rate occurs at about 950 °C for initially microstructure I and III. Barriobero-Vila et al. [14] found that the expansion in β lattice is more significant when the fraction of β phase is low. With the increase of β fraction, the phase transformation has less effect on the composition of β phase. The expansion of β lattice is more sluggish. Meanwhile, the rapid phase transition may cause significant volume contraction. Thus, the expansion rate drops dramatically. As the α-β phase transition was dominant in the four stages of β-α-β transformation, the obvious peak (Figure 8a) in the curves of derivative of length change corresponded with the starting and ending of the transformation from α to β phase.

The phase transformation sequence in continuous heating is summaried in Table 2. Microstructures I and III undergo a four stage phase transformation: the decomposition of residual β phase to the tiny secondary α laths, the transformation of tiny α laths, the transformation of coarse α laths and the transformation of equiaxed α. The secondary and third stages overlap for microstructure II because the existed α laths are thin. The four stages of phase transformation overlap with increasing heating rate.

Table 2. Phase transformation sequence in continuous heating of TA15 alloy.

Microstructure	Phase Transition
I	residual β-secondary α, secondary α-β, lamellar α-β, equaixed α-β
II	residual β-secondary α, secondary α/lamellar α-β, equaixed α-β
III	residual β-secondary α, secondary α-β, lamellar α-β, equaixed α-β

Dilatometry and DSC can be used to quantify the kinetics of phase transformations [21,22]. In this paper, the dilatometric methods have been adopted. Commonly, the transformed rate increases with temperature in a sigmoidal way. So the results are not presented in this work. In fact, the transformation rate was not measured by the volume contraction in α to β transformation but evaluated by the rapid increase in lattice parameter of β phase, as an increase in expansion rate during transformation is observed at all heating rates. Due to the nonlinearity in lattice expansion and the volume contraction by transformation, it may be concluded that the dilatometry can not be used to estimate the transformation kinetics in TA15 titanium alloy.

3.3. *Transformation Kinetics*

The kinetics measured for different initial structures showed that the effect of initial microstructure is negligible especially when the transformation of equiaxed α becomes dominant, as shown in Figure 9. For each initial structure, the volume fraction of β phase increases slowly with the temperature before 800 °C and the increases sharply until all the α phase was transformed. At the heating rate of 5 °C/min, the volume fraction of β phase around 940 °C was the highest for microstructure I. With increasing heating rate, the specimen with microstructure I has the lowest fraction of β phase.

In this temperature range, there still exists a large fraction of lamellar α in specimens with initially microstructure II and III (Figure 10). On the contrast, only the equiaxed α phase is left for microstructure I. At a low heating rate, the phase fraction is more close to the equilibrium value. The phase fraction is more affected by the chemical compositions of the constituent phases. The content of β stabilizers in α lamellae is higher than that in equiaxed α. In the presence of lamellar α, the fraction of α phase is larger according to the lever rule. At a higher heating rate, the transformation rate is more determined by the diffusion of the β stabilizer (Mo and V for the TA15 alloy). The α lamellae are thin and distributed in the residual β matrix. Along with the high content of β stabilizer, the phase transformation is faster when there are large amount of α lamellae.

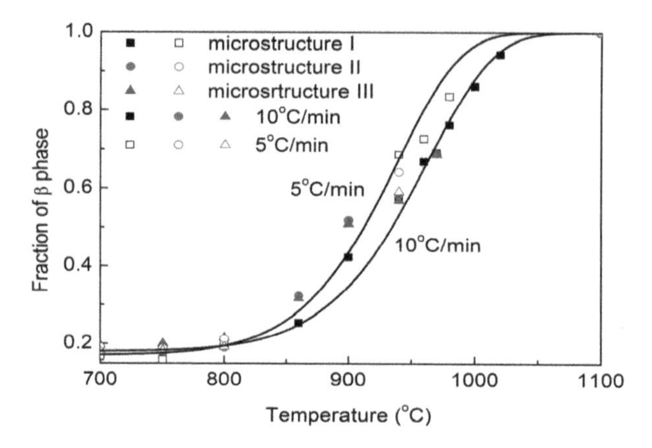

Figure 9. Measured fraction β with temperature at different heating rates.

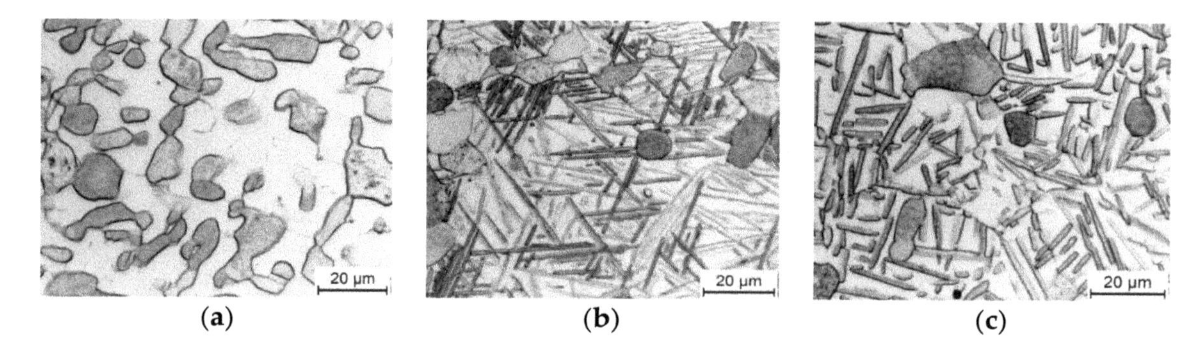

Figure 10. Optical microscopy micrographs of specimens after heated up to 940 °C at 5 °C/min: (**a**) equiaxed structure; (**b**) bimodal structure; (**c**) trimodal structure.

The phase transformation kinetics in isothermal condition is often depicted by the Johnson–Mehl–Avrami (JMA) approach:

$$f(t) = 1 - \exp(-(kt)^n) \tag{1}$$

where $f(t)$ is the transformed fraction at a specific time t, n is the JMA exponent, and k is a rate constant. In non-isothermal conditions, k is taken to be a function of temperature which is given by:

$$k = \mathrm{k}_0 \exp\left(-\frac{Q}{RT}\right) \tag{2}$$

where k_0 is a constant, Q is the activation energy of the transformation, R is the gas constant, and T is the absolute temperature. Elmer et al. [15] suggested that a unique activation energy under non-isothermal conditions is not always possible because the nucleation and growth of the β phase are simultaneously operating. Their contributions are temperature and rate dependent and influenced by the initial structure. However, the α-to-β transformation in titanium alloys often has a large growth component and it is possible to simplify it by using the activation energy of growth. It has been found in the continuous cooling of TA15 titanium alloy that Mo diffusion through β matrix controls the epitaxial growth of primary α due to its slowest diffusivity [23]. Assuming the diffusion of V in β phase controls the transformation and the mobility of the α-β interface is high, Elmer [15] suggested to use the activation energy for the diffusion of V to represent the activation energy for phase transformation in a Ti-6Al-4V alloy. These assumptions are also valid in the TA15 titanium alloy [23].The activation energy for diffusion of Mo is determined to be 154 kJ/mol [24]. It is used as a starting value for parameter identification. The JMA exponent n also has physical significance. Elmer et al. [15] used a value of

$n = 4$ to represent a transformation mechanism involving both growth and nucleation. For diffusion controlled transformation under isothermal condition, the value should be larger than 2.5. The value of 4 is also employed as a starting value.

Using the numerical procedure in Ref. [18], the optimized parameters are given in Table 3. The calculated JMA exponent is near 1. This means the phase transformation is controlled by the growth of existing large β phase. This is in accordance with the microstructural developments in the current work. Elmer [15] reported a much higher value of JMA exponent. In their work, the heating rate is hundreds of times larger than the current work, which may change the mechanism of phase transformation.

Table 3. Calculated Johnson–Mehl–Avrami (JMA) parameters for the continuous heating of TA15 titanium alloy.

ln (k_0)	n	Q (kJ/mol)
26.9	0.75	333

4. Conclusions

Experimental study was carried out on the effect of initial structure on microstructure evolution in continuous heating of a near-α TA15 titanium alloy. The following conclusions were drawn:

(1) A four stage phase transformation occurs during continuous heating of the TA15 titanium alloy for microstructure I and III: the decomposition of residual β phase, the transformation of tiny α laths, the transformation of coarse α laths and the transformation of equiaxed α. The secondary and third stages overlap for microstructure II. The four stages of phase transformation overlap with increasing heating rate.

(2) The α-to-β transformation is accompanied by the coarsening of secondary α laths for microstructure II, resulting in significant changes in the size and volume fraction of α laths.

(3) The phase transformation kinetics is not affected by the initial structure when the transformation of equiaxed α becomes dominant. In the presence of lamellar α, the specimen with initially microstructure I has the highest transformed fraction under slow heating but the lowest transformed fraction under rapid heating.

(4) The transformed fraction increases with temperature in a sigmoidal way which can be fitted by the JMA model. The determined JMA exponent is close to 1, suggesting phase transformation is controlled by the growth of existing large β phase.

Acknowledgments: This work is supported by the National Natural Science Foundation of China (No. 51575449), Research Fund of the State Key Laboratory of Solidification Processing (NWPU), China (No. 104-QP-2014), and 111 Project (B08040).

Author Contributions: Xiaoguang Fan conceived and designed the experiments and interpret the data, Qi Li and Xiaoguang Fan wrote the paper, Anming Zhao analyzed the data and Yuguo Shi performed the experiments, Wenjia Mei collected the literatures.

References

1. Lütjering, G.; Williams, J.C. *Titanium*; Springer: Berlin, Germany, 2007.
2. Beranoagirre, A.; Lacalle, L.N.L. Grinding of gamma TiAl intermetallic alloys. *Procedia Eng.* **2013**, *63*, 489–498. [CrossRef]
3. Beranoagirre, A.; Olvera, D.; Lacalle, L.N.L. Milling of gamma titanium–aluminum alloys. *Int. J. Adv. Manuf. Technol.* **2012**, *62*, 83–88. [CrossRef]
4. Zhou, Y.G.; Zeng, W.D.; Yu, H.Q. An investigation of a new near-beta forging process for titanium alloys and its application in aviation components. *Mater. Sci. Eng. A* **2005**, *393*, 204–212. [CrossRef]

5. Tang, B.; Kou, H.; Zhang, X.; Gao, P.; Li, J. Study on the formation mechanism of α lamellae in a near β titanium alloy. *Prog. Nat. Sci. Mater. Int.* **2016**, *26*, 385–390. [CrossRef]
6. Sun, Z.; Guo, S.; Yang, H. Nucleation and growth mechanism of α-lamellae of Ti alloy TA15 cooling from an α + β phase field. *Acta Mater.* **2013**, *61*, 2057–2064. [CrossRef]
7. He, D.; Zhu, J.C.; Zaefferer, S.; Raabe, D.; Liu, Y.; Lai, Z.L.; Yang, X.W. Influences of deformation strain, strain rate and cooling rate on the Burgers orientation relationship and variants morphology during $\beta \rightarrow \alpha$ phase transformation in a near α titanium alloy. *Mater. Sci. Eng. A* **2012**, *549*, 20–29. [CrossRef]
8. Zhao, Z.B.; Wang, Q.J.; Hu, Q.M.; Liu, J.R.; Yu, B.B.; Yang, R. Effect of β (110) texture intensity on α-variant selection and microstructure morphology during β/α phase transformation in near α titanium alloy. *Acta Mater.* **2017**, *126*, 372–382. [CrossRef]
9. Kherrouba, N.; Bouabdallah, M.; Badji, R.; Carron, D.; Amir, M. β to α transformation kinetics and microstructure of Ti-6Al-4V alloy during continuous cooling. *Mater. Chem. Phys.* **2016**, *181*, 462–469. [CrossRef]
10. Liu, H.H.; Niinomi, M.; Nakai, M.; Cho, K.; Fujii, H. Deformation-induced omega-phase transformation in a beta-type titanium alloy during tensile deformation. *Scr. Mater.* **2017**, *130*, 27–31. [CrossRef]
11. Jonas, J.J.; Aranas, C.; Fall, A.; Jahazi, M. Transformation softening in three titanium alloys. *Mater. Des.* **2017**, *113*, 305–310. [CrossRef]
12. Wang, Y.H.; Kou, H.C.; Chang, H.; Zhu, Z.; Su, X.; Li, J.; Zhou, L. Phase transformation in TC21 alloy during continuous heating. *J. Alloys Compd.* **2009**, *472*, 252–256. [CrossRef]
13. Sha, W.; Guo, Z.L. Phase evolution of Ti–6Al–4V during continuous heating. *J. Alloys Compd.* **1999**, *290*, L3–L7. [CrossRef]
14. Barriobero-Vila, P.; Requena, G.; Buslaps, T.; Alfeld, M.; Boesenberg, U. Role of element partitioning on the α–β phase transformation kinetics of a bi-modal Ti–6Al–6V–2Sn alloy during continuous heating. *J. Alloys Compd.* **2015**, *626*, 330–339. [CrossRef]
15. Elmer, J.W.; Palmer, T.A.; Babu, S.S.; Zhang, W.; DebRoy, T. Phase transformation dynamics during welding of Ti–6Al–4V. *J. Appl. Phys.* **2004**, *95*, 8327–8339. [CrossRef]
16. Guo, L.G.; Zhu, S.; Yang, H.; Fan, X.G.; Chen, F.L. Quantitative analysis of microstructure evolution induced by temperature rise during (α + β) deformation of TA15 titanium alloy. *Rare Met.* **2016**, *35*, 223–229. [CrossRef]
17. Chen, F.; Xu, G.; Zhang, X.; Zhou, K. Exploring the phase transformation in β-quenched Ti-55531 alloy during continuous heating via, dilatometric measurement, microstructure characterization, and diffusion analysis. *Metall. Mater. Trans. A* **2016**, *47*, 5383–5394. [CrossRef]
18. Malinov, S.; Guo, Z.; Sha, W.; Wilson, A. Differential scanning calorimetry study and computer modeling of $\beta \Rightarrow \alpha$ phase transformation in a Ti-6Al-4V alloy. *Metall. Mater. Trans. A* **2001**, *32*, 879–887. [CrossRef]
19. Motyka, M.; Kubiak, K.; Sieniawski, J.; Ziaja, W. Phase transformations and characterization of α + β titanium alloys. In *Comprehensive Materials Processing*; Hashmi, S., Ed.; Elsevier: Amsterdam, The Netherlands, 2014; Volume 2, pp. 7–36.
20. Elmer, J.W.; Palmer, T.A.; Babu, S.S.; Specht, E.D. In situ observations of lattice expansion and transformation rates of α and β phases in Ti–6Al–4V. *Mater. Sci. Eng. A* **2005**, *391*, 104–113. [CrossRef]
21. Guo, Z.; Keong, K.G.; Sha, W. Crystallisation and phase transformation behaviour of electroless nickel phosphorus platings during continuous heating. *J. Alloys Compd.* **2003**, *358*, 112–119. [CrossRef]
22. Liu, Y.C.; Sommer, F.; Mittemeijer, E.J. Abnormal austenite–ferrite transformation behavior in substitutional Fe-based alloys. *Acta Mater.* **2003**, *51*, 507–519. [CrossRef]
23. Meng, M.; Yang, H.; Fan, X.G.; Yan, S.L.; Zhao, A.M.; Zhu, S. On the modeling of diffusion-controlled growth of primary α in heat treatment of two-phase Ti-alloys. *J. Alloys Compd.* **2016**, *691*, 67–80. [CrossRef]
24. Semiatin, S.L.; Lehner, T.M.; Miller, J.D.; Doherty, R.D.; Fueere, D.U. α/β heat treatment of a titanium alloy with a nonuniform microstructure. *Metall. Mater. Trans. A* **2007**, *38*, 910–921. [CrossRef]

2

Improving the Mechanical Properties of a β-type Ti-Nb-Zr-Fe-O Alloy

Vasile Danut Cojocaru [1], Anna Nocivin [2,*], Corneliu Trisca-Rusu [3], Alexandru Dan [1], Raluca Irimescu [1], Doina Raducanu [1] and Bogdan Mihai Galbinasu [4]

1 Materials Science and Engineering Faculty, University Politehnica of Bucharest, 060042 Bucharest, Romania; dan.cojocaru@upb.ro (V.D.C.); alexandru_dan_ro@yahoo.com (A.D.); raluca.irimescu@stud.sim.upb.ro (R.I.); doina.raducanu@mdef.pub.ro (D.R.)
2 Mechanical, Industrial and Maritime Faculty, Ovidius University of Constanța, 900527 Constanța, Romania
3 National Institute for Research and Development in Micro-technologies, 077190 Bucharest, Romania; corneliu.trisca@nano-link.net
4 Dental Medicine Faculty, University of Medicine and Pharmacy "Carol Davila" Bucharest, 020021 Bucharest, Romania; bogdan.galbinasu@yahoo.com
* Correspondence: anocivin@univ-ovidius.ro

Abstract: The influence of complex thermo-mechanical processing (TMP) on the mechanical properties of a Ti-Nb-Zr-Fe-O bio-alloy was investigated in this study. The proposed TMP program involves a schema featuring a series of severe plastic deformation (SPD) and solution treatment (STs). The purpose of this study was to find the proper parameter combination for the applied TMP and thus enhance the mechanical strength and diminish the Young's modulus. The proposed chemical composition of the studied β-type Ti-alloy was conceived from already-appreciated Ti-Nb-Ta-Zr alloys with high β-stability by replacing the expensive Ta with more accessible Fe and O. These chemical additions are expected to better enhance β-stability and thus avoid the generation of ω, α′, and α″ during complex TMP, as well as allow for the processing of a single bcc β-phase with significant grain diminution, increased mechanical strength, and a low elasticity value/Young's modulus. The proposed TMP program considers two research directions of TMP experiments. For comparisons using structural and mechanical perspectives, the two categories of the experimental samples were analyzed using SEM microscopy and a series of tensile tests. The comparison also included some already published results for similar alloys. The analysis revealed the advantages and disadvantages for all compared categories, with the conclusions highlighting that the studied alloys are suitable for expanding the database of possible β-Ti bio-alloys that could be used depending on the specific requirements of different biomedical implant applications.

Keywords: β-Titanium alloys; thermo-mechanical processing; SEM; mechanical properties

1. Introduction

Good biomedical materials for orthopedic implants with long-term service need a combination of a low Young's modulus, close proximity to the human bone, and high strength to avoid the known "stress shielding effect" [1–5]. Among many investigated biocompatible materials, non-cytotoxic β-Ti alloys have been developed in recent decades [6–13], as these alloys have the most attractive combination of a low Young's modulus, high mechanical strength, and high ductility. This property combination can also ensure good processability (i.e., plastic deformation, machinability, welding, etc.) and good tribological characteristics, all of which are necessary for the long-term dimensional stability of an implant [14]. Compared to other possible Ti-bio-alloys, such as the frequently used (α) or (α + β) types, the special advantage of β-type Ti-alloys is their lower Young's modulus (45–65 GPa) [2,4,15–17]. Beta-Ti alloys are also reported to exhibit high fracture toughness [18] and a good response to heat treatment if the alloying is achieved using suitable β-stabilizing chemical elements [5,19–21], such as non-toxic Nb, Ta, Mo, and Zr [22–24]. Ti-Nb-Ta-Zr (TNTZ) alloys are highly regarded [25–30] because of their suitable combination of functional properties, both mechanical and biomedical. However, they exhibit some disadvantages, such as a much higher melting point of Ta compared to other related alloying elements ($T_{top}{}^{Ta}$ = 2996 °C, $T_{top}{}^{Nb}$ = 2468 °C, $T_{top}{}^{Zr}$ = 1855 °C, $T_{top}{}^{Ti}$ = 1660 °C), which decreases the casting alloy characteristics. Even though Nb also has a high melting point compared to Ti and Zr, it is still indispensable due to its strong β-stabilizing characteristics. In addition, Ta is an expensive chemical element [6]. Therefore, efforts have been made to replace Ta with more accessible Fe and O [31–36]. Fe is selected not only for economic considerations but also because it is a strong β stabilizer [37,38]. It was shown that the Young's modulus (E) can be reduced to 91 GPa by adding Fe [38], which is slightly lower than the E value for commonly used α/β Ti-64 (~110 GPa [39]) but higher than the E of cortical bone (~30 GPa) [1]. Similar results were reported in [40–44]. Oxygen, despite not being a β stabilizing element [20,21], may hamper the formation of α″ [1,2]. Recent research found that by adding oxygen to β-Ti alloys, both strength and ductility can be improved simultaneously [43,44]. Moreover, Fe and interstitial oxygen are beneficial for improving mechanical properties through solid-solution strengthening [18,31–33]. Therefore, both Fe and O are attractive elements for the development of high-strength and low-Young's-modulus β-Ti alloys.

Apart from the chemical composition being an important factor for enhancing the mechanical properties in β-Ti alloys, thermo-mechanical processing (TMP) is also a very important way to decrease the Young's modulus and increase the mechanical strength of β-Ti alloys [14,45]. The applied TMP can involve a combination of severe plastic deformations (SPD) and solution treatments (STs) to achieve grain refinement featuring ultra-fine or even nano-meter grain dimensions. Through this, the mechanical strength can increase significantly, and the Young's modulus can decrease. For example, Bertrand et al. [46] reported a very low Young's modulus for a Ti-25Ta-25Nb alloy (55 GPa, one of the lowest values for a β-Ti alloy) developed via TMP; the objective of the applied treatment (cold rolling + solution treatment) was to restore a fully recrystallized β phase microstructure from the cold rolled state using a decreased β grain size and a low Young's modulus.

Previous studies on TNTZ-Fe-O alloys [31–33,47,48] provided valuable information for developing novel alloys by tuning the composition and processing parameters to obtain the best mechanical and biomedical properties suitable for medical implants. These reports refer to alloys in a cold-rolling (CR) state or after a CR + ST combination, for which the Young's modulus can vary between 60 and 107 GPa, and the ultimate tensile strength-UTS can vary between 903 and 1370 MPa. The intent of the present work is also to elaborate a complex TMP program applied to a particular chemical composition from the TNTZ-Fe-O family-alloy (but without Ta and with Fe and O) to find a better combination of TMP parameters coupled with a suitable chemical composition and ultimately obtain better mechanical properties than those reported previously. Thus, a chemical composition of the Ti-Nb-Zr-Fe-O alloy was proposed for investigation, with sufficient β-stabilizing elements to obtain a stable β-Ti alloy. On the other hand, the proposed TMP program contains two distinct directions: One is formed from a CR series with a gradual increase in the applied deformation degree, coupled with an ST series featuring variable parameters; the second one combines a series of severe plastic deformation (SPD) and ST, both in the effort to achieve evident grain refinement. Thus, both previously reported modalities for increasing the strength of β-Ti alloys will be applied [49–51], including the addition of Fe and O and a complex TMP program with variable processing parameters. The experimental results will be compared with those reported in [31–33,47,48] by highlighting the advantages and disadvantages of the results obtained for the proposed alloy and corresponding TMP parameters. At the same time, the proposed investigations seek to describe the relationship between the microstructure and mechanical properties of this alloy under different heat treatment conditions. Alongside this main objective, the intention is also to enlarge the database of possible β-Ti bio-alloys, depending on the specific requirements of different implant applications.

2. Materials and Methods

2.1. *Synthesis of the Studied Alloy*

The nominal new chemical composition for the studied alloy, Table 1, is as follows:

Table 1. The chemical composition of the studied alloy (wt.%).

The Chemical Composition of the Studied Alloy (wt.%)	Ti	Nb	Zr	Fe	O
Ti-Nb-Zr-Fe-O	57.25	34.10	7.59	0.90	0.16

This alloy was obtained using a levitation induction melting furnace FIVE CELES-MP25 with a nominal power of 25 kW and a melting capacity of 30 cm^3 in a high vacuum of 10^{-4}–10^{-5} mbar. The alloy synthesis was conducted with an intense agitation of the melted alloy. The ingots were re-melted twice to achieve a high degree of chemical homogeneity.

2.2. *Thermomechanical Processing Program of the New β-Ti Alloy*

After the alloy synthesis and before the complex TMP program, the as-cast alloy was treated to obtain a quality-homogenised precursor (Figure 1). The applied treatment consisted of (a) Cold Rolling (CR) with a relative reduction of $\varepsilon = 20\%$ using a Mario di Maio LQR120AS rolling-mill (Mario di Maio Inc., Milano, Italy) with a 3 m/min rolling speed and no lubrication; before the CR process, the sample was cleaned using an ultrasonic bath at 60 °C in ethylic alcohol; (b) a Homogenization Treatment (HT) at 1223 K/950 °C (above the β-transus temperature) with a holding time of 20 min. and water quenching (w.q.) using a GERO SR 100 × 500 type oven (Carbolite-Gero Inc., Neuhausen, Germany) under a high vacuum. The final obtained homogenised alloy was named "the initial alloy" and was processed by a complex TMP program.

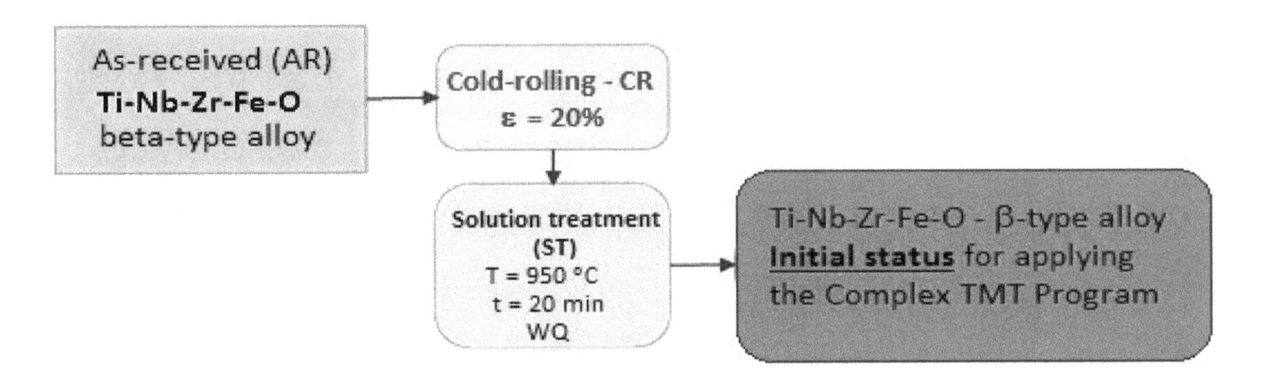

Figure 1. Schema for processing the as-cast sample to obtain a quality homogenised precursor named the initial alloy.

This program was structured and applied to two different research directions on a batch/group of samples obtained from the initial alloy (Figures 2 and 3). These two categories of samples were processed with different thermo-mechanical parameters for the final analysis, comparison, and recommendations for potential selection/application.

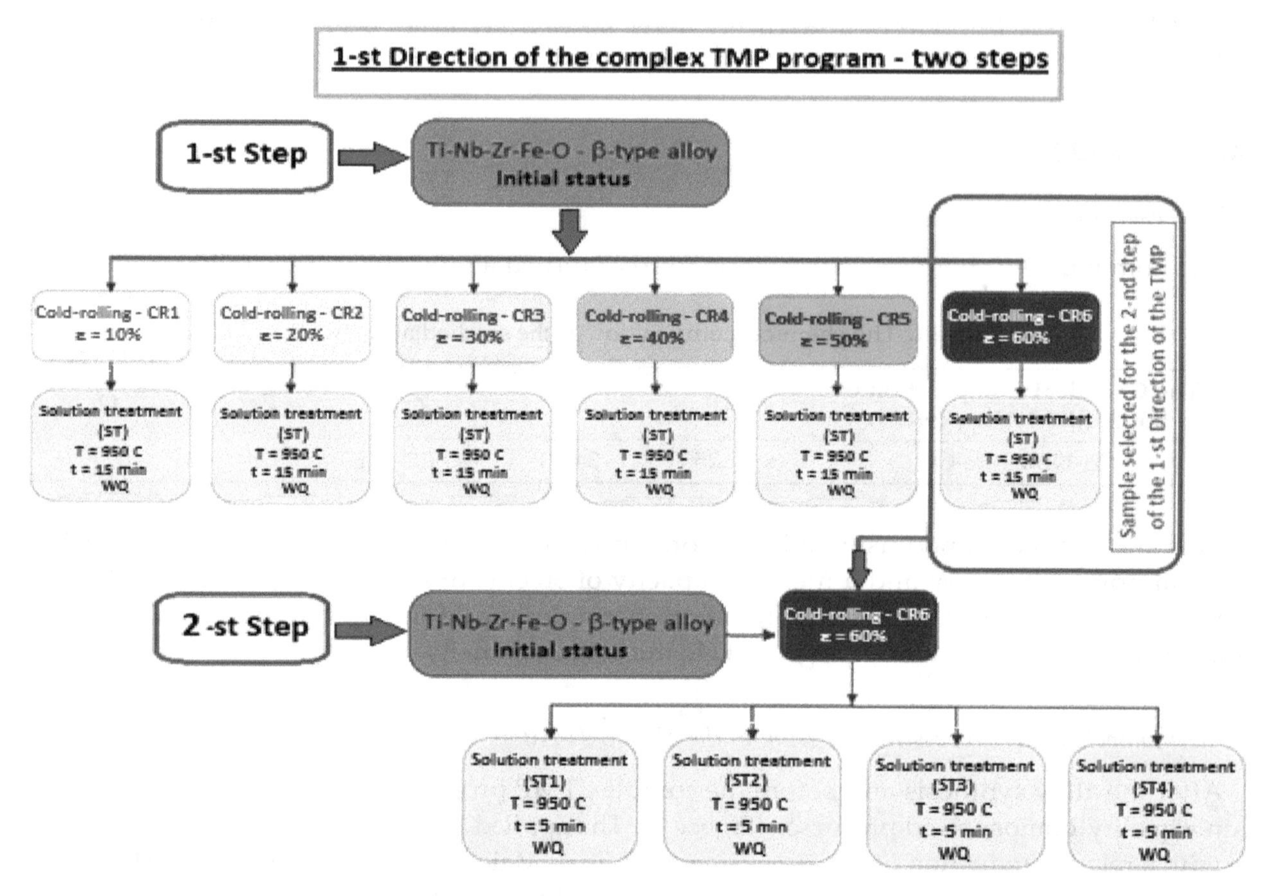

Figure 2. The 1st Direction of the complex TMP Program applied to the studied β-type Ti-Nb-Zr-Fe-O alloy.

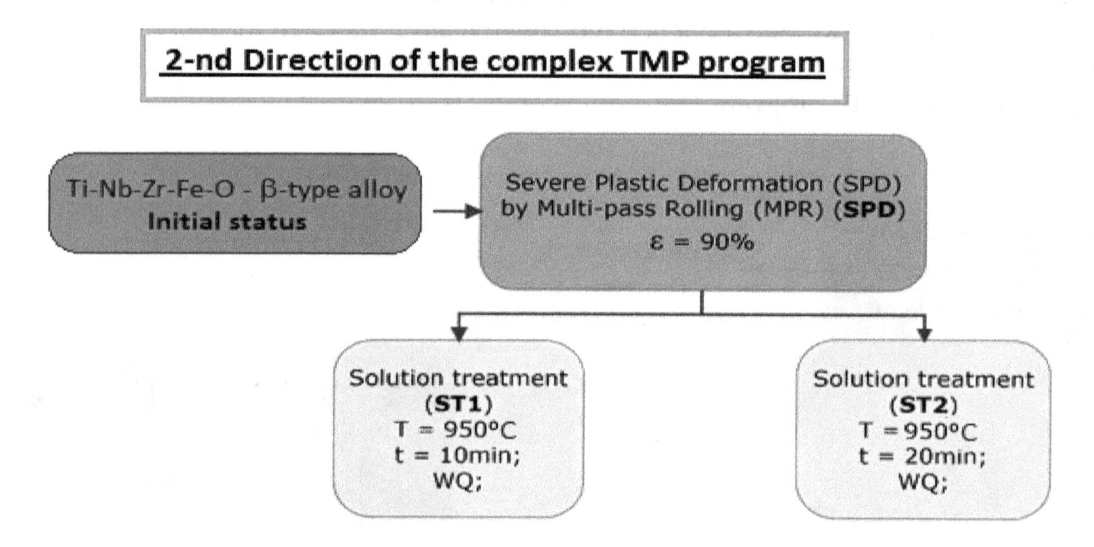

Figure 3. The 2nd Direction of the complex TMP Program applied to the studied β-type Ti-Nb-Zr-Fe-O alloy.

The first direction (Figure 2) included a two-step process: (1) The first step consisted of various CR cycles followed by identical STs for all CR samples—1223 K/15 min./w.q. CR was applied to six distinct samples with the initial state using six different total deformation degrees: ε_{tot} = 10%—1 pass; ε_{tot} = 20%—2 passes; ε_{tot} = 30%—3 passes; ε_{tot} = 40%—4 passes; ε_{tot} = 50%—5 passes; ε_{tot} = 60%—6 passes. (2) For the second step, from the six variants that were experimented on, the one with ((ε_{tot} = 60%) + ST) was selected, as it presented the smallest grain size.

Four STs were applied to this selected sample after cold-rolling (ε_{tot} = 60%—6 passes) for different times of 5, 10, 15, and 20 min, while the other parameters remained unchanged.

The second direction (Figure 3) includes severe plastic deformation (SPD) using the multi-pass rolling (MPR) method, followed by two different STs. MPR processing was performed using 10 rolling passes with ε_{tot} = 90%. The two variants of the ST consisted of the same heating temperature (1223 K/950 °C) and cooling medium (water) but two distinct holding times (10 min. and 20 min.). All STs were performed using a GERO SR 100 × 500-type oven with a high vacuum—the same as that used for the initial treatment.

2.3. Micro-Structural and Mechanical Analysis of the Alloy Samples

A Metkon MICRACUT 200 type machine (Metkon Instruments Inc., Bursa, Turkey) with diamond cutting disks was used for cutting. The specimens were then fixed on a specific epoxy resin of a Buehler Sampl-Kwick type, abraded with 1200 grit SiC paper using a Metkon Digiprep ACCURA machine (Metkon Instruments Inc., Bursa, Turkey), and then mechanically polished using 6, 3, and 1 μm diamond paste and 0.03 μm colloidal silica on a Buehler VibroMet2 machine (Buehler Ltd., Lake Bluff, IL, USA).

The SEM images and analysis were realized using a scanning electron microscope—a TESCAN VEGA II—XMU (Tescan Orsay Holding, a.s., Brno, Czech Republic). The CR- and SPD-processed samples were examined mainly in the RD–ND cross-section (RD—rolling direction; ND—normal direction) to observe the grain deformation/texture degree evolution. To highlight the main microstructural characteristics, only the most representative images were selected.

The tensile tests, performed on the final states from both research directions, were achieved in the RD using a Gatan MicroTest-2000N-type machine (Gatan Inc., Pleasanton, CA, USA) with a strain rate of 1×10^{-4} s^{-1}. Based on the obtained data, the following average values of the mechanical characteristics were determined: the ultimate tensile strength (σ_{UTS}); yield strength ($\sigma_{0.2}$); the elongation to fracture (ε_f); and the elastic modulus (E). The standard deviation (SD) was also calculated.

3. Results

3.1. Micro-Structural Analysis of the As-Cast and Homogenized Sample, Named the Initial State

The structures corresponding to the cast sample and the homogenized sample (by applying CR + HT) are indicated in Figure 4. Due to consistent β-stabilizing alloying elements (~42% in total, excluding oxygen), the alloy's structure was formed from only β-phase grains with an equiaxe shape. The measured average dimensions of the β grains were 121 μm for the as-cast sample and slightly larger for the homogenized sample (145 μm) due to grain growing during the applied heat treatment.

In metastable β-type Ti alloys, it is already known that the ω, α′, and α″ phases can be produced in the β matrix as a result of cold plastic deformation [52–55]. These phases are detrimental to the formability of the material, making it difficult to process. Therefore, the interest lies in designing β-Ti-type alloys containing a sufficient quantity of alloying elements with β-stabilizing characteristics. For that purpose, the main β-stabilizing element—Nb—should be no less than 35–38% to ensure good stability of the β phase in binary Ti-Nb alloys (see Hu et al. [56]). For this study, besides high Nb content (34.1 wt.%), the addition of Fe and Zr was considered, as both elements are reported to be effective for solid solution strengthening, while also playing a very important role from the perspective of β-phase stability by suppressing the generation of α′ and ω and by shifting the martensite starting temperature (Ms) to a lower one [47]. Fe also is a strong β stabilizer [37], and Zr, despite being considered by some reports to be a neutral element from a β-stabilization perspective [57], has been demonstrated by others to be beneficial for β-Ti-alloys because it suppresses the nucleation of isothermal ω [6] and also decreases the Ms temperature [58]. Therefore, it can be presumed that the absence of the ω, α′, and α″ phases may be a consequence of the combined stabilization of the β-phase under all the added alloying elements (~42%), a fact also reported in [4,59,60]. It can be also presumed that the design of the chemical composition of the studied alloy achieved the initially stated goal—obtaining a single β-phase for safe and easy mechanical processing. However, this assertion should be taken only as probably because, to prove the absence of the above secondary phases, a detailed XRD analysis should be performed, which represents a future objective for the following stages of this research work.

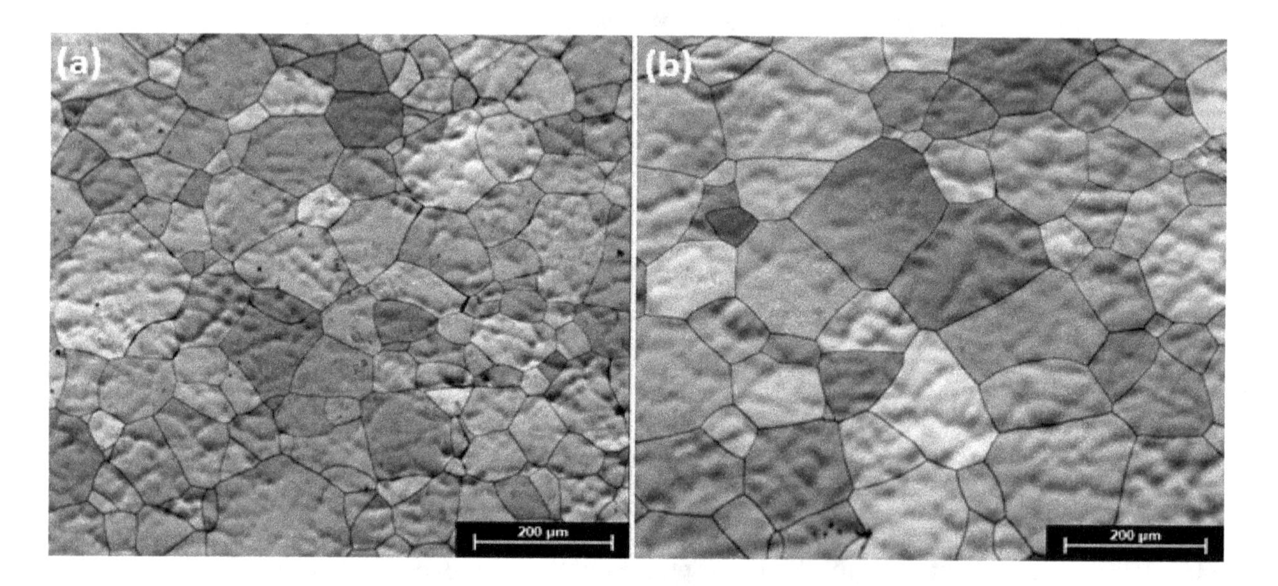

Figure 4. The SEM images of the Ti-Nb-Zr-Fe-O alloy microstructures in: (**a**) the as-cast state; (**b**) after the homogenisation treatment—named the initial state.

3.2. Micro-Structural and Mechanical Analysis of the Samples Processed by the First Direction of the TMP Program

Following the first direction schema of the TMP Program, a series of samples was processed by CR with different deformation degrees from $\varepsilon_{tot} = 10\%$ to $\varepsilon_{tot} = 60\%$ (Figure 2). Figure 5 shows the obtained microstructures. It can be observed that the β grains lengthened in the rolling direction step-by-step from sample to sample due to the increase in the applied total deformation degree. It can also be observed that the deformation products became increasingly visible and pronounced with an increase of the deformation degree (Figure 5a–c). These deformation products seemed to be kink bands, twin bands, or even shear bands [4], depending on the opening/value of the misorientation angle, which were smaller or larger between the β matrix and these deformation products. For example, if the misorientation angles between the kink bands and the β matrix for the Ti-22.4Nb-0.73Ta-2Zr-1.34O alloy during its compression straining, as reported by Yang et al. [59], are between 10–30°, the visible deformation products of the band-types in Figure 5a (CR at 10%) with similar misorientation angles can also be considered kink bands. Further, in Figure 5b,c, the misorientation angle increases between 50–55°. Since the misorientation angle between the {332} <113> β-mechanical twins and the β matrix is 50.5° in the <110> β direction [61], it can be presumed that, for the present case, the deformation products are evolving gradually from kink to twin bands (white arrows on the images). Concerning increases in the misorientation angle with an increase in plastic deformation, similar research results were reported for other β-Ti alloys [52,53]. Generally, it is already known that the twining phenomenon facilitates subsequent grain fragmentation/refinement during plastic deformation process development [62–65]. For this study, Figure 5b,c show the gradual grain refinement as a function of an increase in the deformation degree. In addition to this grain fragmentation, the structure acquires a clear textured appearance (Figure 5d,f); these last three images show the formation of shear bands due to severe deformation degrees, which were also reported in [59,64] for other similar β-type Ti-alloys.

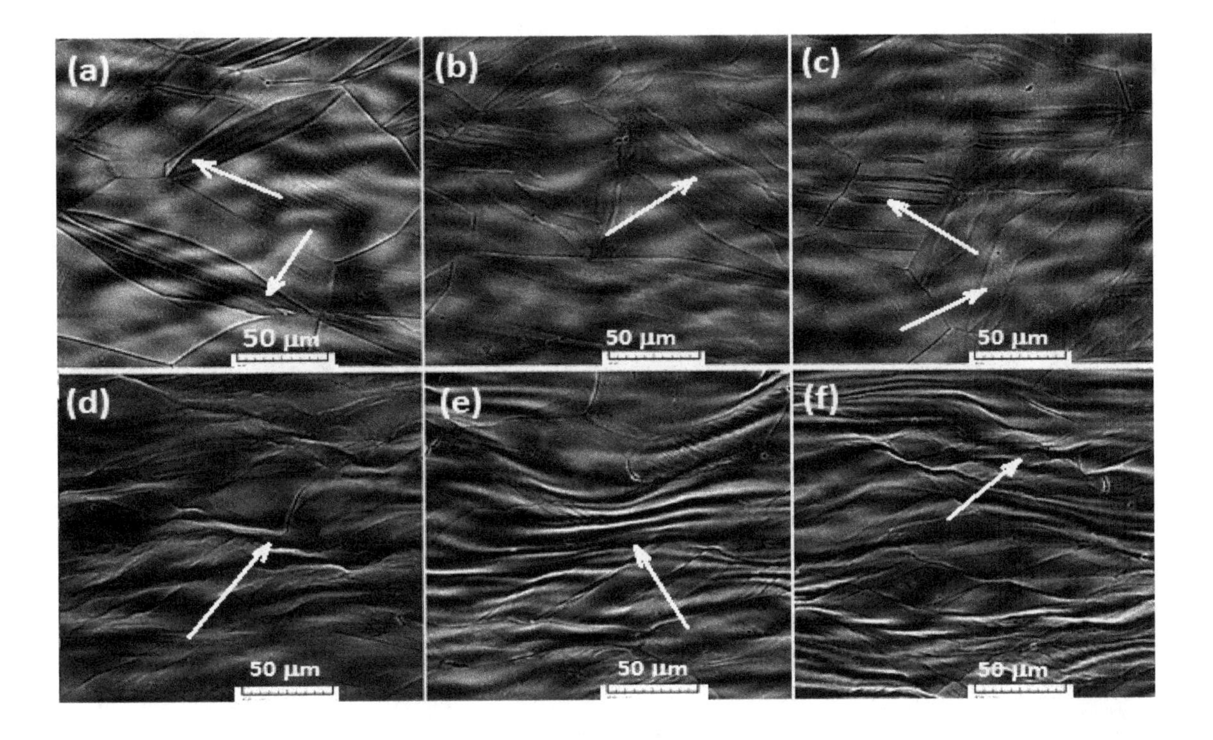

Figure 5. The SEM images of the Cold Rolling (CR) processed samples from the 1st step of the 1st direction of the TMP Program: (**a**) CR—ε_{tot} = 10%; (**b**) CR—ε_{tot} = 20%; (**c**) CR—ε_{tot} = 30%; (**d**) CR—ε_{tot} = 40%; (**e**) CR—ε_{tot} = 50%; (**f**) CR—ε_{tot} = 60%.

From Figure 5, it can also be observed that the shear bands are not homogeneously distributed and have a wavy shape. This wavy shape of the slip bands is reported to be typical for bcc metals [38], as in our case. Shear bands are considered to be formed from rotated and severely distorted crystals due to the localization/accumulation of a mass of shear stress on a slip plane [59].

In this study, the process starting with kinking, passing through {332} <113> β-mechanical twinning, and arriving to shear band formations provides the most probable deformation products that will gradually facilitate alloy deformation and texture during CR as the degree of deformation increases. However, this assertion will also have to be proven by a future XRD/TEM analysis, as noted previously.

After applying the same ST for all six CRed samples (Figure 2), the β-grains again became equiaxe (Figure 6), but with the grain size decreasing steadily depending on the ascending degree of the deformation that was previously applied. This decreasing phenomenon of the grain size is proven once again in Figure 7, which shows the measured average values of the obtained β-grains from 138 µm corresponding to the sample (CR1 + ST) and those at 75 µm corresponding to the sample (CR6 + ST). In this way, the β-grain dimensions became almost two times smaller than the initial dimensions (138/75 = 1.84) by applying a deformation degree that was six times greater (from 10% to 60%).

At this stage of the experiments, the sample (CR6 + ST) with the smallest β grains (75 µm) was selected to be further processed, following the 2nd step of the 1st direction of the TMP program. This second step (Figure 2) requires that for the sample selected (the one with the smallest β grains), after applying CR with the same ε_{tot} = 60%, the ST treatments will be varied/diversified by applying four different holding times of 5 min, 10 min, 15 min, and 20 min (same 1223 K/water quenching). The resulteding microstructures are shown in Figure 8, and the measured average values of the obtained β-grains are shown in Figure 9.

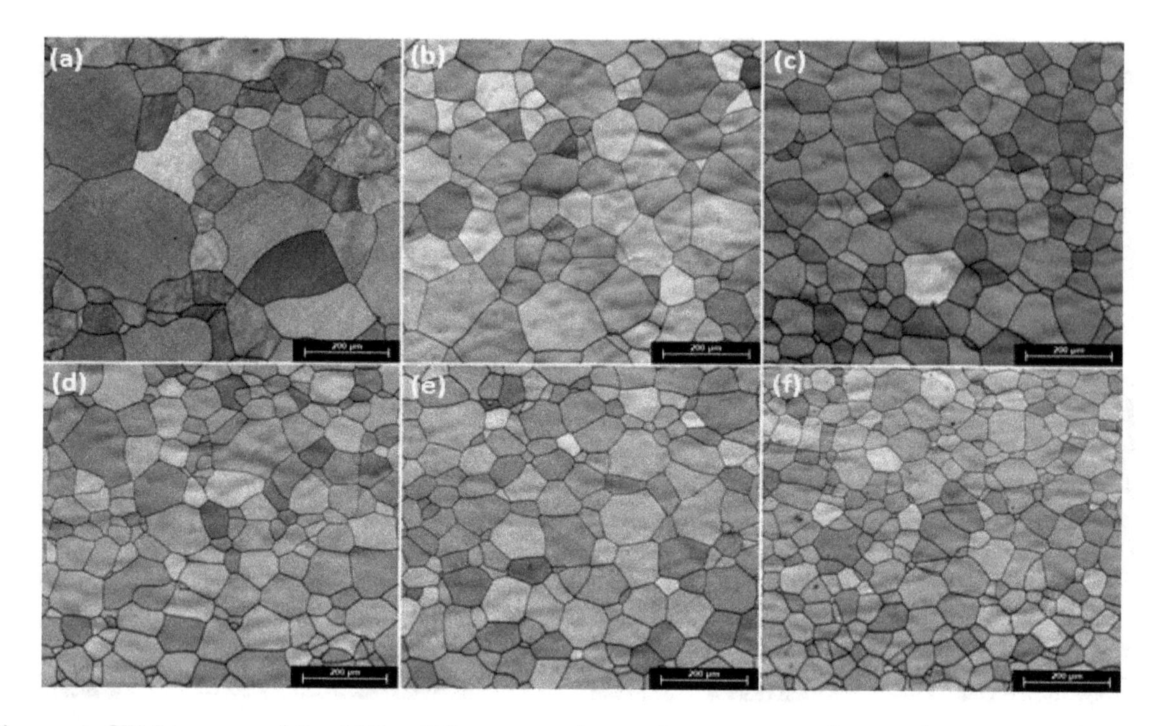

Figure 6. SEM images of the (CR + ST) processed samples corresponding to the 1st step of the 1st direction of the TMP Program, with various ε_{tot} values for CR and similar STs for all six samples: 1223 K/15 min./w.q.: (**a**) CR(ε_{tot} = 10%) + ST; (**b**) CR(ε_{tot} = 20%) + ST; (**c**) CR(ε_{tot} = 30%) + ST; (**d**) CR(ε_{tot} = 40%) + ST; (**e**) CR(ε_{tot} = 50%) + ST; (**f**) CR(ε_{tot} = 60%) + ST.

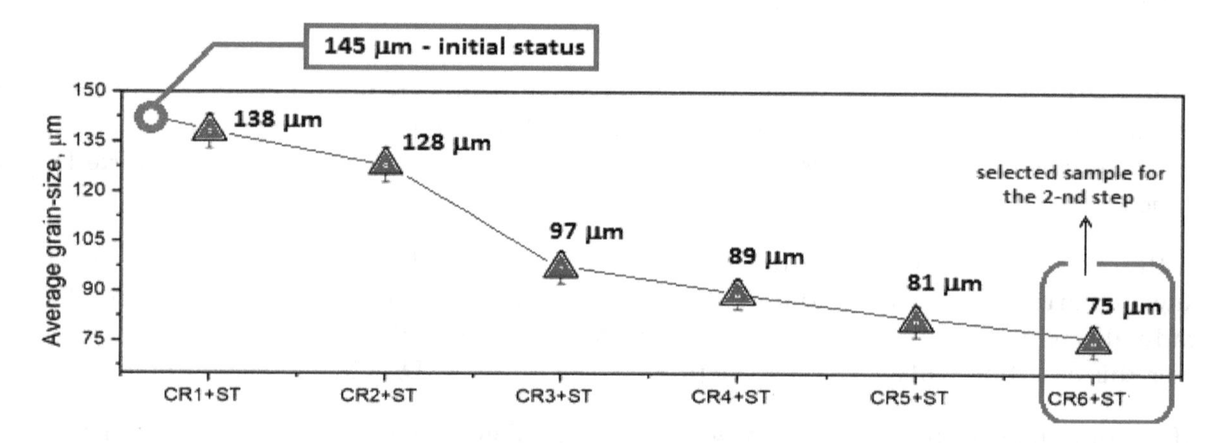

Figure 7. The evolution of the β grain average dimension as a function of the applied TMP variant corresponding to the 1st step of the 1st direction of the TMP Program.

The microstructures from Figure 8 also represent a series of β grains like those in Figure 6 but with visibly smaller grain dimensions. The grains are homogeneous and equiaxial without any traces of deformation products. Considering that the applied CR here was identical for all four samples (ε_{tot} = 60%), the variation of the β grain size was due to the use of a shorter or longer time for grain growing during the different holding times applied to the STs: The smallest grains correspond to the smallest holding time of 5 min., and vice versa. However, the resulting grains' dimensions (Figure 9) are much smaller than those in the anterior 1st step of the TMP program. This time, the smallest grains of 55 μm were obtained for the sample with CR (ε_{tot} = 60%) and ST1 (1223 K/5 min./w.q.).

Furthermore, tensile tests were applied to all four resulting samples. Figure 10 shows the strain–stress curves for the following samples: (a) the sample in the initial state; (b) the sample after applying the CR (ε_{tot} = 60%); and (c) the sample CR (ε_{tot} = 60%) + ST3 (1223 K/15 min./w.q.). For reasons of space, the inclusion of curves was waived for the other three CR+ST samples. Nevertheless, the mechanical characteristics were determined for all four samples. Table 2 indicates the average values for the ultimate tensile strength (σ_{UTS}); yield strength ($\sigma_{0.2}$); elongation to fracture (ε_f); and elastic modulus (E). Standard deviation is also included.

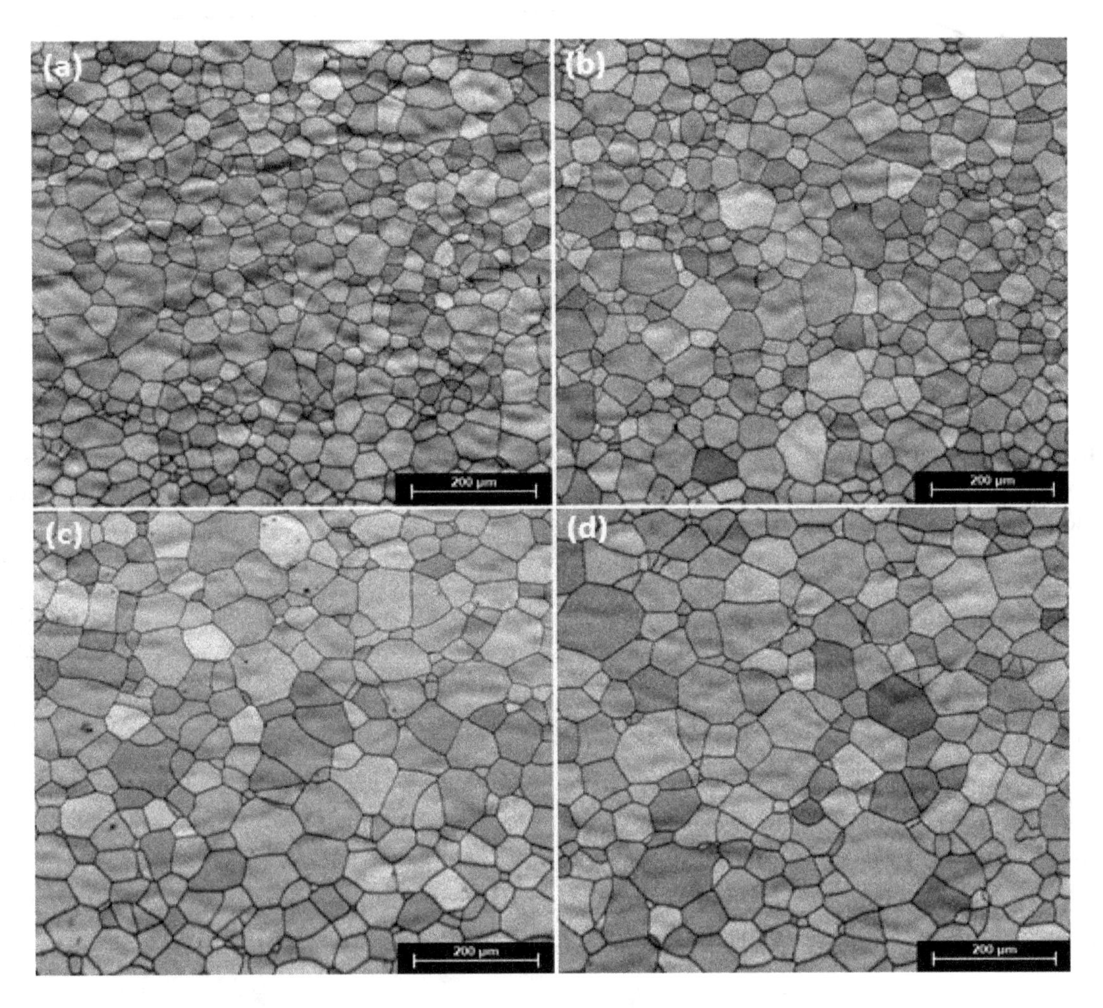

Figure 8. The SEM images of the (CR + solution treatment (ST) processed samples corresponding to the 2nd step of the 1st direction of the TMP Program with the same ε_{tot} = 60% for the applied CR but with various STs: (**a**) CR + ST1 (1223 K/5 min./w.q.); (**b**) CR + ST2 (1223 K/10 min./w.q.); (**c**) CR + ST3 (1223 K/15 min./w.q.); (**d**) CR + ST4 (1223 K/20 min./w.q.).

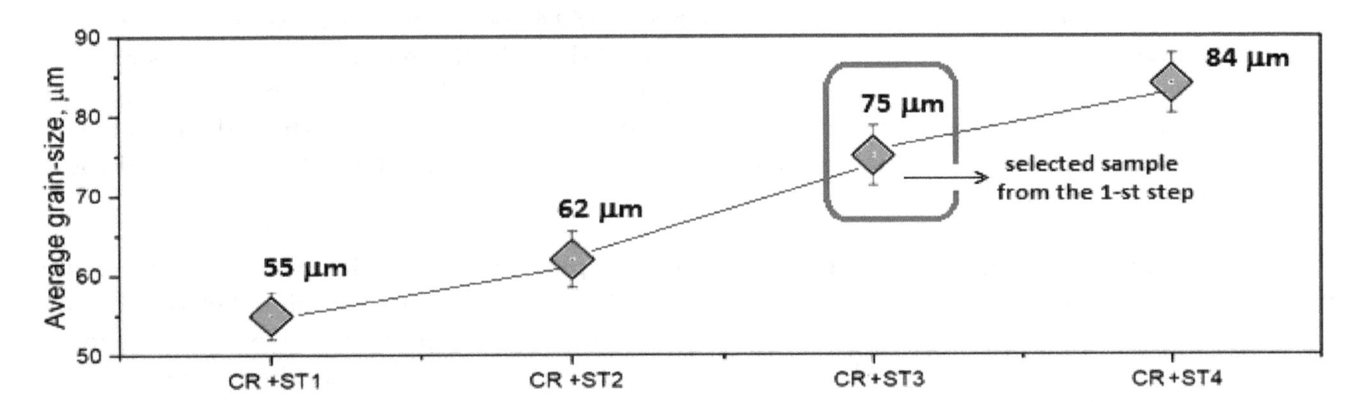

Figure 9. The evolution of the β grain average dimension as a function of the applied TMP variant, corresponding to the 2nd step of the 1st direction of the TMP Program.

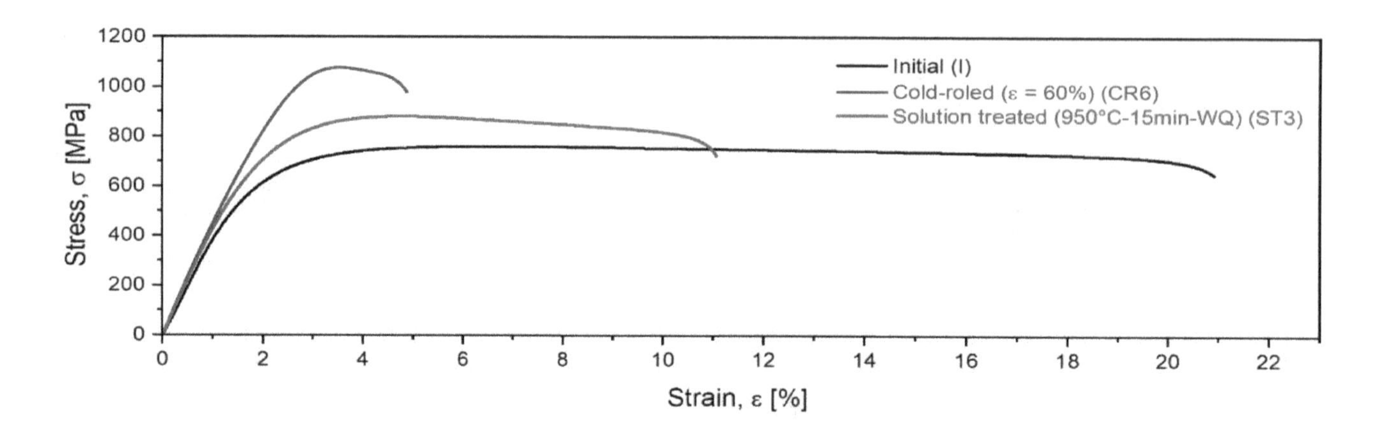

Figure 10. The stress–strain curves of the samples corresponding to the 2nd step of the TMT program (sample selected from the 1st step).

Table 2. Mechanical properties of the studied alloy at the end of the 1st direction of experiments from the TMP Program: Ultimate Tensile Strength (σ_{UTS}); Yield Strength ($\sigma_{0.2}$); Elongation to Fracture (ε_f); Elastic Modulus (E); SD—Standard Deviation.

Structural State	Mechanical Properties			
	σ_{UTS} (SD) (MPa)	$\sigma_{0.2}$ (SD) (MPa)	ε_f (SD) (%)	E (SD) (GPa)
Initial state (I)	759.7 (12.1)	546.1 (10.3)	20.9 (0.8)	49.1 (2.1)
CR—ε_{tot} = 60%	1076.1 (15.3)	808.1 (13.8)	4.9 (0.1)	48.8 (1.8)
CR + ST1 (1223 K-5 min-w.q.)	962.8 (14.5) (↑26.7%)	703.7 (12.6) (↑28.8%)	6.9 (0.2) (↓66.9%)	48.3 (1.6) (↓1.6%)
CR + ST2 (1223 K-10 min-w.q.)	901.7 (15.1) (↑18.7%)	631.4 (10.2) (↑15.6%)	8.2 (0.2) (↓60.7%)	47.6 (0.9) (↓3.0%)
CR + ST3 (1223 K-15 min-w.q.)	879.5 (13.7) (↑15.7%)	601.6 (10.4) (↑10.2%)	11.1 (0.4) (↓46.9%)	47.8 (1.1) (↓2.6%)
CR + ST4 (1223 K-20 min-w.q.)	809.0 (13.3) (↑6.5%)	574.4 (10.4) (↑5.2%)	15.9 (0.4) (↓23.9%)	49.3 (2.2) (↑0.4%)

A general characteristic for strain–stress curves, observed for bcc β-Ti alloys with soluble substitution elements and interstitial oxygen, is a flat zone lacking classic double yielding [60,66,67]. This characteristic is also visible in Figure 10. The flat zone indicates that the yield strength is very close to the ultimate tensile strength. Moreover, the stress values remain constant by increasing the strain. The explanation of this phenomenon reported in [30,42,60,68,69] relates to the heterogeneity of the deformation process for β-Ti alloys: dislocation sliding, β-phase twining, or "stress-induced martensitic transformation" β→α" (SIM) depending on the alloy's chemical composition, electronic parameters, deformation temperature, or even oxygen content. For the presently studied alloy, all these details regarding the deformation mechanisms are intended to be analysed in future research through TEM analyses specifically dedicated to this issue.

As shown in Table 2, it can be observed that the samples with CR + ST, compared to the initial state, feature increased values for UTS, between 6.5 and 26.7%, and YS provides values between 5.2 and 28.8%. The values of Young's modulus are almost the same (a very small decrease), but the elongation to fracture decreased significantly to between 23.9 and 66.9%. Considering that the grain dimension decreased from 145 μm (the initial state) to 55 μm (by about 62%), a preliminary conclusion here is that a combination of the applied parameters increased the UTS and YS for this first direction of the experiments without losing the acceptable level of a low Young's modulus (~48 GPa).

3.3. Micro-Structural and Mechanical Analysis of the Samples Processed by the 2nd Direction of the TMP Program

Following the 2nd Direction of the proposed experiments from the complex TMP Program, two samples of the initial state were processed by SPD-MPR with the same deformation degree of ε_{tot} = 60% (Figure 3). Figure 11 shows the obtained microstructures.

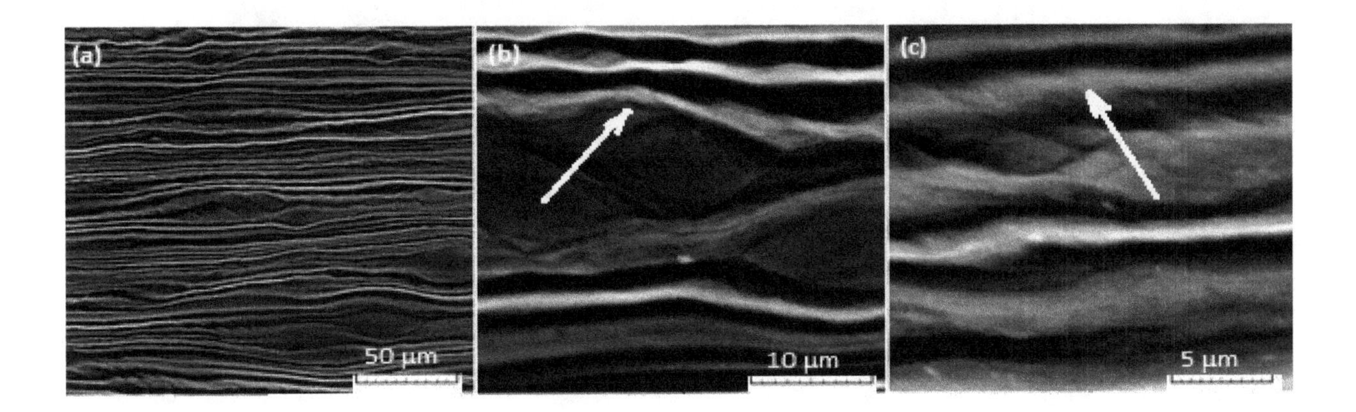

Figure 11. (**a–c**) SEM images of the sample processed by SPD-MPR (ε_{tot} = 60%) corresponding to the 2nd direction of the TMP Program (**a–c**—same sample, with gradual increasing magnifications).

It can be observed that the β grains were strongly textured on the rolling direction. From the images with greater magnification, it can also be observed that the mechanism of deformation was similar to the mechanisms from the first category of the experimental samples, as discussed above. Here, due to the very high deformation degree being applied, the resulting deformation products are more like shear bands with very large misorientation angles, indicated by white arrows in Figure 11b. Moreover, the absence of any other phases can be observed—phases that, alongside β-grains, could have been formed during SPD (e.g., the ω, α′, and α″ phases). This means that the amount of the alloying elements with β-stabilizing effects was sufficient to secure the stability of the β phase and suppress the formation of secondary phases.

After SPD processing, both samples were subjected to ST using the same heating temperature and quenching medium (1223 K and water, respectively) as the first category of experiments but with two different holding times: 10 min and 20 min, respectively. The resulting microstructures are indicated in Figure 12, with the measured average β-grain dimensions indicated directly on the images. The first sample (SPD-60% + ST1(1223 K-10 min-w.q.)) had a similar β grain dimension (77 µm) to the sample (CR-60% + ST3(1223 K-15 min-w.q.)) from the first category of experiments (75 µm). The second sample (SPD-60% + ST2(1223 K-20 min-w.q.)) had a similar β-grain dimension (91 µm) to the sample (CR-40% + ST (1223 K/15 min/w.q.)) of 89 µm.

To assess the mechanical properties, tensile tests were applied to both the resulting SPD + ST samples. Similar to Figure 10, Figure 13 shows the strain–stress curves for initial, severely deformed, and solution-treated samples. Determination of the mechanical characteristics was also performed for both samples. Table 3 indicates the average values for the ultimate tensile strength (σ_{UTS}); yield strength ($\sigma_{0.2}$); elongation to fracture (ε_f); and elastic modulus (E).

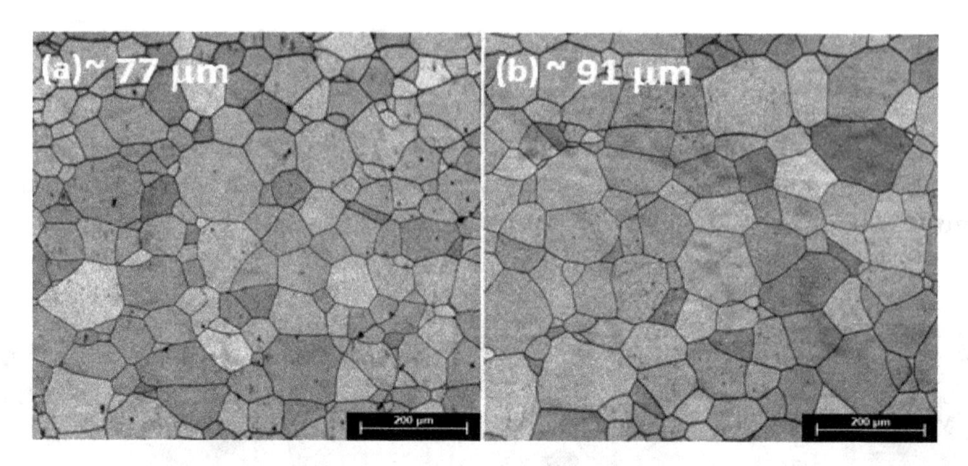

Figure 12. The SEM images of the two samples corresponding to the 2nd direction of the TMP Program: (**a**) SPD-MPR (ε_{tot} = 60%) + ST1 (1223 K-10 min-w.q.); (**b**) SPD-MPR (ε_{tot} = 60%) + ST2 (1223 K-20 min-w.q.).

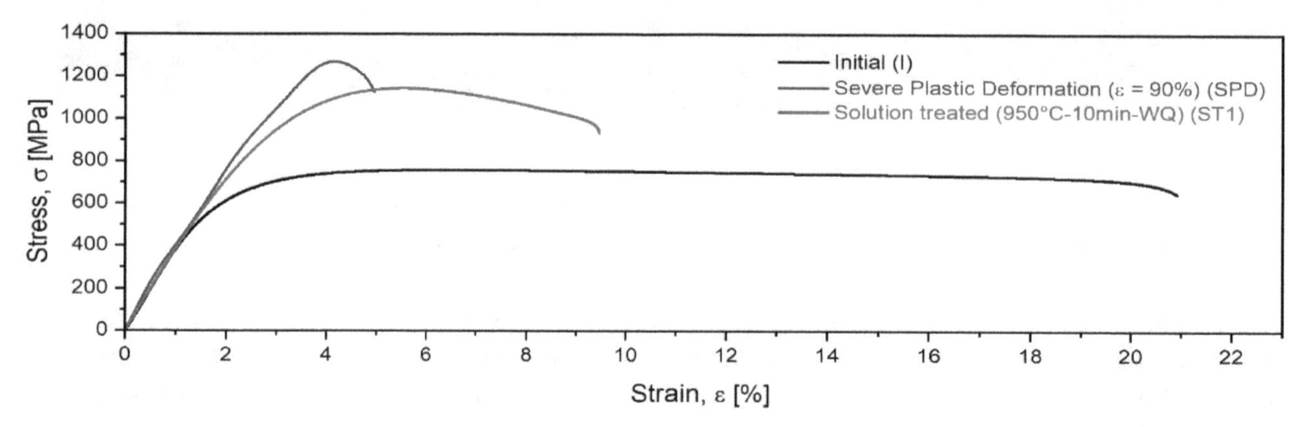

Figure 13. The stress–strain curve of the samples processed by the 2nd direction of the TMP program: SPD-MPR (ε_{tot} = 60%) + ST1 (1223 K-10 min-WQ).

Table 3. Mechanical properties of the studied alloy at the end of the 2nd direction of experiments: Ultimate Tensile Strength (σ_{UTS}); Yield Strength ($\sigma_{0.2}$); Elongation to Fracture (ε_f); Elastic Modulus (E); SD (Standard Deviation).

Structural State	Mechanical Properties			
	σ_{UTS} (SD) (MPa)	$\sigma_{0.2}$ (SD) (MPa)	ε_f (SD) (%)	E (SD) (GPa)
Initial state (I)	759.7 (13.3)	546.1 (10.6)	20.9 (0.6)	49.1 (1.6)
SPD by MPR—(ε_{tot} = 60%)	1270.7 (16,4)	1011.2 (11.2)	4.9 (0.1)	50.1 (1.4)
SPD + ST1 (1223 K-10 min-w.q.)	1143.8 (16.8) (↑50.6%)	772.6 (11.4) (↑41.5%)	9.5 (0.3) (↓54.5%)	48.7 (0.9) (↓0.8%)
SPD + ST2 (1223 K-20 min-w.q.)	1035.5 (15.9) (↑36.3%)	701.1 (10.1) (↑28.4%)	6.9 (0.2) (↓66.9%)	49.2 (1.1) (↑0.2%)

In Table 3, it can be observed that the two samples with SPD + ST provide much greater UTS and YS values (almost 50%) compared to results from the first category of experiments (CR + ST) and the initial state. This time, the UTS increased by 36.3% and 50.6%, respectively, while the YS increased by 28.4% and 41.5%, respectively. At the same time, the values of Young's modulus remained almost the same, and the elongation to fracture decreased significantly (by 54.5% and 66.9%, respectively). The obtained β-grain dimensions (77 µm and 91 µm, respectively) are situated in between the 1st step of the 1st experiments, at the bottom of the obtained values (138 → 128 → 97 → 89 → 81 → 75 µm—Figure 7), and those obtained from the 2nd steps of the 1st experiments, at the top of the obtained values (55 → 62 → 75 → 84 µm, Figure 9).

Comparing the mechanical properties obtained from the two research directions of the experimental program (Table 2 versus Table 3), the following observations can be underlined. The first direction of the experiments involved more processing steps than the second direction (two series of CR with two series of ST), but the result was a smaller β-grain dimension (55 μm versus 77 μm). The results for the Young's modulus were similar for both experimental directions (about 48–49 GPa), indicating a very good achievement. The results for YS and UTS were slightly higher (about 150–200 MPa) for the samples processed by SPD-MPR compared to those processed two times by CR. Nevertheless, the ε_f values were better/higher for the samples processed by CR. Thus, both processing schemas can provide good results, depending on the final specific application requirements.

3.4. Comparison Between the Experimental Alloy and Other Published Results for Similar Alloys

To better appreciate the obtained experimental results, this analysis includes not only a presentation and discussion of the obtained results but also a comparison with other published data for similar alloys. These data are shown in Table 4.

Table 4. Mechanical properties of the studied alloy and those of other similar alloys.

Studied Alloy: Ti-34.1Nb-7.59Zr-0.9Fe-0.16O	σ_{UTS} (MPa)	$\sigma_{0.2}$ (MPa)	ε_f (%)	E (GPa)
Samples from 1st direction of the TMP program:	-	-	-	-
CR + ST1-(1223 K-5 min-w.q.)	962.8	703.7	6.9	48.3
CR + ST2-(1223 K-10 min-w.q.)	901.7	631.4	8.2	47.6
CR + ST3-(1223 K-15 min-w.q.)	879.5	601.6	11.1	47.8
CR + ST4-(1223 K-20 min-w.q.)	809.0	574.4	15.9	49.3
Samples from 2nd direction of the TMP program:				
SPD + ST1-(1223 K-10 min-w.q.)	1143.8	772.6	9.5	48.7
SPD + ST2-(1223 K-20 min-w.q.)	1035.5	701.1	6.9	49.2
Other reported data:	-	-	-	-
Furuta et al. [47]: Ti-32Nb-2Ta-3Zr-0.5O (ST)	1370	-	12	55
Hussein et al. [31]: Ti-24.96Nb-17.15Ta-0.88Fe-0.25O (CR-50%)	851.1	-	11.1	60
M.A.-H. Gepreel [48]: Ti-20Zr-10Nb-3Ta-1Fe-1O (CR-90%) Ti-20Zr-10Nb-3Ta-1Fe-1O (ST)	- -	1198 784	- -	65 50
Strasky et al. [32]: Ti-35.3Nb-5.7Ta-7.3Zr-2Fe-0.4O (forged) Ti-35.3Nb-5.7Ta-7.3Zr-0.4O (forged) Ti-35.3Nb-5.7Ta-7.3Zr-0.7O (forged)	1130 903 1217	817 860 1017	28 16 21	107 81 80

By highlighting the positive achievements and possible negative points, the following observations can be underlined. The first observation, common to all anterior reported alloys, relates to the presence of expensive Tantalum, which is absent in the alloy that was studied here. The reported alloys were in an ST condition (like the present alloy) in only two studies—Furuta et al. [47] and M.A.-H. Gepreel [48]. Compared to these two cases, the obtained results are similar, i.e., they have very good limits. It can be observed that the modulus is even lower for the reported results (~48 GPa compared to 50–55 GPa). Further, compared with Hussein et al. [31] and Strasky et al. [32], who reported similar alloys (but in CR or forged states with Ta), the studied alloy in an ST state (both experimental research directions) provides superior characteristics, i.e., the ST state achieved almost the same positive values as those of the reported CR or forged alloys. Usually, β-type alloys in an ST state are expected to have lower values for YS and UTS than alloys in a severely deformed state; here, the obtained values were very close to similar alloys in a severely deformed state, proving that the addition of Fe and O is beneficial for improving mechanical properties [18,31–33]. In addition, for the studied alloy, the Young's modulus was much lower (~48 GPa) compared to the reported results (60, 80, or even 107 GPa).

As a conclusion, the mechanical parameters presented in Table 4 show that our experimental results are in good agreement with the values reported in the literature.

4. Conclusions

1. A stable β-Ti alloy–Ti-34.1Nb-7.59Zr-0.9Fe-0.16O—Was subjected to two TMP processing directions with variable parameters to obtain mechanical properties suitable for orthopedic implants with long-term service.
2. The amount of the β-stabilizing alloying elements was sufficient to obtain a single β-phase, without any subsequent generation of the secondary phases that are possible to be formed during complex TMP and can decrease the processability of the alloy.
3. The deformability of the studied alloy in the monophasic bcc β-status allowed the application of high deformation degrees for the CR process, as well as for SPD.
4. The ST parameters in both TMP variants led to the restoration of a fully recrystallized β phase microstructure from the cold rolled state, with a small β grain size and a reduced Young's modulus.
5. The structural deformation mechanisms during cold-rolling were predicted by SEM analysis, but these mechanisms must be proven by future detailed analyses.
6. The selected chemical composition and TMP parameters can obtain promising mechanical properties with high levels of YS (about 600–770 MPa) and UTS (about 900–1140 MPa), and low values for the Young's modulus (around 48 GPa, very close to the 30 GPa of the cortical bone), comparable to or even better than other data reported for similar alloys.
7. Both TMP methods applied to the studied alloy provided good final results, including a low β-grain size, low modulus, and high mechanical strength, all for a single β-phase, which can ensure adequate subsequent processability for the final implant shape.
8. In these obtained experimental results, the database for the mechanical properties of Ti-Nb-Zr-Fe-O family alloys was extended to a beneficial application as biomaterials for human implants with long-term service.

Author Contributions: Conceptualization, V.D.C.; methodology, A.N.; software, A.D.; validation, D.R. and A.N.; formal analysis, B.M.G.; investigation, C.T.-R.; writing—original draft preparation, A.N. and D.R.; writing—review and editing, A.N.; visualization, R.I.; supervision, V.D.C.; project administration, D.R. All authors have read and agreed to the published version of the manuscript.

References

1. Niinomi, M.; Yi, L.; Nakai, M.; Liu, H.; Hua, L. Biomedical titanium alloys with Young's moduli close to that of cortical bone. *Regener. Biomater.* **2016**, *3*, 173–185. [CrossRef] [PubMed]
2. Gepreel, M.A.-H.; Niinomi, M. Biocompatibility of Ti-alloys for long-term implantation. *J. Mech. Behav. Biomed. Mater.* **2013**, *20*, 407–415. [CrossRef] [PubMed]
3. Fu, Y.; Wang, J.; Xiao, W.; Zhao, X.; Ma, C. Microstructure evolution and mechanical properties of Ti–8Nb–2Fe-0.2O alloy with high elastic admissible strain for orthopedic implant applications. *Prog. Nat. Sci.-Mater. Int.* **2020**, *30*, 100–105. [CrossRef]
4. Ozan, S.; Lin, J.; Zhang, Y.; Li, Y.; Wen, C. Cold rolling deformation and annealing behavior of a β-type Ti–34Nb–25Zr titanium alloy for biomedical applications. *J. Mater. Res. Technol.* **2020**, *92*, 2308–2318. [CrossRef]
5. Raducanu, D.; Cojocaru, V.D.; Nocivin, A.; Cinca, I.; Serban, N.; Cojocaru, E.M. Surface Modifications of a Biomedical Gum-Metal Type Alloy by Nano Surface-Severe Plastic Deformation. *JOM* **2019**, *71*, 4114–4124. [CrossRef]

6. Kolli, R.P.; Devaraj, A. A Review of Metastable Beta Titanium Alloys. *Metals* **2018**, *8*, 506. [CrossRef]
7. Chen, W.; Li, C.; Feng, K.; Lin, Y.; Zhang, X.; Chen, C.; Zhou, K. Strengthening of a Near β-Ti Alloy through β Grain Refinement and Stress-Induced α Precipitation. *Materials* **2020**, *13*, 4255. [CrossRef] [PubMed]
8. Du, Z.; Ma, Y.; Liu, F.; Xu, N.; Chen, Y.; Wang, X.; Chen, Y.; Gong, T.; Xu, D. The Influences of Process Annealing Temperature on Microstructure and Mechanical Properties of near β High Strength Titanium Alloy Sheet. *Materials* **2019**, *12*, 1478. [CrossRef]
9. Wang, W.; Xu, X.; Ma, R.; Xu, G.; Liu, W.; Xing, F. The Influence of Heat Treatment Temperature on Microstructures and Mechanical Properties of Titanium Alloy Fabricated by Laser Melting Deposition. *Materials* **2020**, *13*, 4087. [CrossRef]
10. Lan, C.; Wu, Y.; Guo, L.; Chen, H.; Chen, F. Microstructure, texture evolution and mechanical properties of cold rolled Ti-32.5Nb-6.8Zr-2.7Sn biomedical β-Ti alloy. *J. Mater. Sci. Technol.* **2018**, *34*, 788–792. [CrossRef]
11. Dong, R.; Li, J.; Kou, H.; Fan, J.; Tang, B. Dependence of mechanical properties on the microstructure characteristics of a near β-Ti alloy Ti-7333. *J. Mater. Sci. Technol.* **2019**, *35*, 48–54. [CrossRef]
12. Lei, X.; Dong, L.; Zhang, Z.; Liu, Y.; Hao, Y.; Yang, R.; Zhang, L.-C. Microstructure, Texture Evolution and Mechanical Properties of VT3-1 Titanium Alloy Processed by Multi-Pass Drawing and Subsequent Isothermal Annealing. *Metals* **2017**, *7*, 131. [CrossRef]
13. Ma, Y.; Du, Z.; Cui, X.; Cheng, J.; Liu, G.; Gong, T.; Liu, H.; Wang, X.; Chen, Y. Effect of cold rolling process on microstructure and mechanical properties of high strength β titanium alloy thin sheets. *Prog. Nat. Sci. Mater. Int.* **2018**, *28*, 711–717. [CrossRef]
14. Mohammed, M.T.; Khan, Z.A.; Siddiquee, A.N. Beta Titanium Alloys: The Lowest Elastic Modulus for Biomedical Applications: A Review; World Academy of Science Engineering and Technology. *Int. J. Chem. Mol. Nucl. Mater. Metall. Eng.* **2014**, *8*, 822–827.
15. Gupta, A.; Khatirkar, R.K.; Kumar, A.; Parihar, M.S. Investigations on the effect of heating temperature and cooling rate on evolution of microstructure in an α + β titanium alloy. *Mater. Res. Soc.* **2018**, *33*, 946–957. [CrossRef]
16. Padmalatha, T.S.R.V.; Chakkingal, U. The effect of heat treatment and the volume fraction of the alpha phase on the workability of Ti-5Al-5Mo-5V-3Cr alloy. *J. Mater. Eng. Perform.* **2019**, *28*, 5352–5360. [CrossRef]
17. Xu, P.; Zhou, L.; Han, M.; Wei, Z.; Liang, Y. Flash-butt welded Ti6242 joints preserved base-material strength and ductility. *Mater. Sci. Eng. A* **2020**, *774*, 138915. [CrossRef]
18. Terlinde, G.; Fischer, G. Beta titanium alloys. In *Titanium and Titanium Alloys—Fundamentals and Applications*; Leyens, C., Peters, M., Eds.; Wiley-VCH: Weinheim, Germany, 2003; pp. 37–59.
19. Cordeiro, J.M.; Beline, T.; Ribeiro, A.L.R.; Rangel, E.C.; da Cruz, N.C.; Landers, R.; Faverani, L.P.; Vaz, L.G.; Fais, L.M.; Vicente, F.B.; et al. Development of binary and ternary titanium alloys for dental implants. *Dent. Mater.* **2017**, *33*, 1244–1257. [CrossRef]
20. Mohammed, M.T. Development of a new metastable beta titanium alloy for biomedical applications. *Karbala Int. J. Mod. Sci.* **2017**, *3*, 224–230. [CrossRef]
21. Niinomi, M.; Nakai, M.; Hieda, J. Development of new metallic alloys for biomedical applications. *Acta Biomater.* **2012**, *8*, 3888–3903. [CrossRef]
22. Mantri, S.A.; Banerjee, R. Microstructure and micro-texture evolution of additively manufactured β-Ti alloys. *Addit. Manuf.* **2018**, *23*, 86–98. [CrossRef]
23. Hafeez, N.; Liu, J.; Wang, L.; Wei, D.; Tang, Y.; Lu, W.; Zhang, L.C. Superelastic response of low-modulus porous beta-type Ti-35Nb-2Ta-3Zr alloy fabricated by laser powder bed fusion. *Addit. Manuf.* **2020**, *34*, 101264. [CrossRef]
24. Ozan, S.; Lin, J.; Li, Y.; Wen, C. New Ti-Ta-Zr-Nb alloys with ultrahigh strength for potential orthopedic implant applications. *J. Mech. Behav. Biomed. Mater.* **2017**, *75*, 119–127. [CrossRef]
25. Liu, H.; Niinomi, M.; Nakai, M.; Cho, K. β-Type titanium alloys for spinal fixation surgery with high Young's modulus variability and good mechanical properties. *Acta Biomater.* **2015**, *24*, 361–369. [CrossRef]
26. Zheng, Y.; Williams RE, A.; Nag, S.; Banerjee, R.; Fraser, H.L.; Banerjee, D. The effect of alloy composition on instabilities in the β phase of titanium alloys. *Scr. Mater.* **2016**, *116*, 49–52. [CrossRef]
27. Li, Y.; Yang, C.; Zhao, H.; Qu, S.; Li, X.; Li, Y. New Developments of Ti-Based Alloys for Biomedical Applications. *Materials* **2014**, *7*, 1709. [CrossRef]
28. Kalaie, M.A.; Zarei-Hanzaki, A.; Ghambari, M.; Dastur, P.; Málek, J.; Farghadany, E. The effects of second

phases on super elastic behavior of TNTZ bio alloy. *Mater. Sci. Eng. A* **2017**, *703*, 513–520. [CrossRef]
29. Wu, C.; Zhan, M. Microstructural evolution, mechanical properties and fracture toughness of near β titanium alloy during different solution plus aging heat treatments. *J. Alloys Compd.* **2019**, *805*, 1144–1160. [CrossRef]
30. Nakai, M.; Niinomi, M.; Akahori, T.; Tsutsumi, H.; Ogawa, M. Effect of Oxygen Content on Microstructure and Mechanical Properties of Biomedical Ti-29Nb-13Ta-4.6Zr Alloy under Solutionized and Aged Conditions. *Mater. Trans.* **2009**, *50*, 2716–2720. [CrossRef]
31. Hussein, A.H.; Gepreel, M.A.-H.; Gouda, M.K.; Hefnawy, A.M.; Kandil, S.H. Biocompatibility of new Ti–Nb–Ta base alloys. *Mater. Sci. Eng. C* **2016**, *61*, 574–578. [CrossRef]
32. Strasky, J.; Harcuba, P.; Vaclavova, K.; Horvath, K.; Landa, M.; Srba, O.; Janecek, M. Increasing strength of a biomedical Ti-Nb-Ta-Zr alloy by alloying with Fe, Si and O. *J. Mech. Behav. Biomed. Mater.* **2017**, *71*, 329–336. [CrossRef]
33. Nocivin, A.; Cojocaru, V.D.; Raducanu, D.; Cinca, I.; Angelescu, M.L.; Dan, I.; Serban, N.; Cojocaru, M. Finding an Optimal Thermo-Mechanical Processing Scheme for a Gum-type Ti-Nb-Zr-Fe-O Alloy. *J. Mater. Eng. Perform.* **2017**, *26*, 4373–4380. [CrossRef]
34. Biesiekierski, A.; Lin, J.; Li, Y.; Ping, D.; Yamabe-Mitarai, Y.; Wen, C. Investigations into Ti-(Nb,Ta)-Fe Alloys for Biomedical Applications. *Acta Biomater.* **2016**, *32*, 336–347. [CrossRef]
35. Kopova, J.; Strasky, P.; Harcuba, M.; Landa, M.; Janecek, M.; Bacakova, L. Newly developed Ti-Nb-Zr-Ta-Si-Fe biomedical beta titanium alloys with increased strength and enhanced biocompatibility. *Mater. Sci. Eng. C* **2016**, *60*, 230–238. [CrossRef]
36. Zhang, D.C.; Mao, Y.F.; Li, Y.L.; Li, J.J.; Yuan, M.; Lin, J.G. Effect of ternary alloying elements on microstructure and superelasticity of Ti–Nb alloys. *Mater. Sci. Eng. A* **2013**, *559*, 706–710. [CrossRef]
37. Li, Q.; Miao, P.; Li, J.; He, M.; Nakai, M.; Niinomi, M.; Chiba, A.; Nakano, T.; Liu, X.; Zhou, K.; et al. Effect of Nb Content on Microstructures and Mechanical Properties of Ti-xNb-$_2$Fe Alloys. *J. Mater. Eng. Perform.* **2019**, *28*, 5501–5508. [CrossRef]
38. Ehtemam, H.S.; Attar, H.; Okulov, I.V.; Dargusch, M.S.; Kent, D. Microstructural evolution and mechanical properties of bulk and porous low-cost Ti–Mo–Fe alloys produced by powder metallurgy. *J. Alloys Compd.* **2020**, *853*, 156768. [CrossRef]
39. Park, C.H.; Park, J.W.; Yeom, J.T.; Chun, Y.S.; Lee, C.S. Enhanced mechanical compatibility of submicrocrystalline Ti-13Nb-13Zr alloy. *Mater. Sci. Eng. A* **2010**, *527*, 4914–4919. [CrossRef]
40. Hsu, H.C.; Hsu, S.K.; Wu, S.C.; Lee, C.J.; Ho, W.F. Structure and mechanical properties of as-cast Ti–$_5$Nb–xFe alloys. *Mater. Charact.* **2010**, *61*, 851–858. [CrossRef]
41. Cui, W.F.; Guo, A.H. Microstructure and properties of biomedical TiNbZrFe β-titanium alloy under aging conditions. *Mater. Sci. Eng. A* **2009**, *527*, 258–262. [CrossRef]
42. Gordin, D.M.; Ion, R.; Vasilescu, C.; Drob, S.I.; Cimpean, A.; Gloriant, T. Potentiality of the "Gum Metal" titanium-based alloy for biomedical applications. *Mater. Sci. Eng. C* **2014**, *44*, 362–370. [CrossRef] [PubMed]
43. Lei, Z.; Liu, X.; Wu, Y.; Wang, H.; Jiang, S.; Wang, S.; Hui, X.; Wu, Y.; Gault, B.; Kontis, P.; et al. Enhanced strength and ductility in a high-entropy alloy via ordered oxygen complexes. *Nature* **2018**, *563*, 546–550. [CrossRef]
44. Liu, H.; Niinomi, M.; Nakai, M.; Cong, X.; Cho, K.; Boehlert, C.J.; Khademi, V. Abnormal Deformation Behavior of Oxygen-Modified β-Type Ti-29Nb-13Ta-4.6Zr Alloys for Biomedical Applications. *Metall. Mater. Trans. A* **2016**, *48*, 139–149. [CrossRef]
45. Zafari, A.; Ding, Y.; Cui, J.; Xia, K. Achieving fine beta grain structure in a metastable beta titanium alloy through multiple forging-annealing cycles. *Metal. Mater. Trans. A* **2016**, *47*, 3633–3648. [CrossRef]
46. Bertrand, E.; Gloriant, T.; Gordin, D.M.; Vasilescu, E.; Drob, P.; Vasilescu, C.; Drob, S.I. Synthesis and characterisation of a new superelastic Ti–25Ta–25Nb biomedical alloy. *J. Mech. Behav. Biomed. Mater.* **2010**, *3*, 559–564. [CrossRef]
47. Furuta, T.; Kuramoto, S.; Hwang, J.; Nishino, K.; Saito, T.; Niinomi, M. Mechanical Properties and Phase Stability of Ti-Nb-Ta-Zr-O Alloys. *Mater. Trans.* **2007**, *48*, 1124–1130. [CrossRef]
48. Gepreel, M.A.-H. Improved elasticity of new Ti-Alloys for Biomedical Applications. *Mater. Today Proceed.* **2015**, *2*, S979–S982. [CrossRef]
49. Ozan, S.; Lin, J.; Li, Y.; Ipek, R.; Wen, C. Development of Ti–Nb–Zr alloys with high elastic admissible strain

for temporary orthopedic devices. *Acta Biomater.* **2015**, *20*, 176–187. [CrossRef]
50. Biesiekierski, A.; Lin, J.X.; Munir, K.; Ozan, S.; Li, Y.C.; Wen, C.E. An investigation of the mechanical and microstructural evolution of a TiNbZr alloy with varied ageing time. *Sci. Rep.* **2018**, *8*, 5737. [CrossRef] [PubMed]
51. Shekhar, S.; Sarkar, R.; Kar, S.K.; Bhattacharjee, A. Effect of solution treatment and aging on microstructure and tensile properties of high strength β titanium alloy Ti-5Al-5V-5Mo-3Cr. *Mater. Des.* **2015**, *66*, 596–610. [CrossRef]
52. Ozan, S.; Lin, J.; Li, Y.; Zhang, Y.; Munir, K.; Jiang, H.; Wen, C. Deformation mechanism and mechanical properties of a thermo-mechanically processed β Ti–28Nb–35.4Zr alloy. *J. Mech. Behav. Biomed. Mater.* **2018**, *78*, 224–234. [CrossRef]
53. Ozan, S.; Li, Y.C.; Lin, J.X.; Zhang, Y.W.; Jiang, H.W.; Wen, C.E. Microstructural evolution and its influence on the mechanical properties of a thermos-mechanically processed beta Ti-32Zr-30Nb alloy. *Mater. Sci. Eng. A-Struct. Mater. Prop. Microstruct. Process* **2018**, *719*, 112–123. [CrossRef]
54. Tang, B.; Chu, Y.; Zhang, M.; Meng, C.; Fan, J.; Kou, H.; Li, J. The ω phase transformation during the low temperature aging and low rate heating process of metastable β titanium alloys. *Mater. Chem. Phys.* **2020**, *239*, 122–125. [CrossRef]
55. Ali, T.; Wang, L.; Cheng, X.; Liu, A.; Xu, X. Omega phase formation and deformation mechanism in heat treated Ti-5553 alloy under high strain rate compression. *Mater. Lett.* **2019**, *236*, 163–166. [CrossRef]
56. Hu, Q.-M.; Li, S.-J.; Hao, Y.-L.; Yang, R.; Johansson, B.; Vitos, L. Phase stability and elastic modulus of Ti alloys containing Nb, Zr, and/or Sn from first-principles calculations. *Appl. Phys. Lett.* **2008**, *93*, 121902. [CrossRef]
57. Correa, D.R.N.; Vicente, F.B.; Donato, T.A.G.; Arana-Chavez, V.E.; Buzalaf, M.A.R.; Grandini, C.R. The effect of the solute on the structure, selected mechanical properties, and biocompatibility of Ti–Zr system alloys for dental applications. *Mater. Sci. Eng. C* **2014**, *34*, 354–359. [CrossRef]
58. Banerjee, D.; Williams, J.C. Perspectives on Titanium Science and Technology. *Acta Mater.* **2013**, *61*, 844–879. [CrossRef]
59. Yang, Y.; Wu, S.Q.; Li, G.P.; Li, Y.L.; Lu, Y.F.; Yang, K.; Ge, P. Evolution of deformation mechanisms of Ti-22.4Nb-0.73Ta-2Zr-1.34Oalloy during straining. *Acta Mater.* **2010**, *58*, 2778–2787. [CrossRef]
60. Besse, M.; Castany, P.; Gloriant, T. Mechanisms of deformation in gum metal TNTZ-O and TNTZ titanium alloys: A comparative study on the oxygen influence. *Acta Mater.* **2011**, *59*, 5982–5988. [CrossRef]
61. Furuta, T.; Kuramoto, S.; Hwang, J.; Nishino, K.; Saito, T. Elastic deformation behavior of multi-functional Ti-Nb-Ta-Zr-O alloys. *Mater. Trans.* **2005**, *46*, 3001–3007. [CrossRef]
62. Tan, M.H.C.; Baghi, A.D.; Ghomashchi, R.; Xiao, W.; Oskouei, R.H. Effect of niobium content on the microstructure and Young's modulus of Ti-xNb-7Zr alloys for medical implants. *J. Mech. Behav. Biomed. Mater.* **2019**, *99*, 78–85. [CrossRef]
63. Ozan, S.; Lin, J.; Weng, W.; Zhang, Y.; Li, Y.; Wen, C. Effect of thermomechanical treatment on the mechanical and microstructural evolution of a β-type Ti-40.7Zr-24.8Nb alloy. *Bioact. Mater.* **2019**, *4*, 303–311. [CrossRef]
64. Wang, L.; Lu, W.; Qin, J.; Zhang, F.; Zhang, D. Microstructure and mechanical properties of cold-rolled TiNbTaZr biomedical β-titanium alloy. *Mater. Sci. Eng. A* **2008**, *490*, 421–426. [CrossRef]
65. Okulov, I.V.; Wendrock, H.; Volegov, A.S.; Attar, H.; Kühn, U.; Skrotzki, W.; Eckert, J. High strength beta titanium alloys: New design approach. *Mater. Sci. Eng. A* **2015**, *628*, 297–302. [CrossRef]
66. Xie, K.Y.; Wang, Y.; Zhao, Y.; Chang, L.; Wang, G.; Chen, Z.; Cao, Y.; Liao, X.; Lavernia, E.J.; Valiev, R.B.; et al. Nanocrystalline β-Ti alloy with high hardness, low Young's modulus and excellent in vitro biocompatibility for biomedical applications. *Mater. Sci. Eng. C* **2013**, *33*, 3530–3536. [CrossRef]
67. Tahara, M.; Kim, H.Y.; Inamura, T.; Hosoda, H.; Miyazaki, S. Role of interstitial atoms in the microstructure and non-linear elastic deformation behavior of Ti–Nb alloy. *J. Alloys Compd.* **2013**, *577*, S404–S407. [CrossRef]
68. Wei, L.S.; Kim, H.Y.; Miyazaki, S. Effects of oxygen concentration and phase stability on nano-domain structure and thermal expansion behavior of Ti–Nb–Zr–Ta–O alloys. *Acta Mater.* **2015**, *100*, 313–322. [CrossRef]
69. Wei, Q.; Wang, L.; Fu, Y.; Qin, J.; Lu, W.; Zhang, D. Influence of oxygen content on microstructure and mechanical properties of Ti–Nb–Ta–Zr alloy. *Mater. Des.* **2011**, *32*, 2934–2939. [CrossRef]

3

Reconstruction of Complex Zygomatic Bone Defects using Mirroring Coupled with EBM Fabrication of Titanium Implant

Khaja Moiduddin [1,*], Syed Hammad Mian [1], Usama Umer [1], Naveed Ahmed [1,2], Hisham Alkhalefah [1] and Wadea Ameen [1]

[1] Advanced Manufacturing Institute, King Saud University, Riyadh 11421, Saudi Arabia; syedhammad68@yahoo.co.in (S.H.M.); uumer@ksu.edu.sa (U.U.); anaveed@ksu.edu.sa (N.A.); halkhalefah@ksu.edu.sa (H.A.); wqaid@ksu.edu.sa (W.A.)

[2] Department of Industrial and Manufacturing Engineering, University of Engineering and Technology, Lahore 54000, Pakistan

* Correspondence: khussain1@ksu.edu.sa

Abstract: Reconstruction of zygomatic complex defects is a surgical challenge, owing to the accurate restoration of structural symmetry as well as facial projection. Generally, there are many available techniques for zygomatic reconstruction, but they hardly achieve aesthetic and functional properties. To our knowledge, there is no such study on zygomatic titanium bone reconstruction, which involves the complete steps from patient computed tomography scan to the fabrication of titanium zygomatic implant and evaluation of implant accuracy. The objective of this study is to propose an integrated system methodology for the reconstruction of complex zygomatic bony defects using titanium comprising several steps, right from the patient scan to implant fabrication while maintaining proper aesthetic and facial symmetry. The integrated system methodology involves computer-assisted implant design based on the patient computed tomography data, the implant fitting accuracy using three-dimensional comparison techniques, finite element analysis to investigate the biomechanical behavior under loading conditions, and finally titanium fabrication of the zygomatic implant using state-of-the-art electron beam melting technology. The resulting titanium implant has a superior aesthetic appearance and preferable biocompatibility. The customized mirrored implant accurately fit on the defective area and restored the tumor region with inconsequential inconsistency. Moreover, the outcome from the two-dimensional analysis provided a good accuracy within 2 mm as established through physical prototyping. Thus, the designed implant produced faultless fitting, favorable symmetry, and satisfying aesthetics. The simulation results also demonstrated the load resistant ability of the implant with max stress within 1.76 MPa. Certainly, the mirrored and electron beam melted titanium implant can be considered as the practical alternative for a bone substitute of complex zygomatic reconstruction.

Keywords: zygomatic bone; customized reconstruction; electron beam melting; 3D comparison; titanium alloy; finite element analysis

---

1. Introduction

Advancements in the field of biomaterials, fabrication techniques, and computer-assisted technologies have been taking place owing to the huge demand for medical implants [1]. Biomaterials are natural or artificial materials that are used to enhance or replace any tissue, organ, or biological structure in order to improve the quality of human life [2]. Titanium and its alloys have a unique combination of high strength to low weight ratio, low density and better corrosion resistance, are biocompatible, and have non-magnetic properties. They are one of the few biomaterials that naturally match the requirement of bone tissue replacement in the human body. Among them, commercial pure titanium (grade 2), Ti-6Al-4V (grade 5), and Ti-6Al-4V ELI (Extra low interstitial) (grade 23) are widely used biomaterial in medical application [3]. In addition, a significant rise in the fabrication of custom-built titanium implants with better design and biomechanical properties has been realized in clinical applications. The facilitation and application of tailor-made effective implants are crucial to improve the quality of the patient's life. Moreover, it is of utmost importance to study the implant concerning its design, fitting accuracy, biomechanical properties, and fabrication method.

Zygomatic bone reconstruction is indeed a challenging procedure for maxillofacial surgeons due to its unique position which is close to the orbital rim that houses the eyeball [4]. The integrity of the zygomatic reconstruction can be achieved through the maintenance of the person's aesthetic appearance, its functionality, and the prominence of the cheek [5]. The zygomatic bone is highly robust and variable in shape. It is located at the crucial junction between the zygomatic arc, orbital wall, and tooth bearing snout. It consists of multiple bone tissues types, namely the cancellous bone which is surrounded by cortical bone shell. The cortical bone was found to be thickest (5.0 mm) in the upper zygomatic bone and thinnest (1.1 mm) at the anterior of maxillary sinus region [6]. In addition, the cortical bone has greater volume fraction in the zygomatic bone than the cancellous bone [7]. Studies indicate that the young modulus of zygomatic bone is in the range from 10.4 GPa to 19.6 GPa, respectively [8]. Titanium implants are commonly used for zygomatic bony reconstruction because they provide reliable support for the orbital contents [9]. The first zygomatic implant reconstruction was developed by Branemark without grafting procedures for maxillectomized patients [10]. Different surgical approaches and treatments have been employed since then, for successful zygomatic complex reconstruction, including osteotomy, autologous bone graft, and synthetic implants [11]. Among all, the autologous bone graft is considered gold standards; however, limited bone availability, volatile resorption rate, and deformities remain serious challenges. Hence, various alloplastic implants, including polymers [12], silicone [13], metals [14], and hydroxyapatite based products [15] were also used to replace the autologous bone graft.

Long-term success and effectiveness of the implants depend on several factors, including the design, implant material, accuracy, biomechanical study, fabrication process, and, of course, the skills of the surgeon [4]. In all, the mirror reconstruction technique is one of the widely used implant design techniques for medical applications [16,17]. It can replace the defective bony region with a healthy region, thus maintaining the symmetry and ideal anatomical structure [18]. Titanium alloy (Ti-6Al-4V ELI) is one of the most favored biocompatible materials owing to its high strength to weight ratio, excellent corrosion resistance, and mechanical properties [19]. Titanium and its alloys have excellent biocompatibility when compared to other metallic biomaterials such as stainless steel and cobalt chromium alloys. The most widely used titanium alloy (Ti-6Al-4V ELI) used for bone replacement has a modulus of elasticity (110 GPa) almost half that of stainless steel (200 GPa) and Cobalt chromium alloy (210 GPa) [20]. Development of new titanium alloys with shape memory is a growing attraction for biomedical application with superior elasticity and properties closer to that of bone [21]. One of the important properties that differentiate titanium and its alloys from other biomaterials is the natural formation of thin oxide film on its surface which is responsible for good chemical stability and biocompatibility [22]. The biomechanical study involving stress analysis on the implant and its surrounding bone is also critical to study the orthopedic mechanical failure [23]. It helps to evaluate different designs virtually during functional loading, thus reducing material usage, physical prototyping, and testing methods [24]. Implant accuracy evaluation is also a mandatory procedure to analyze its fitting accuracy. The implant–bone interface contact is necessary for immediate restoration as well as to avoid any damage to the vital structures [25].

Fabrication of an implant from the patient computer tomography (CT) scan involves several steps encompassing data collection and processing, design template, virtual simulation, and fabrication utilizing three-dimensional (3D) printing. 3D printing also known as rapid prototyping and additive manufacturing is a technique of producing physical 3D objects from computer-aided-design (CAD) files in a successive material layering. Electron beam melting (EBM) is one of the widely used metal 3D printing processes with a US Federal Drug Administration (FDA) approval [26]. It has been widely adopted by surgeons at an impressive rate and is used in a large variety of medical applications [27,28]. The 3D printed models can also be used for mock surgeries, pre-operative planning, surgical guides, and educational training purposes [29].

It is indeed very difficult to reconstruct a complex anatomical structure with proper aesthetic and facial symmetry [30]. Any deviation or aberration in structural alignment between the implant and bone contours may most likely lead to functional disturbance and implant failure. In this study, an integrated methodology involving custom design using the mirror reconstruction technique has been proposed. It also consists of biomechanical evaluation for the custom built implant, implant accuracy assessment using 3D comparison technique, and, finally, the three-dimensional (3D) printing of the titanium implant using state-of-the-art EBM technology. Certainly, the objective is to produce a zygomatic titanium implant, which is reliable based on its functionality, appearance, and mechanical strength. The methodology adopted in this work involves multidisciplinary fields including custom-built design, biomechanical analysis, implant accuracy, and titanium-based EBM fabrication. Most often, the clinicians or engineers, owing to the vastness or intricacies of the field, overlook some important step or assessment test. Henceforth, in this work, the different steps that are pertinent in zygomatic rehabilitation as well as implant realization are comprehensively described.

2. Materials and Methods

The methodology as presented in Figure 1 involved the integration of several technologies to realize the customized zygomatic titanium implant. This approach also comprised of the formal meeting and communication between the engineering department and the medical field in each step, to evaluate and verify the design model for the enhanced aesthetic outcome, and to reduce implant revision and failure. Indeed, a multidisciplinary approach involving biomedical engineers, surgeons, and researchers are vital for the successful fabrication and implementation of implants [31]. Studying implant failure would help the clinicians and engineers in understanding the failure factors and to develop better prosthesis. In this study, a zygomatic implant was first produced using conventional techniques as illustrated in Figure 1a. Later, a new computer-aided custom design implant was produced as shown in Figure 1b and compared with the conventionally produced implant. Some of the drawbacks of conventional techniques in implant design are the excessive time consumption, inaccurate fitting of implant, diagnostic limitations, and lack of surgical planning. From the past decade, the development of computer-assisted implant design and surgical planning has become extremely important in orthopedic surgeries.

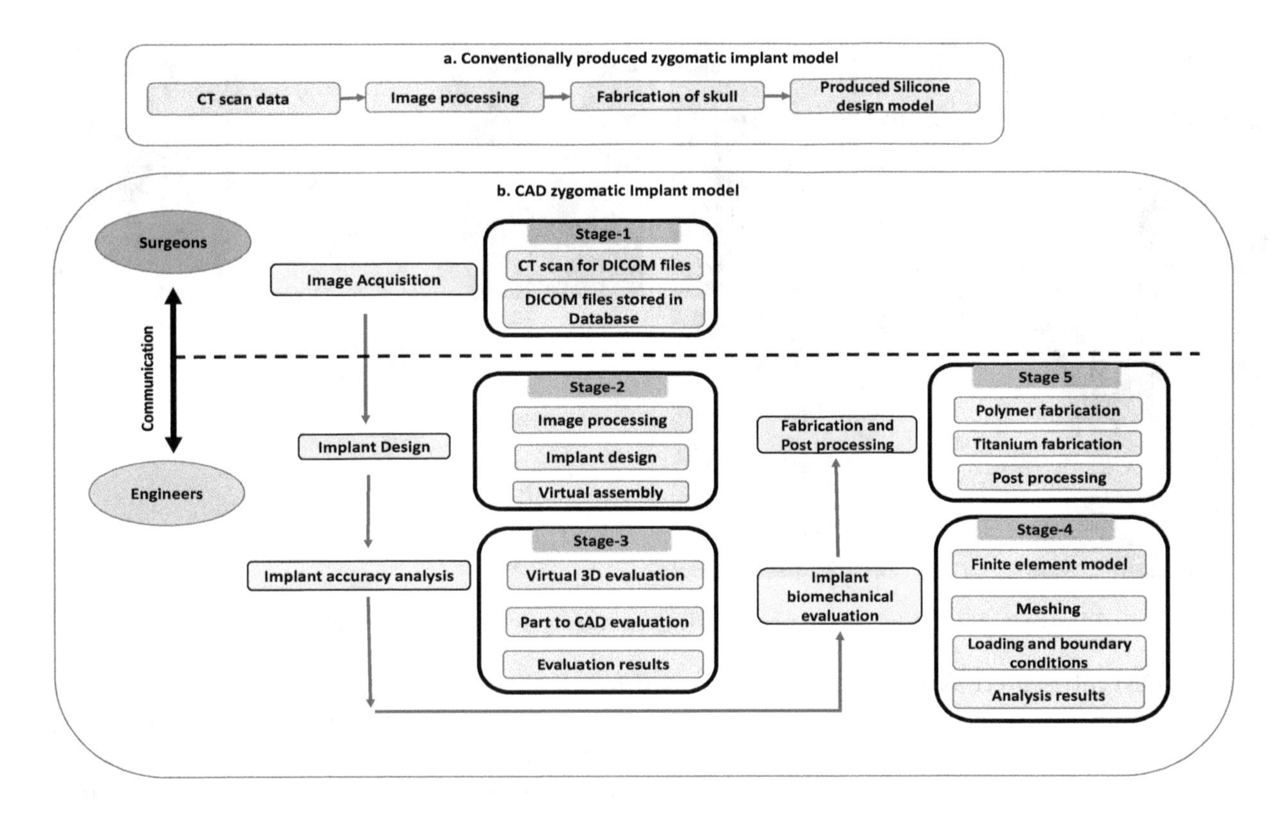

Figure 1. Methodology employed in the fabrication of (**a**) conventionally produced zygomatic implant model and (**b**) computer-aided-design zygomatic implant model.

2.1. Implant Customization and Fabrication

A cone-beam computed tomography (CBCT) scan was performed using Promax 3D (Planmeca, Helsinki, Finland) on a 34-year-old male patient who was suffering from painful cheek swelling. The series of two-dimensional (2D) images obtained from the CT scan were stored as a Digital Imaging and Communications in Medicine (DICOM) file in a database. The DICOM file was then imported into Mimics® (version, Materialise, Leuven, Belgium), a medical modeling software, which distinctly converted the series of 2D images into a 3D model as shown in Figure 2. The thresholding function from Mimics® with a thresholding value of 226 to 3071 Hounsfield Units (HU) was used to segregate the hard and soft tissues. The HU is a universally acceptable dimensionless unit that is used to express CT numbers. The Segmentation and region growing functions using Mimics® were used to divide and subdivide the image into several regions until the region of interest–Skull (green) was obtained. The obtained 3D skull model was then saved as a Standard Translation Language (STL) file and imported into a polymer 3D printer-fused deposition modeling (FDM) for fabrication.

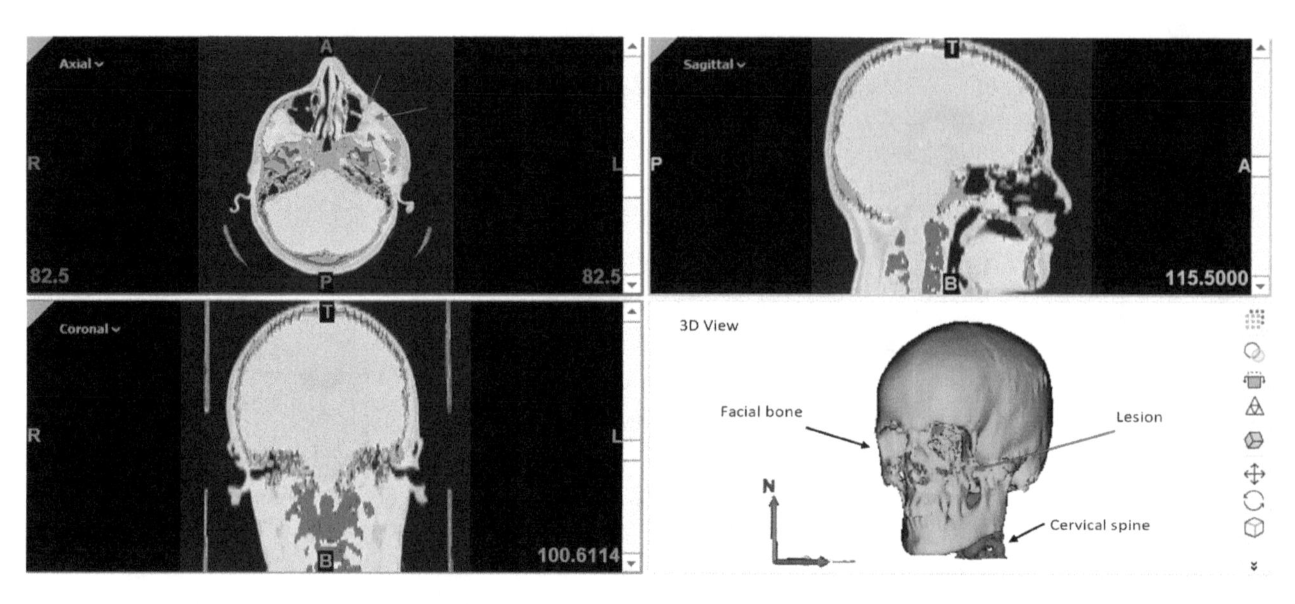

Figure 2. 3D model of the patient skull anatomy revealing the tumor location.

The FDM fabricated skull 3D model was used for designing the zygomatic implant. A silicone impression material was placed over the defective area and gently molded to obtain the shape of the zygomatic bone as shown in Figure 3a. As the process was done manually, a lot of contours and rough surfaces could be observed on the posterior end of the implant (Figure 3c). The conventionally designed implant (Figure 3b) did not adapt to the recipient defect perfectly, and hence it was rejected based on the formal meeting with the medical clinicians. On further discussion with the surgeons, it was decided to have a new CAD implant model using reconstruction techniques.

There are two types of implant reconstruction techniques widely used in custom designs [32]. The first is a mirror reconstruction technique and the other is an anatomical reconstruction design based on a curve based and refinement approach [4]. The mirroring technique provides better facial symmetry and is more accurate for medium and large complex tumors, whereas the anatomical reconstruction technique is a lengthy process, requires lots of human expertise, and provides good results for smaller tumors [33,34]. One of the major differences between these two techniques is that mirroring techniques can be applied only in symmetrical regions, whereas anatomical reconstruction can be applied to both symmetrical and asymmetrical parts.

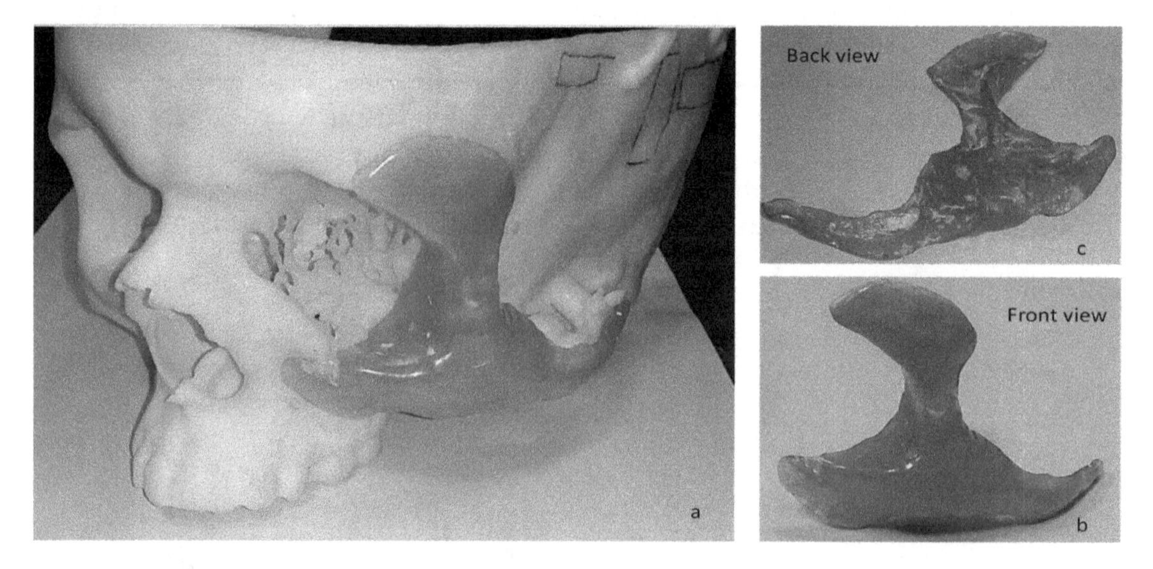

Figure 3. (**a**) Fused deposition Modeling fabricated skull model with the attached silicone implant; (**b**) silicone implant front view; and (**c**) silicone implant back view.

Based on the inputs from the medical clinicians, an implant mirror reconstruction technique was utilized to replace the defective region with the healthier bony region. The steps involved in the mirroring reconstruction technique are illustrated in Figure 4. Primarily, the STL defective model (Figure 4a) was first resected into two halves (Figure 4b) using Magics® (Materialise, Leuven, Belgium). The defective side was removed (Figure 4c) and replaced by the healthy right side (Figure 4d) using the mirror reconstruction technique. The next step was to join the two error-free sides through merging operation (Figure 4e). Any voids or gaps were removed by wrapping operations (Figure 4f). Boolean subtraction process (Figure 4g) was performed between the error-free model (Figure 4f) and the tumor model (Figure 4a) to obtain the zygomatic implant template (Figure 4h), which was then saved as an STL file for subsequent fabrication.

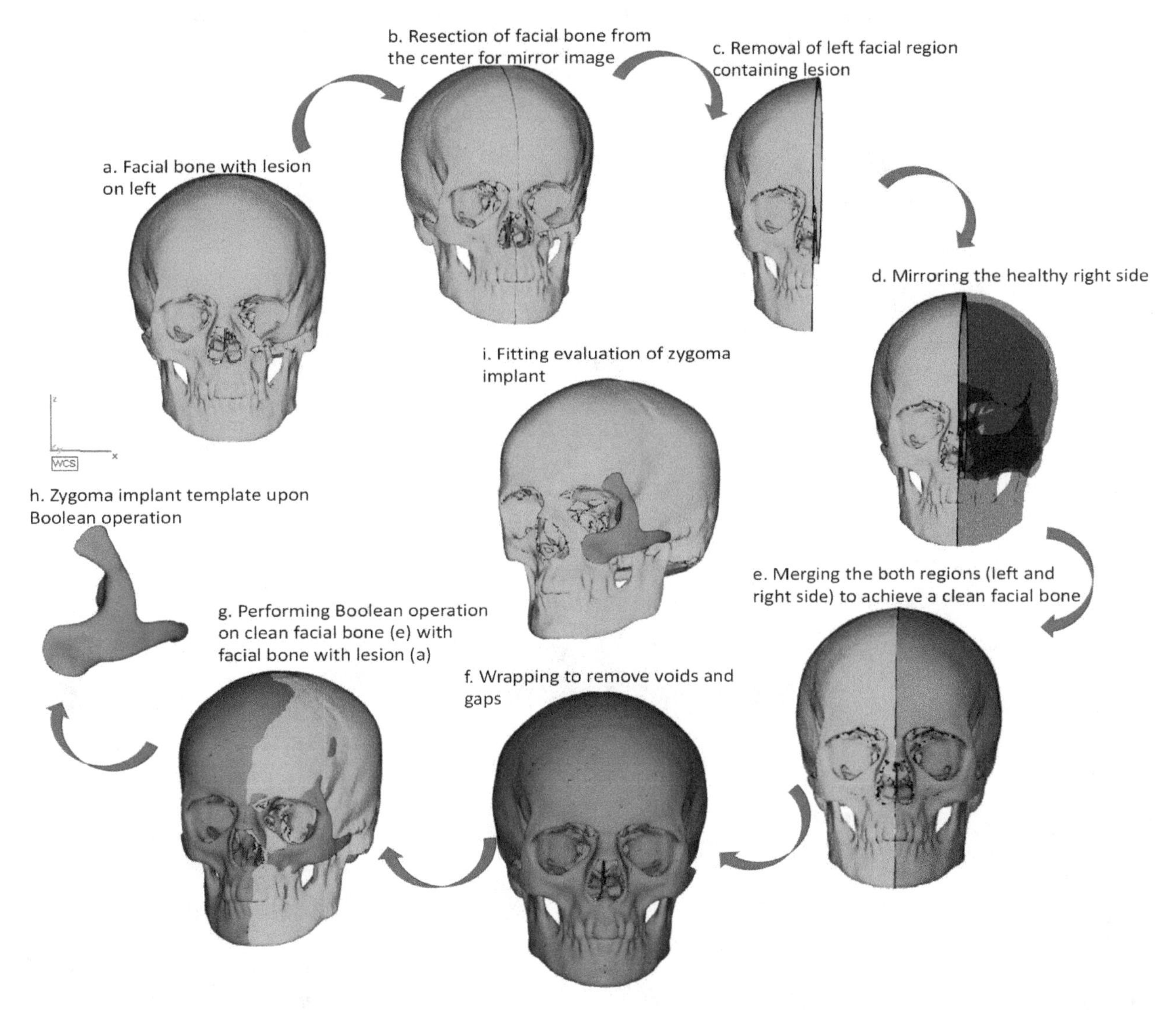

Figure 4. Computer-aided customized zygomatic implant process cycle. (**a**) facial bone with lesion; (**b**) resection of facial bone for mirror image; (**c**) removal of left facial region with lesion; (**d**) mirroring the healthy right side; (**e**) merging both left and right regions to obtain clean facial bone; (**f**) wrapping operation to remove gaps; (**g**) boolean operation between clean facial bone and facial bone with lesion; (**h**) obtained zygomatic bone template upon Boolean operation; (**i**) fitting evaluation of zygoma implant.

Figure 5 illustrates the FDM fabricated zygomatic implant designed using a mirror reconstruction technique as well as its counterpart that is a conventionally produced implant model. On visual observation, it can be clearly seen that the manually produced implant model has lots of curves and rough surfaces when compared to the mirror implant model produced through FDM. Figure 6 demonstrates the precise fitting of the mirror implant model onto the skull model. The polymer produced FDM models also assisted the medical clinicians in comprehensive surgical planning and rehearsal in addition to precise drilling of screw holes.

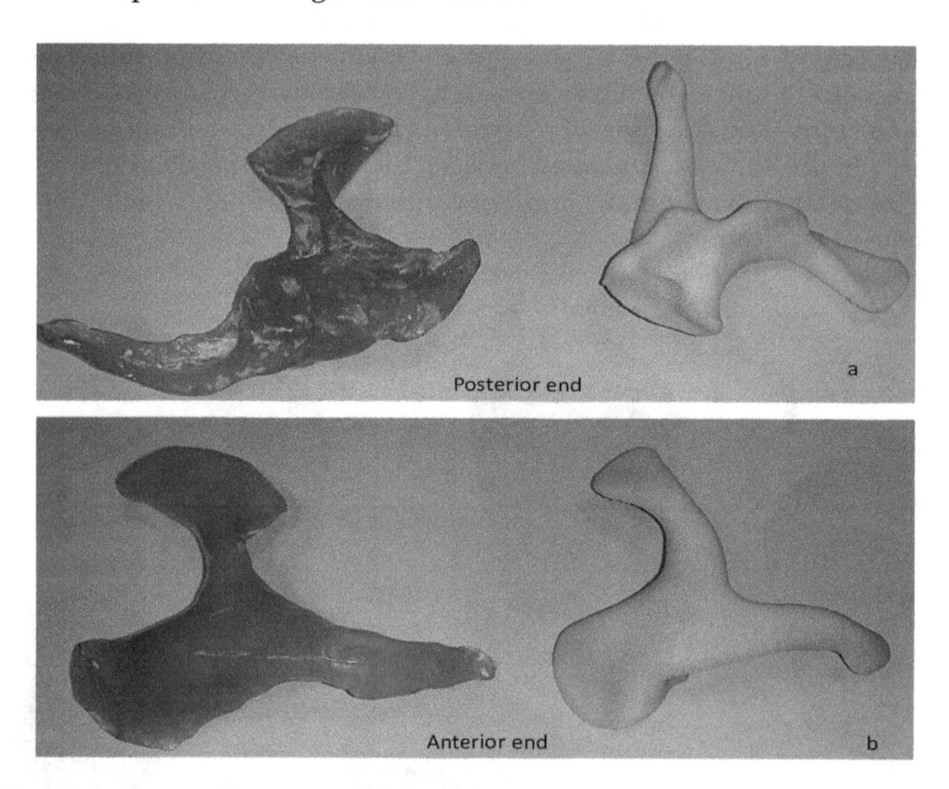

Figure 5. Silicone based zygomatic implant produced manually without CAD technique (**a**) and FDM fabricated zygomatic implant by CAD mirroring technique (**b**).

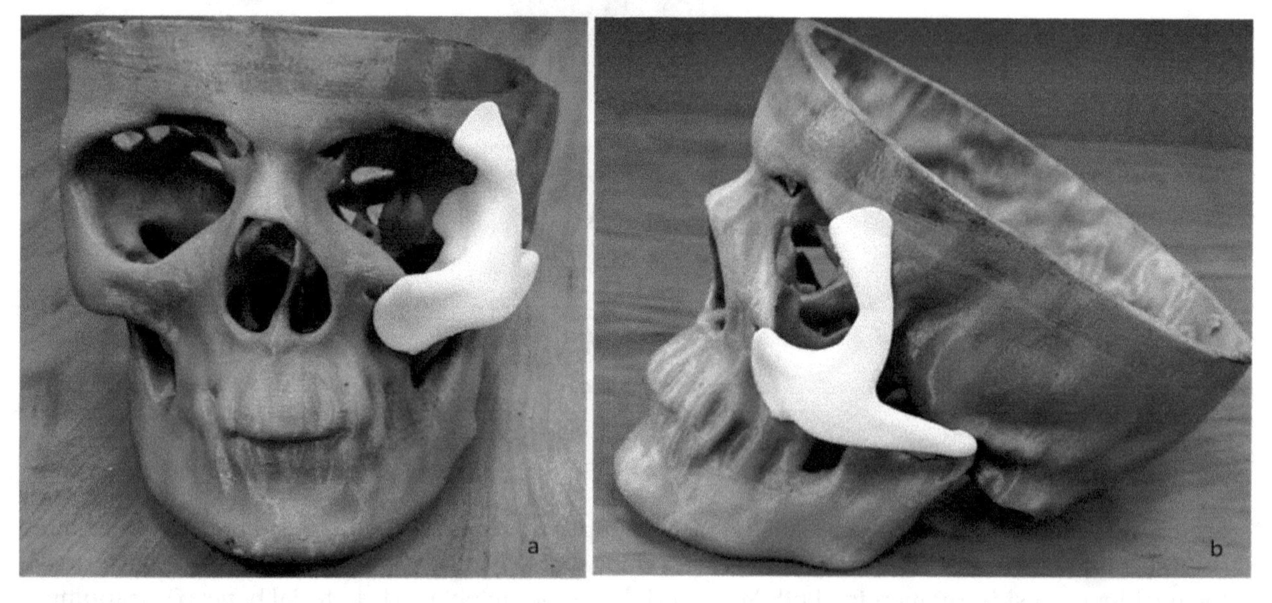

Figure 6. (**a**) FDM fabricated customized zygomatic implant fitted onto the skull model (front view); (**b**) side view illustration.

After the successful rehearsal and fitting operation of the FDM produced a zygomatic implant onto the skull model and based on the approval from the clinicians, the mirror implant model was fabricated using EBM technology. Arcam's EBM technology offers a new method of rapid manufacturing of near net shaped titanium products, thus eliminating the time, cost and challenges in machining and casting [35]. Arcam EBM A2 machine (Arcam AB, Mölndal, Sweden) as illustrated in Figure 7b was used for the fabrication of titanium zygomatic implant. The accuracy of the EBM machine is in the range of 0.13 mm to 0.20 mm for shorter to long range build parts and the highest resolution is 50 μm [36,37]. The EBM process cycle is illustrated in Figure 7c where the filament (1), when heated to a temperature of above 2500 °C, accelerates a beam of electron (5) through the series of magnetic lenses including astigmatism lens, focus lens, and deflection lens before hitting the titanium powder (Ti-6Al-4V ELI) (8). The first magnetic lens (2) generates a circular beam of electrons and corrects astigmatism while the second magnetic lens (3) focuses the electron beam to the desired diameter, and the third magnetic lens (4) deflects the focused beam to the desired position. The powder hoppers (6) continuously feed the titanium powder (8) onto the start plate (10) inside the build platform. A mechanical raking blade (7) spreads the titanium powder evenly onto the build platform (9). Initially, a high-speed beam of electrons scans the titanium powder to preheat the powder to a sintered state. After preheating, the melting of powder takes place at slower beam scans. On completion of each melting cycle, the build platform is lowered by one-layer thickness. The entire EBM build process takes place under vacuum and under elevated temperature. This is done to prevent reactions between reactive metals (titanium) with oxygen and to prevent residual stresses [38].

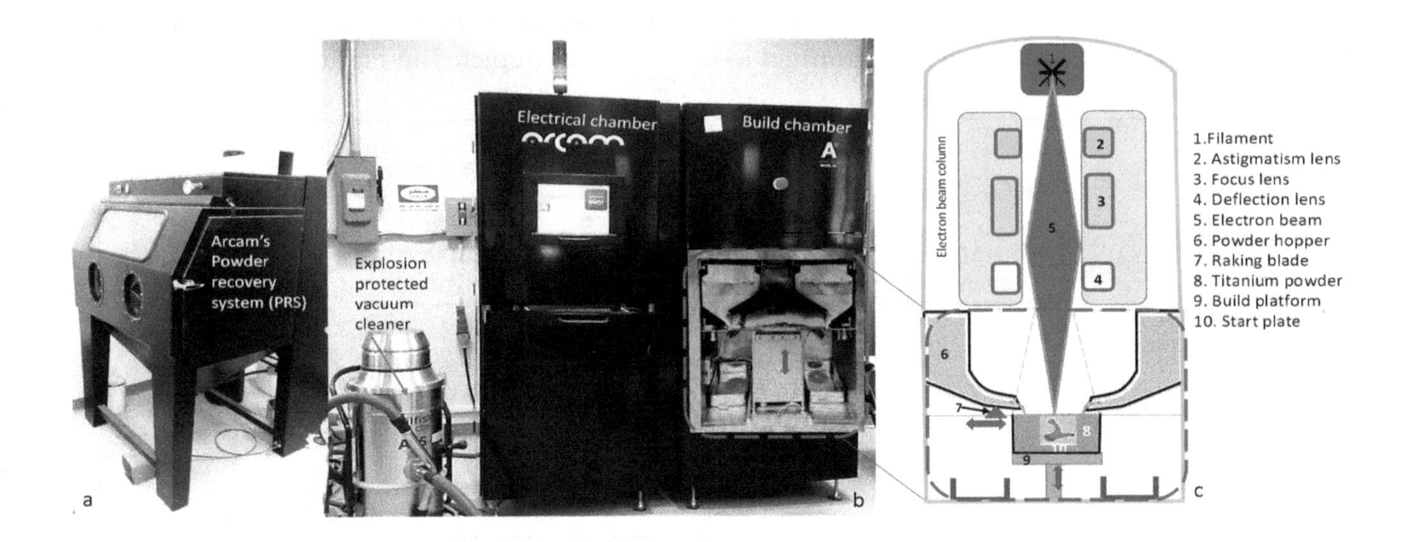

Figure 7. (**a**) ARCAM powder recovery system unit to remove the sintered powder attached to the build part, (**b**) ARCAM A2 electron beam melting machine, and (**c**) the schematic working diagram of electron beam melting machine.

Figure 8a illustrates the fabricated titanium zygomatic implant with support structures. The titanium implant after fabrication was placed inside a powder recovery system (PRS) to remove the semi-sintered powder attached to the implant. The support structures were removed manually using pliers. The titanium zygomatic implant was then fixed onto the polymer skull model for fitting evaluation (Figure 8b).

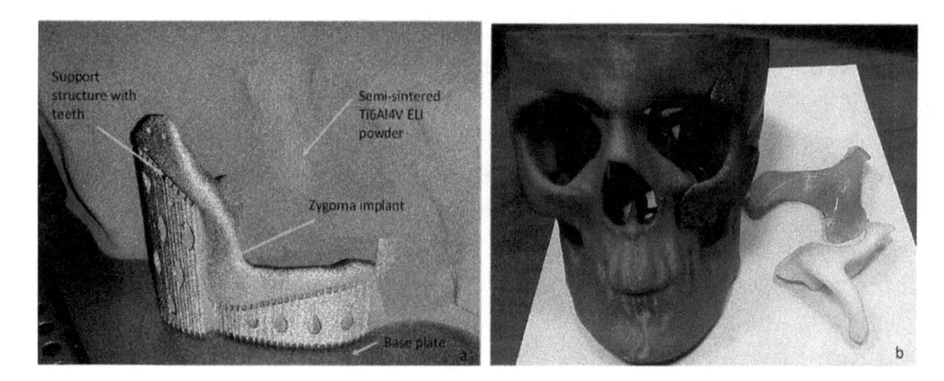

Figure 8. (**a**) EBM fabricated titanium zygomatic implant with support structures inside the powder recovery system and (**b**) titanium zygomatic implant fitted to the skull model after support removal.

2.2. Evaluation

To study the accurate fitting of the implant onto the skull region, an accuracy analysis of the implant was executed in Geomagics Control® (3D systems, Rock Hill, SC, USA). It is one of the most dynamic and comprehensive techniques to estimate the deviation between the test and reference CAD object [39]. It was carried out to quantify the error between the implant and the face model. The implant accuracy analysis was performed in two stages as depicted in Figure 9.

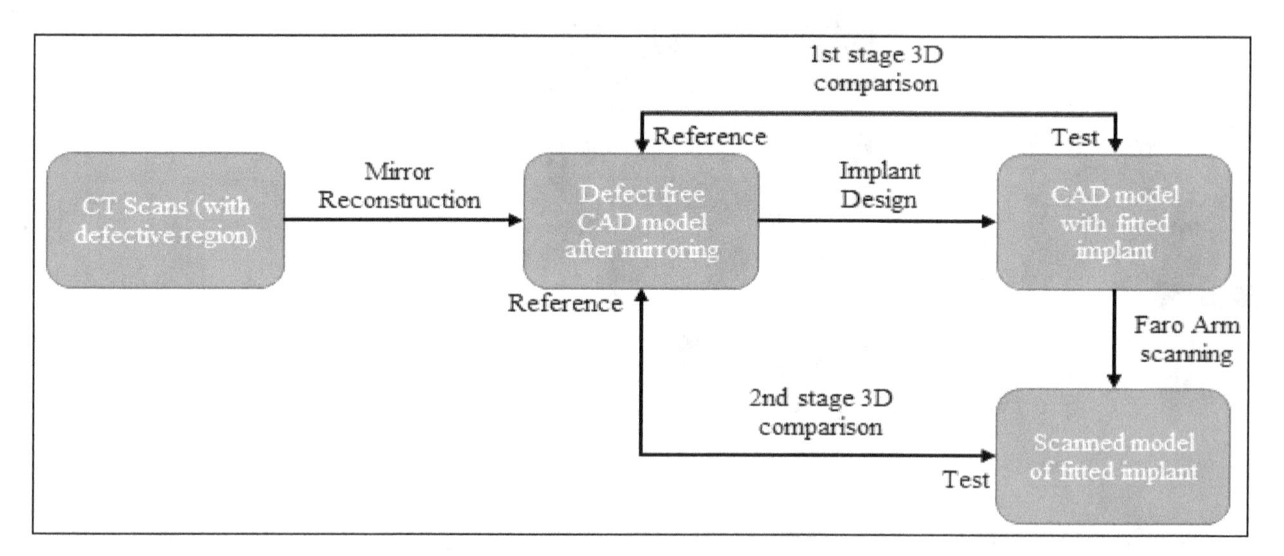

Figure 9. Evaluation of designed implant using comparison approach in Geomagics®.

- First stage comparison before fabrication. In this stage, the tumor-free model after mirroring was taken as a reference and the skull model with the implant was considered as a test file. It was done to quantify the error from the mirroring stage.
- The second stage after fabrication. The 3D model obtained using the Faro platinum arm scanner (FARO, Lake Mary, FL, USA) was used to inspect the implant with the skull model. The 3D model obtained from a Faro arm scanner was considered as a test file and the tumor-free model after mirroring was fixed as the reference file. The test was aligned and superimposed on the error-free mirror model to analyze the deviation error.

The Finite Element Analysis (FEA) was also conducted to investigate the biomechanical behavior of the zygomatic implant and its supporting bone. Typical occlusal loads were applied and stresses on the implants and its surrounding bones were examined. The STL models of Skull and zygomatic implant were first transformed to Solid B-Rep models before analysis. The finite element model (FEM) was developed using ABAQUS/CAE (Version 6.14, Dassault Systemes, Veliez-Villacublai, France).

ABAQUS/CAE is an interactive, graphical environment for Abaqus software. It allows models to be created quickly and easily by producing or importing the geometry of the structure to be analyzed and decomposing the geometry into meshable regions. Once the model is complete, ABAQUS/CAE can submit, monitor, and control the analysis jobs. The Visualization module can then be used to interpret the results. The attributes of cortical bone were assigned to the skull model, and titanium alloy (Ti-6Al-4V ELI) was assigned to the zygomatic implant [40,41]. The material properties designated in the FEA model are presented in Table 1.

Table 1. Material properties utilized in the finite element analysis model [40,41].

Materials	Young's Modulus (GPa)	Poisson's Ratio	Yield Strength (MPa)
Cortical bone	13.7	0.3	122
Zygomatic implant (Ti-6Al-4V ELI)	120	0.3	930

The FEM with load and boundary conditions is shown in Figure 10. The skull was fixed around the neck area and a force of 50 Newton was applied to the zygomatic implant over an area of 500 mm^2. In previous studies, researchers have also applied similar loading conditions of 5.5 kg on zygomatic under the mastication process [42]. The joints between the skull and zygomatic implant were modeled using mesh independent fastener available in ABAQUS/STANDARD. ABAQUS/STANDARD (Version 6.14, Dassault Systemes, Veliez-Villacublai, France) is a general-purpose analysis product that can solve a wide range of linear and nonlinear problems involving the static, dynamic, thermal, electrical, and electromagnetic response of components. Abaqus/Standard solves a system of equations implicitly at each solution increment. Mesh independent fasteners provide a point-based connection between surfaces similar to spot-weld connections. A total of 12 fasteners were utilized across the three joint areas of the zygomatic implant (red-colored) as shown in Figure 10.

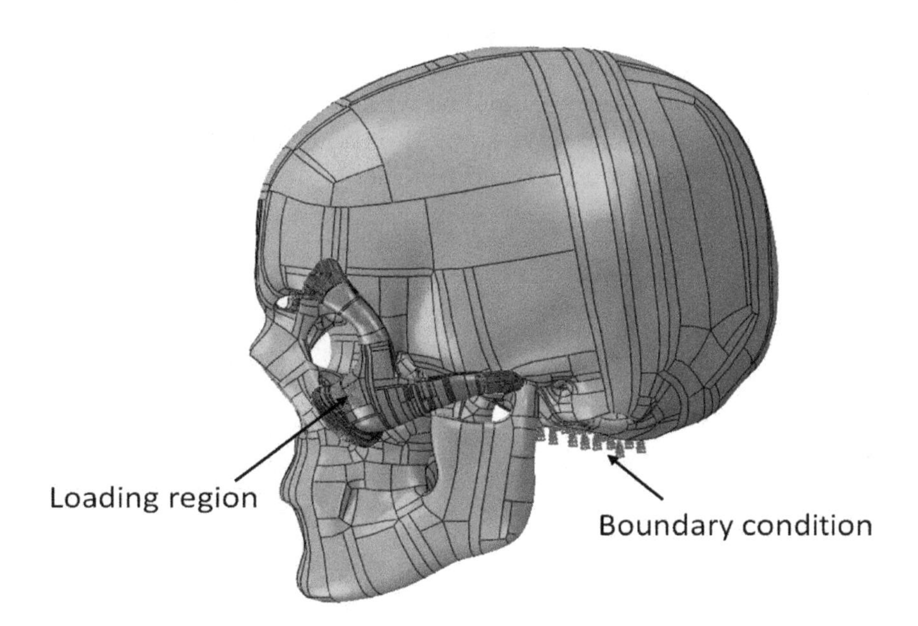

Figure 10. Loading and boundary conditions on the skull and the zygomatic implant.

3D stress quadratic tetrahedron elements (C3D10) with 10 nodes were selected for the FEM as shown in Figure 11. To save computational time, an optimum size mesh is selected that resulted in 149,966 elements in the whole model. To model interaction between implant and skull surfaces, a general contact algorithm has been selected along with frictionless tangential behavior.

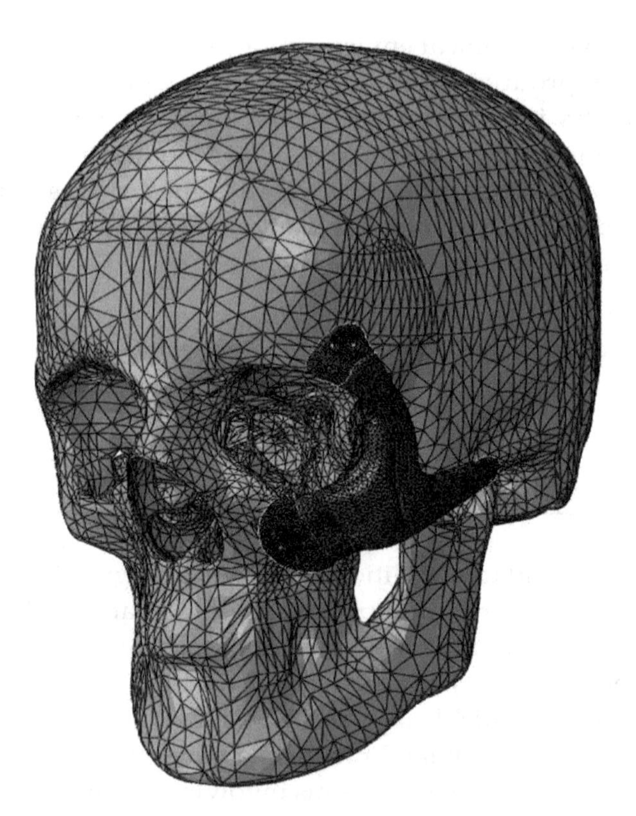

Figure 11. Finite element mesh of the skull and zygomatic implant.

3. Results and Discussion

The tumor-free skull model obtained from the mirror reconstruction technique (Figures 4f and 12b) was used as a reference and the skull model with the attached implant (Figure 12c) was taken as a test file. The test file was aligned and superimposed onto the reference file to obtain the deviation error, which reflected both the mirroring as well as implant shape effect as shown in Figure 12a. In addition to 3D analysis, the 2D comparison was also conducted which quantified the implant shape effect. As it can be observed, there is an inappreciable deviation onto the zygomatic implant region, which states that the obtained zygomatic implant model fitted precisely and replaced the tumor region without much deviation.

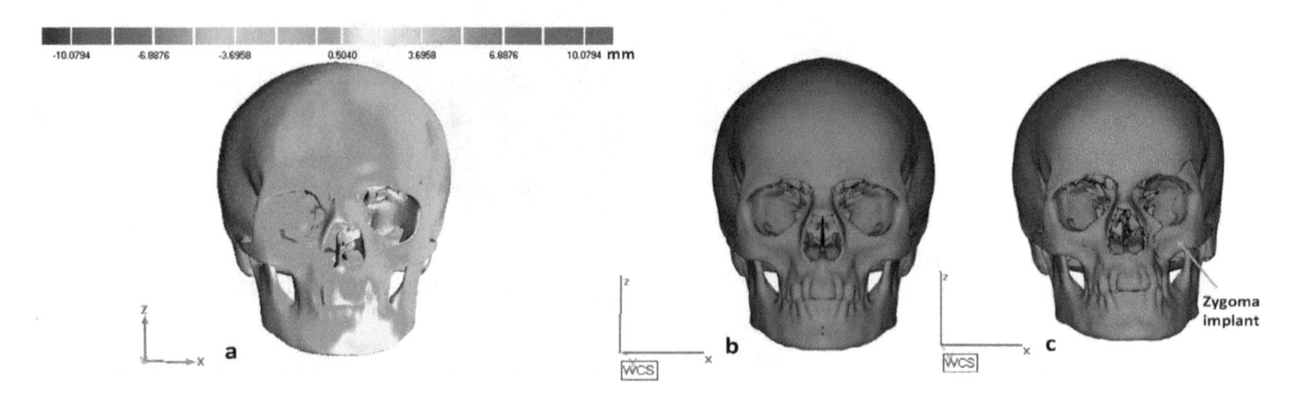

Figure 12. (**a**) 3D inspection results of model deviation between the (**b**) mirror reconstruction model and (**c**) zygomatic implant model.

The 2D comparison results between the reference and the test model are illustrated in Figure 13. In 2D comparison, a cross-sectional plane was created onto the test and reference model in the zygomatic region. Figure 13a–c illustrate the different cross-sectional views of the superimposition of the test file (Figure 13e) onto the reference file (Figure 13d). The color spectrum in the whisker deviation (Figure 13f) provides the deviation error between them. The 3D and 2D comparison results obtained through virtual model inspection in 1st stage are provided in Table 2.

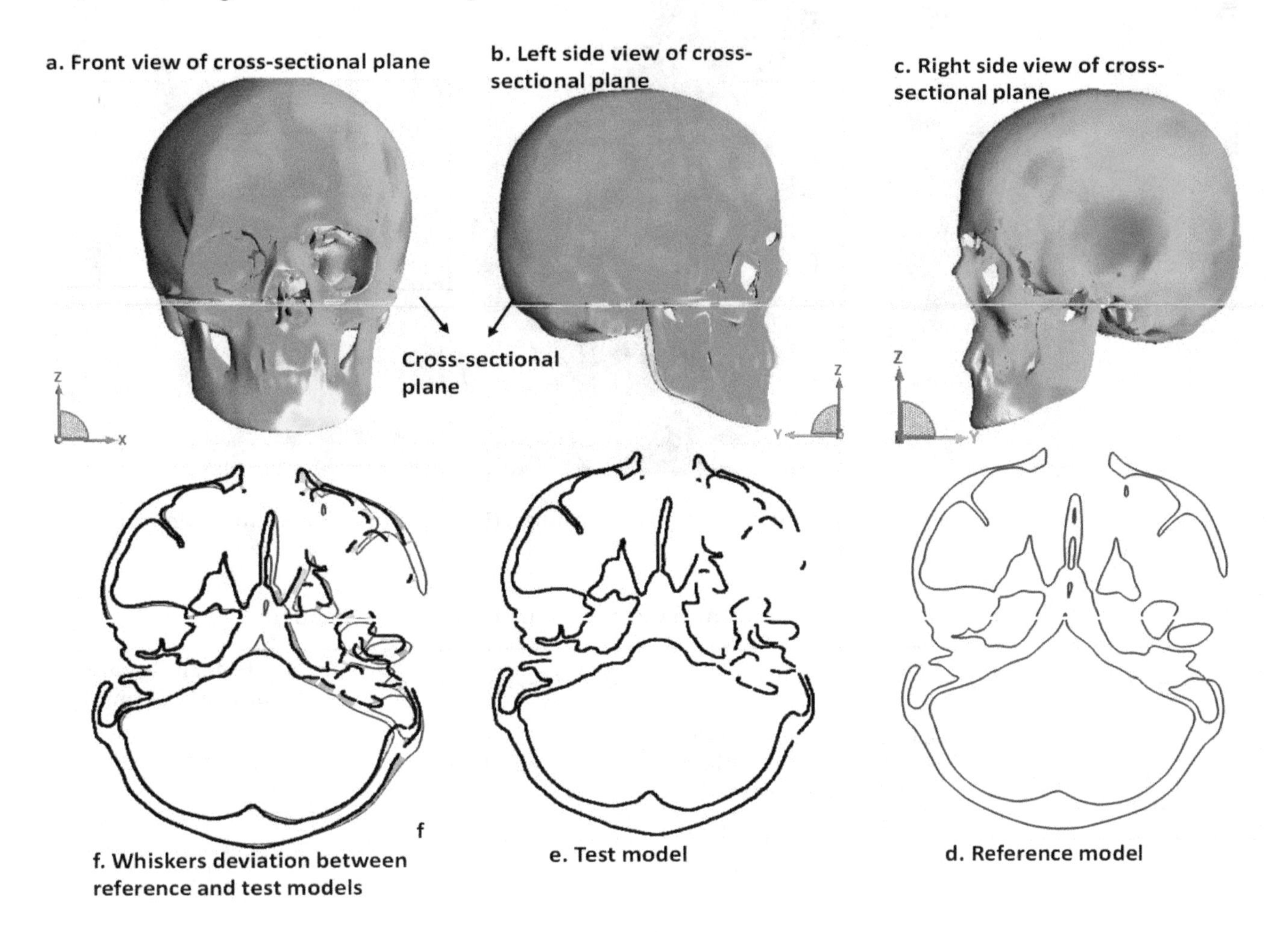

Figure 13. Different views (**a**–**c**) of cross-sectional plane on the implant region. The 2D comparison results (**f**) illustrating the deviation between the (**d**) reference model and the (**e**) test model.

The statistics used to investigate the implant accuracy were the average (AVG) deviation in the positive and negative directions and the root mean square error (RMSE) as illustrated in Equation (1). The average deviation was used as they approximate the differences between the test and reference files in the inward and outward direction. The RMSE represents the average magnitude of the error between two data sets or models. It also quantifies the overall accuracy of the models:

$$\mathrm{RMSE} = \frac{1}{\sqrt{n}}\sqrt{\sum_{i=1}^{n}(X_{1,i} - X_{2,i})^2}, \quad (1)$$

where $X_{1,i}$ is the measurement point of i in the reference data, $X_{2,i}$ is the measurement point of i in the test data, and n is the number of measuring points.

Similarly, the Faro arm scanning and its deviation analysis were carried out on the physical models as shown in Figure 14. The faro arm scanner (Figure 14a) was used to scan the physical facial model attached with an EBM fabricated zygomatic implant. The scanned model was taken as a test file, whereas the mirror reconstruction CAD model was taken as a reference file as illustrated in Figure 14b. The results obtained on the superimposition of test file over a reference file in the 3D comparison technique are shown in Table 2

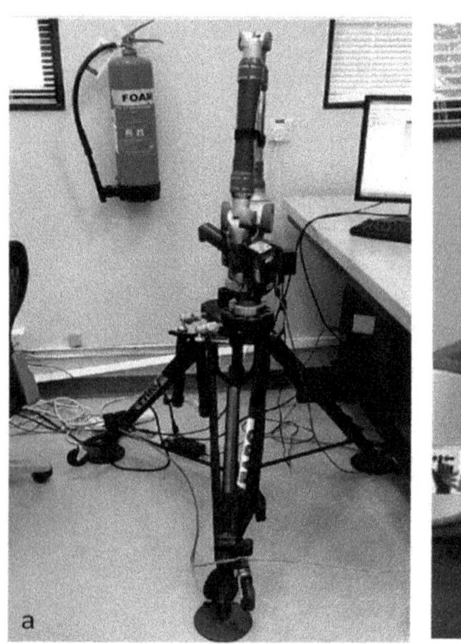

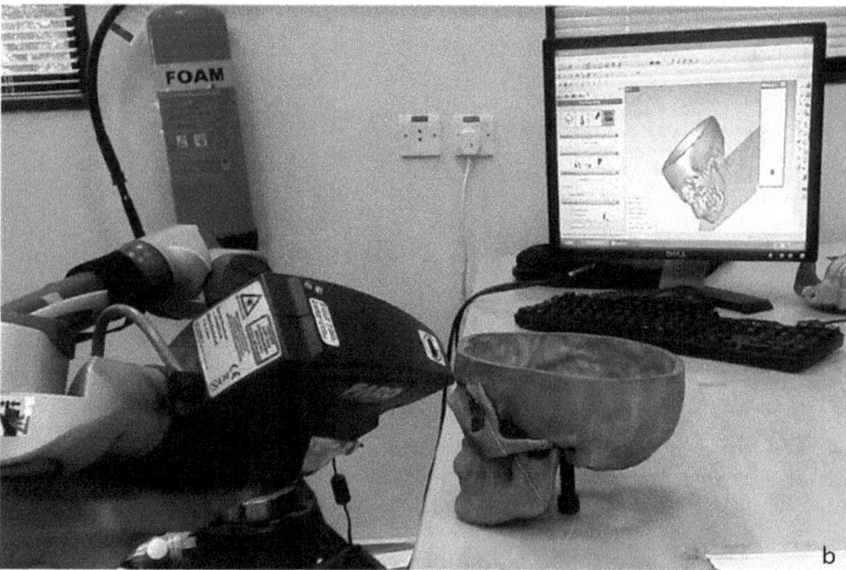

Figure 14. 3D measurement (**a**) Faro platinum arm with scanner; (**b**) scanning of the skull model fixed with titanium zygomatic implant.

Table 2. 3D and 2D comparison results for implant accuracy evaluation.

Comparison	Models	AVG Deviation (mm)	Root Mean Square Error (RMSE) (mm)
1st Stage (virtual evaluation using Geomagics®)	Zygomatic mirror and zygomatic with implant (3D comparison)	1.65/–1.55	2.38
	Zygomatic mirror and zygomatic with implant (2D comparison on Implant region)	0.86/–0.97	1.28
2nd Stage (physical model evaluation using Faro arm scanner)	Zygomatic mirror and zygomatic with implant (3D comparison)	1.82/–3.86	4.25
	Zygomatic mirror and zygomatic with implant (2D comparison on Implant region)	1.40/–1.79	1.96

The results from the 1st stage accuracy analysis, where the remodeled or reconstructed face was analyzed with the mirrored face virtually, the average deviation was in the range of 1.65 mm in the outside direction and –1.55 mm in the inward of the face with RMSE as 2.38 mm. In the second stage, when the remodeled physical face model was analyzed with the mirrored face, the average deviation exhibits in the range of 1.82 mm in the outside direction and –3.86 mm in the inward of the face along with RMSE of 4.25 mm. The deviation (or error) was slightly higher in the physical evaluation when compared to virtual evaluation due to the inclusion of several uncertainties from fabrication as well as the scanning procedure performed by the Faro Arm scanner (FARO, Lake Mary, FL, USA). The lesser RSME in 2D comparison of virtual and physical evaluation further validated the implant accuracy for fitting and aesthetic appearance. The 2D comparison was implemented to evaluate the test model at the defect region where the implant was fixed. It did not consider the entire model, but only the region of interest for comparison. Notice that, at this point, it is difficult to establish or compare these results due to the unavailability of similar studies.

This study is the first of its kind where authors have made an effort to quantify the fitting accuracy of the zygomatic implant on the human face. However, the authors along with this analysis also physically anchored the part onto the face to visualize the aesthetics as well as the facial symmetry. The results were satisfactory and acceptable. In addition, the 2D average deviation of physical model with 1.40 mm in the outward direction and −1.79 mm in the inward direction also justify the results—that there is not much deviation at the zygomatic implant. In addition, the authors will carry out similar studies in the future with zygomatic defects to prepare the data for fitting accuracy and determine their acceptable values. As part of the future studies, an in vivo study will also be performed with the zygomatic implant to confirm its fitting performance.

The FEA outcomes can be realized in Figure 15. The Von Mises stress distribution in the model is shown in Figure 15a. Highly stressed regions can be visualized around the joint and contact areas between the skull and the implant. An inverted image of the zygomatic implant is shown in Figure 15b, which identifies different stress regions. Maximum von Mises stress is found to be 1.76 MPa at one of the fastener positions that is well below the yield strength of implant material—thus ensuring the absence of any plastic deformation. Figure 15c shows the maximum principal strain contours for the zygomatic implant. It is clear that high strained regions comprise interaction and fasteners' areas. Nonetheless, these strains are very low and the maximum value approaches 1.34×10^{-5}. The total displacement pattern on the zygomatic implant due to the applied load is shown in Figure 15d. As depicted, maximum displacement occurs at the load-bearing area. However, maximum deformation has been around 2 microns, which again confirms high stiffness of the implant design.

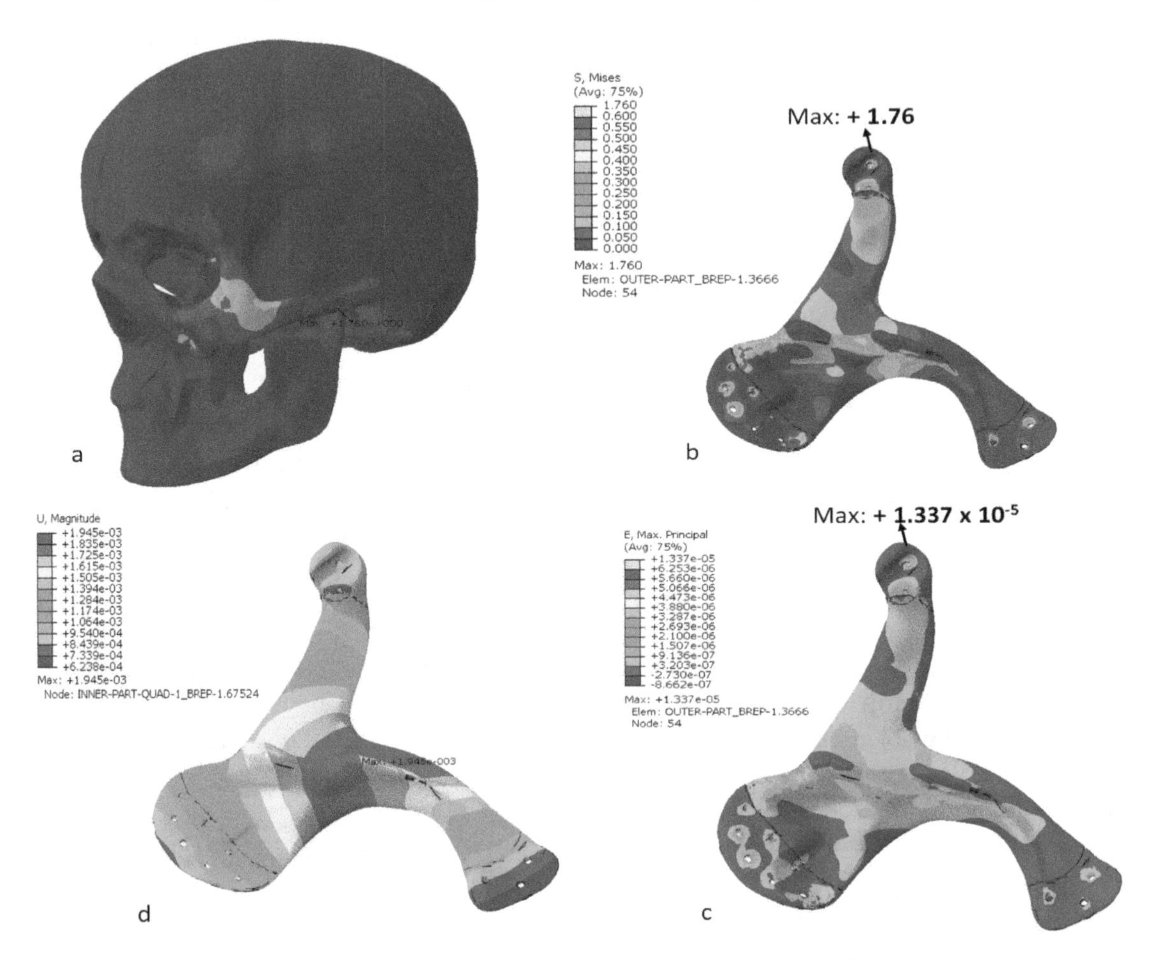

Figure 15. (**a**) Mises stress distributions for the zygomatic implant (MPa), (**b**) Mises stress distributions for the zygomatic implant (MPa), (**c**) maximum principal strain for the zygomatic implant and (**d**) total displacement contour for the zygomatic implant (mm).

4. Conclusions

In this study, a customized zygomatic titanium implant reconstructed using the mirror reconstruction technique was investigated based on the design, biomechanical study, and implant fitting accuracy. The work is particularly important from earlier studies as the authors have identified a technique (2D and 3D comparison) to quantify the implant fitting accuracy or fitting error. Initially, the zygomatic implant was designed manually using silicone, which could not adapt to the bone contours perfectly. Later, a computer-aided custom design implant was constructed using a computer-aided mirroring technique and fabricated using titanium based on state-of-the-art electron beam melting technology. The custom design mirror reconstruction implant precisely fits on the facial region with a maximum deviation error (RMSE) of 2.38 mm in the virtual assembly and 4.25 mm in the physical assembly. Furthermore, the outcomes from the 2D analysis revealed average deviations within 2 mm at the implant region. The implant when practiced on a physical prototype provided a flawless fitting, good symmetry, and pleasant aesthetics. This fitting evaluation study provides crucial information about the implant actualization on the patient's face. This outcome also provides indispensable knowledge for future in vivo or cadaveric implantation studies to further establish the implant's superior performance. The designed implant also successfully withstands the load, with max stress found to be of 1.76 MPa, which is well below the yield strength of implant material (titanium). This proves that the fabricated titanium zygomatic implant possesses the required mechanical strength. Finally, the authors conclude that the EBM fabricated mirror designed titanium implant satisfies the aesthetic, functional, and mechanical properties for efficient zygomatic bone reconstruction. The proposed design methodology can also be applied for other bone reconstruction surgeries. An in vivo or cadaveric based study of the zygomatic titanium implant will be performed as part of future investigation to confirm its fitting performance.

Author Contributions: K.M.: Conceptualization and experiments; K.M. and S.H.M.: Methodology and draft preparation; U.U.: Analysis and Investigation; N.A.: Data curation, resources, and investigation; H.A.: Project administration, revision and funding acquisition, W.A.: Revision and validation.

Acknowledgments: The authors extend their appreciation to the Deanship of Scientific Research at King Saud University for funding this work through Research Group no. RG-1440-034.

References

1. Di, M.P.; Coburn, J.; Hwang, D.; Kelly, J.; Khairuzzaman, A.; Ricles, L. Additively manufactured medical products-the FDA perspective. *3D Print Med.* **2016**, *2*, 1–6.
2. de Viteri, V.S.; Fuentes, E. Titanium and Titanium Alloys as Biomaterials. *Tribol. Fundam. Adv. InTech.* **2013**, 155–181.
3. Balazic, M.; Kopac, J.; Jackson, M.J.; Ahmed, W. Review: Titanium and titanium alloy applications in medicine. *IJNBM* **2007**, *1*, 3. [CrossRef]
4. Parthasarathy, J. 3D modeling, custom implants and its future perspectives in craniofacial surgery. *Ann. Maxillofac. Surg.* **2013**, *4*, 9. [CrossRef]
5. Starch-Jensen, T.; Linnebjerg, L.B.; Jensen, J.D. Treatment of Zygomatic Complex Fractures with Surgical or Nonsurgical Intervention: A Retrospective Study. *Int. J. Oral Maxillofac. Surg.* **2018**, *12*, 377–387. [CrossRef]
6. Lee, H.-S.; Choi, H.-M.; Choi, D.-S.; Jang, I.; Cha, B.-K. Bone thickness of the infrazygomatic crest area in skeletal Class III growing patients: A computed tomographic study. *Imaging Sci. Dent.* **2013**, *43*, 261–266. [CrossRef]
7. Pryor McIntosh, L.; Strait, D.S.; Ledogar, J.A.; Smith, A.L.; Ross, C.F.; Wang, Q.; Opperman, L.A.; Dechow, P.C. Internal Bone Architecture in the Zygoma of Human and Pan. *Anat. Rec.* **2016**, *299*, 1704–1717. [CrossRef]
8. Milne, N.; Fitton, L.C.; Kupczik KFagan, M.J.; O'Higgins, P. The role of the zygomaticomaxillary suture in modulating strain distribution within the skull of Macaca fascicularis. *Homo J. Comp. Hum. Biol.* **2009**, *60*, 281.
9. Foletti, J.M.; Martinez, V.; Haen, P.; Godio-Raboutet, Y.; Guyot LThollon, L. Finite element analysis of

the human orbit. Behavior of titanium mesh for orbital floor reconstruction in case of trauma recurrence. *J. Stomatol. Oral Maxillofac. Surg.* **2019**, *120*, 91–94. [CrossRef]

10. Parel, S.M.; Brånemark, P.I.; Ohrnell, L.O.; Svensson, B. Remote implant anchorage for the rehabilitation of maxillary defects. *J. Prosthet. Dent.* **2001**, *86*, 377–381. [CrossRef]
11. Quatela, V.C.; Chow, J. Synthetic facial implants. *Facial Plast. Surg. Clin. N. Am.* **2008**, *16*, 1–10. [CrossRef] [PubMed]
12. Scolozzi, P. Maxillofacial reconstruction using polyetheretherketone patient-specific implants by 'mirroring' computational planning. *Aesthetic Plast. Surg.* **2012**, *36*, 660–665. [CrossRef] [PubMed]
13. Ivy, E.J.; Lorenc, Z.P.; Aston, S.J. Malar augmentation with silicone implants. *Plast. Reconstr. Surg.* **1995**, *96*, 63–68. [CrossRef] [PubMed]
14. El-Khayat, B.; Eley, K.A.; Shah, K.A.; Watt-Smith, S.R. Ewings sarcoma of the zygoma reconstructed with a gold prosthesis: A rare tumor and unique reconstruction. *Oral Surg. Oral Med. Oral Pathol. Oral Radiol. Endod.* **2010**, *109*, e5–e10. [CrossRef] [PubMed]
15. Hoffmann, J.; Cornelius, C.P.; Groten, M.; Pröbster, L.; Pfannenberg, C.; Schwenzer, N. Orbital reconstruction with individually copy-milled ceramic implants. *Plast. Reconstr. Surg.* **1998**, *101*, 604–612. [CrossRef] [PubMed]
16. Zhang, Y. Orbital Defect Repair and Secondary Reconstruction of Enophthalmos with Mirror-Technique Fabricated Titanium Mesh. *J. Oral Maxillofac. Surg.* **2008**, *66*, 19–20. [CrossRef]
17. Moiduddin, K.; Mian, S.H.; Umer, U.; Alkhalefah, H. Fabrication and Analysis of a Ti6Al4V Implant for Cranial Restoration. *Appl. Sci.* **2019**, *9*, 2513. [CrossRef]
18. Liu, Y.; Xu, L.; Zhu, H.; Liu, S.S.-Y. Technical procedures for template-guided surgery for mandibular reconstruction based on digital design and manufacturing. *Biomed Eng. Online* **2014**, *13*, 63. [CrossRef]
19. Christensen, A.; Kircher, R.; Lippincott, A. Qualification of electron beam melted (EBM) Ti6Al4V-ELI for orthopaedic implant applications. In *Medical Device Materials IV: Proceedings of the Materials and Processes for Medical Devices Conference*; Jeremy, G., Ed.; ASM International: Cleveland, OH, USA, 2007; Volume 6, pp. 48–53.
20. Niinomi, M. Recent research and development in titanium alloys for biomedical applications and healthcare goods. *Sci. Technol. Adv. Mater.* **2003**, *4*, 445. [CrossRef]
21. Ehtemam-Haghighi, S.; Prashanth, K.G.; Attar, H.; Chaubey, A.K.; Cao, G.H.; Zhang, L.C. Evaluation of mechanical and wear properties of Ti xNb 7Fe alloys designed for biomedical applications. *Mater. Des.* **2016**, *111*, 592–599. [CrossRef]
22. Huynh, V.; Ngo, N.K.; Golden, T.D. Surface Activation and Pretreatments for Biocompatible Metals and Alloys Used in Biomedical Applications. *Int. J. Biomater.* **2019**, *2019*, 21. [CrossRef] [PubMed]
23. Coquim, J.; Clemenzi, J.; Salahi, M.; Sherif, A.; Tavakkoli Avval, P.; Shah, S.; Schemitsch, E.H.; Shaghayegh Bagheri, Z.; Bougherara, H.; Zdero, R. Biomechanical Analysis Using FEA and Experiments of Metal Plate and Bone Strut Repair of a Femur Midshaft Segmental Defect. *Biomed. Res. Int.* **2018**, *2018*, 11. [CrossRef] [PubMed]
24. Lemu, H.G. Virtual engineering in design and manufacturing. *Adv. Manuf.* **2014**, *2*, 289–294. [CrossRef]
25. Shah, F.A.; Thomsen, P.; Palmquist, A. Osseointegration and current interpretations of the bone-implant interface. *Acta Biomater.* **2019**, *84*, 1–15. [CrossRef] [PubMed]
26. Wysocki, B.; Maj, P.; Sitek, R.; Buhagiar, J.; Kurzydłowski, K.; Święszkowski, W. Laser and Electron Beam Additive Manufacturing Methods of Fabricating Titanium Bone Implants. *Appl. Sci.* **2017**, *7*, 657. [CrossRef]
27. Li, X.; Wang, C.; Zhang, W.; Li, Y. Fabrication and characterization of porous Ti6Al4V parts for biomedical applications using electron beam melting process. *Mater. Lett.* **2009**, *63*, 403–405. [CrossRef]
28. Moiduddin, K.; Hammad Mian, S.; Alkindi, M.; Ramalingam, S.; Alkhalefah, H.; Alghamdi, O. An in vivo Evaluation of Biocompatibility and Implant Accuracy of the Electron Beam Melting and Commercial Reconstruction Plates. *Metals* **2019**, *9*, 1065. [CrossRef]
29. Hoang, D.; Perrault, D.; Stevanovic, M.; Ghiassi, A. Surgical applications of three-dimensional printing: A review of the current literature & how to get started. *Ann. Transl. Med.* **2016**, *4*, 456.
30. Lee, J.-W.; Fang, J.-J.; Chang, L.-R.; Yu, C.-K. Mandibular Defect Reconstruction with the Help of Mirror Imaging Coupled with Laser Stereolithographic Modeling Technique. *J. Formos. Med. Assoc.* **2007**, *106*, 244–250. [CrossRef]
31. Kashi, A.; Saha, S. Mechanisms of failure of medical implants during long-term use. In *Biointegration of*

Medical Implant Materials; Sharma, C.P., Ed.; Woodhead Publishing: Cambridge, UK, 2010; pp. 326–348.

32. Singare, S.; Shenggui, C.; Sheng, L. The use of 3D printing technology in human defect reconstruction-a review of cases study. *Med. Res. Innov.* **2017**, *1*, 1–4. [CrossRef]
33. Moiduddin, K.; Al-Ahmari, A.; Nasr, E.S.A.; Mian, S.H.; Al Kindi, M. A comparison study on the design of mirror and anatomy reconstruction technique in maxillofacial region. *Technol. Health Care* **2016**, *24*, 377–389. [CrossRef] [PubMed]
34. Jardini, A.L.; Larosa, M.A.; Maciel Filho, R.; de Carvalho Zavaglia, C.A.; Bernardes, L.F.; Lambert, C.S.; Calderoni, D.R.; Kharmandayan, P. Cranial reconstruction: 3D biomodel and custom-built implant created using additive manufacturing. *J. Cranio-Maxillofac. Surg.* **2014**, *42*, 1877–1884. [CrossRef] [PubMed]
35. Lütjering, G.; Williams, J.C. *Titanium*; Springer Science & Business Media: Berlin, Germany, 2007.
36. Arcam, A. Arcam A2 Technical Specification. ARCAM A2 TECHNICAL DATA. 2019. Available online: http://www.arcam.com/wp-content/uploads/Arcam-A2.pdf (accessed on 23 May 2019).
37. New! 50 μm Process for High Resolution and Surface Finish. 21 May 2012. Available online: http://www.arcam.com/new-50-um-process-for-high-resolution-and-surface-finish (accessed on 14 September 2019).
38. Umer, U.; Ameen, W.; Abidi, M.H.; Moiduddin, K.; Alkhalefah, H.; Alkahtani, M.; Al-Ahmari, A. Modeling the Effect of Different Support Structures in Electron Beam Melting of Titanium Alloy Using Finite Element Models. *Metals* **2019**, *9*, 806. [CrossRef]
39. Mian, S.H.; Mannan, M.A.; Al-Ahmari, A.M. The influence of surface topology on the quality of the point cloud data acquired with laser line scanning probe. *Sens. Rev.* **2014**, *34*, 255–265. [CrossRef]
40. El-Anwar, M.I.; Mohammed, M.S. Comparison between two low profile attachments for implant mandibular overdentures. *J. Genet. Eng. Biotechnol.* **2014**, *12*, 45–53. [CrossRef]
41. Arcam, Ti6Al4V ELI Titanium Alloy, Ti6Al4V ELI Titanium Alloy. Available online: http://www.arcam.com/wp-content/uploads/Arcam-Ti6Al4V-ELI-Titanium-Alloy.pdf (accessed on 23 September 2019).
42. Nagasao, M.; Nagasao, T.; Imanishi, Y.; Tomita, T.; Tamaki, T.; Ogawa, K. Experimental evaluation of relapse-risks in operated zygoma fractures. *Auris. Nasus. Larynx.* **2009**, *36*, 168–175. [CrossRef]

4

Corrosion and Tensile Behaviors of Ti-4Al-2V-1Mo-1Fe and Ti-6Al-4V Titanium Alloys

Yanxin Qiao [1], Daokui Xu [2,*], Shuo Wang [1,2,3], Yingjie Ma [2,*], Jian Chen [1], Yuxin Wang [1] and Huiling Zhou [1]

1 School of Materials Science and Engineering, Jiangsu University of Science and Technology, Zhenjiang 212003, China; yxqiao@just.edu.cn (Y.Q.); 1910221@mail.neu.edu.cn (S.W.); jchen496@uwo.ca (J.C.); ywan943@163.com (Y.W.); zhouhl@just.edu.cn (H.Z.)
2 Institute of Metal Research, Chinese Academy of Sciences, Shenyang 110016, China
3 School of Materials Science and Engineering, Northeastern University, Shenyang 110004, China
* Correspondence: dkxu@imr.ac.cn (D.X.); yjma@imr.ac.cn (Y.M.)

Abstract: X-ray diffraction (XRD), scanning electron microscope (SEM), immersion, electrochemical, and tensile tests were employed to analyze the phase constitution, microstructure, corrosion behaviors, and tensile properties of a Ti-6Al-4V alloy and a newly-developed low cost titanium alloy Ti-4Al-2V-1Mo-1Fe. The results showed that both the Ti-6Al-4V and Ti-4Al-2V-1Mo-1Fe alloys were composed of α and β phases. The volume fractions of β phase for these two alloys were 7.4% and 47.3%, respectively. The mass losses after 180-day immersion tests in 3.5 wt.% NaCl solution of these alloys were negligible. The corrosion resistance of the Ti-4Al-2V-1Mo-1Fe alloy was higher than that of the Ti-6Al-4V alloy. The tensile tests showed that the Ti-4Al-2V-1Mo-1Fe alloy presented a slightly higher strength but a lower ductility compared to the Ti-6Al-4V alloy.

Keywords: titanium alloy; corrosion; passive film; mechanical behavior

---

1. Introduction

Titanium and its alloys are widely used in aerospace [1,2], marine [3], chemical [4,5], and biomedical [6–8] fields due to their excellent mechanical properties, high corrosion resistance, and good biocompatibility. Their high corrosion resistance in aggressive environments is ensured by the formation of a compact and chemically-stable oxide film, mainly composed of titanium oxide, TiO_2, which spontaneously covers the metal surface to protect the metal substrate [9–11]. The corrosion behaviors of the commercial Ti-6Al-4V alloy vary with corrosive environments. Yue et al. [12] found that the Ti-6Al-4V alloy exhibited a poor corrosion behavior in a solution with a high Cl^- concentration or in acid environments with a local accumulation of Cl^- ions. Blanco-Pinzon et al. [13] reported that the Ti-6Al-4V alloy was susceptible to corrosion in H_2SO_4 solution but its corrosion resistance could be significantly improved by alloying Pd and Ni. Wang et al. [5] investigated the effects of alloying elements Pd, Mo, and Ni on the corrosion behavior of titanium alloy in H_2SO_4 solution containing fluoride ions, and found that these alloying elements had no influence on the interaction of the F^- ions with titanium matrix, and on the film composition. The addition of Ni accelerated the cathodic reactions while the addition of Mo retarded the anodic process [5]. Newman et al. [14,15] found that Mo located at defect sites preferentially dissolved, leading to the formation of stable Mo oxides to decrease the anodic dissolution rate [16].

Generally, element alloying is an effective method to improve the mechanical properties of titanium and its alloys [17,18]. Since Fe and Al elements are characteristic of low toxicity and cost [19–21], they are added into the titanium and titanium alloys, and the effect of Fe or Al alloying on the mechanical and corrosion properties of titanium alloys has been studied. It is reported that the Ti-4.5Al-3V-2Mo-2Fe alloy has superior mechanical properties to the Ti-6Al-4V alloy [17] due to its microstructural characteristics and element alloying (i.e., Mo and Fe). Lu et al. [22] investigated the mechanical properties and electrochemical corrosion behaviors of Ti-6Al, Ti-6Al-4V, and Ti-6Al-xFe alloys, and found that the Ti-6Al-4Fe alloy possessed the lowest Young's modulus and exhibits the highest strength:modulus ratios, and the Ti-6Al-xFe alloys exhibited a higher corrosion resistance in simulated human body fluid (SBF) than both the Ti-6Al and Ti-6Al-4V alloys. However, the corrosion resistance of the Ti-6Al-xFe alloys decreased with the increasing Fe content, suggesting that the content of Fe added into the titanium alloys needed to be controlled at a lower level to achieve a better corrosion performance. It is consistent with the results conducted by Pimenova et al. [21] and Hsu et al. [23,24]. When the concentration of Al was higher than 15 wt.%, Ti-xAl-yFe alloys underwent severe pitting corrosion due to the precipitates of β phase and uneven distribution of the alloying elements [21]. Thus, more work has to be performed to clarify the effect of Al and Fe alloying on the corrosion and mechanical performances of titanium alloys, especially when Mo is added to improve the resistance of localized corrosion.

In this research, the Ti-4Al-2V-1Mo-1Fe alloy developed by the Institute of Metal Research, Chinese Academy of Sciences (IMR), was investigated. This type of alloy has lower contents of Al and V but higher contents of Mn and Fe. The cost of this alloy is relatively low, providing a potential alternative to the Ti-6Al-4V alloy. The aim of this work is to investigate the corrosion behavior of this newly-developed alloy in a simulated marine environment (3.5 wt.% NaCl solution), and its tensile property. The difference of alloy property between Ti-6Al-4V and Ti-4Al-2V-1Mo-1Fe alloys was systematically studied.

2. Experimental Details

The materials used in the present study were the commercial Ti-6Al-4V alloy, and the Ti-4Al-2V-1Mo-1Fe alloy fabricated in the IMR, Shenyang, China. The chemical compositions (wt.%) of these two alloys are listed in Table 1. Prior to the study, the Ti-6Al-4V and Ti-4Al-2V-1Mo-1Fe alloys were heated to 750 °C for 3 h and then cooled to room temperature in air. Samples for the immersion test were cut into sheets with dimensions of 40 mm × 20 mm × 4 mm. Samples for the electrochemical test and microstructure observation were cut into square sheets (10 mm × 10 mm × 2 mm) and sealed in a mixture of epoxy and polyamide resins with an exposed surface of 1 cm^2. Then each specimen was gradually ground with SiC papers up to 1000 grit, polished with a diamond paste of 0.5 μm, then cleaned in ethanol, and finally dried with hot air.

Table 1. Chemical compositions of the tested alloys (wt.%).

Alloy	Al	V	Mo	Fe	Ti
Ti-6Al-4V	5.95	4.03	-	0.33	Bal.
Ti-4Al-2V-1Mo-1Fe	3.96	2.03	1.05	0.92	Bal

The immersion and electrochemical tests were performed in a 3.5 wt.% NaCl solution at 25 ± 1 °C (controlled by a thermostat water bath). The 3.5 wt.% NaCl solution was prepared using analytical-grade sodium chloride and distilled water. Immersion tests were carried out to investigate the long-term corrosion behaviors of the Ti-6Al-4V and Ti-4Al-2V-1Mo-1Fe alloys. Five samples were prepared for each test solution, with an immersion period of 180 days to ensure the reproducibility. Samples with dimensions of 10 mm × 10 mm × 2 mm were successively ground with abrasive papers to 1000 grit and then were immersed in aerated 3.5% NaCl solution (volume: 1.5 L) without stirring. The solutions were replaced every 10 days. The electrochemical behaviors of the tested alloys were measured using a CS350 (Wuhan Corrtest Instruments Corp., Ltd. Wuhan, China) electrochemical workstation and a three-electrode electrochemical cell, and the method was described in the literature [25]. Potentiodynamic polarization was performed at a scan rate of 0.1667 mV/s from

−500 mV_{SCE} below the open circuit potential (OCP) and terminated at 2500 mV_{SCE}. After immersing in 3.5 wt.% NaCl solution at the OCP for 1 h, electrochemical impedance spectroscopy (EIS) was conducted with a sinusoidal potential perturbation of 10 mV and a frequency range from 10^5 to 10^{-2} Hz. All measurements were repeated at least three times in naturally-aerated 3.5 wt.% NaCl solution without stirring to ensure the reproducibility. Cview and Zview software were used to fit the electrochemical data. Tensile tests of these two alloys were carried out on an Instron-type testing machine with a stain rate of 3 $mm \cdot s^{-1}$ at 25 ± 1 °C [26]. To ensure the reliability of the measured data, at least three repeated measurements were carried out for each time. The standard deviation method was used to analyze the data and obtain the mechanical property parameters.

The crystal structures of the tested alloys were determined using a D/Max 2400 X-ray diffractometer (Rigaku Corporation, Tokyo, Japan) with Cu K_{α} radiation at 10 kV and 35 mA at a step size of 0.02° and a scan rate of 4°/min. The specimens used for the microstructure observation were etched in Kroll reagent (3 mL HF, 9 mL HNO_3 and 88 mL H_2O) for 10 s. The microstructure was observed by a Keyence VHX-700 (Keyence Co. Ltd., Osaka, Japan) (LM) and a scanning electron microscope (XL30-FEG ESEM, FEI, Hillsboro, OR., USA) equipped with EDS. Image-Pro Plus software was used to calculate the phase volume fraction of the tested alloys.

3. Results and Discussion

3.1. Microstructure Characterization

The XRD patterns of the Ti-6Al-4V and Ti-4Al-2V-1Mo-1Fe alloys are shown in Figure 1. The diffraction peaks in Figure 1 corresponded to the peaks of α and β phases, suggesting that both alloys have duplex structure. However, the proportions of these two phases in the Ti-6Al-4V and Ti-4Al-2V-1Mo-1Fe alloys were different. As seen in Figure 1, the peak intensity of α phase was higher, indicating a higher content of α phase in both alloys.

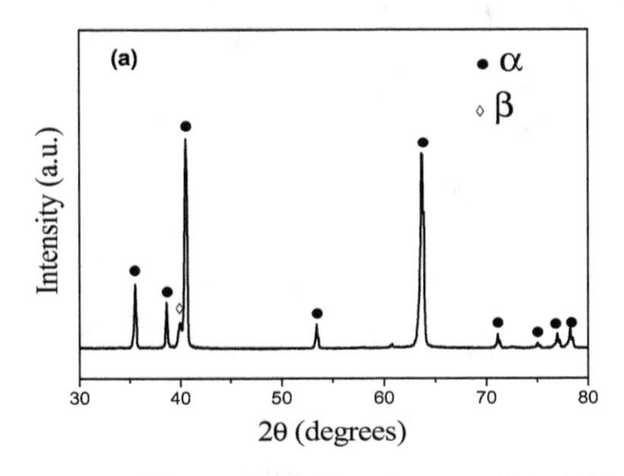

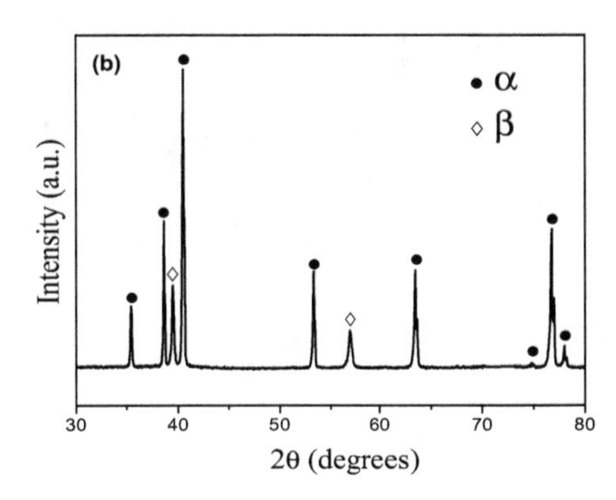

Figure 1. XRD patterns of the (**a**) Ti-6Al-4V and (**b**) Ti-4Al-2V-1Mo-1Fe alloys.

The LM images of the Ti-6Al-4V and Ti-4Al-2V-1Mo-1Fe alloys are shown in Figure 2. It was found that the two tested alloys had a bi-phase structure, consistent with the XRD results in Figure 1. The average grain size of the α phase in the Ti-6Al-4V alloy was ~25 μm, and some grains were larger than 50 μm, as shown in Figure 2a. In comparison, the grain size of the α phase in the Ti-4Al-2V-1Mo-1Fe alloy was smaller, with an average grain size of about ~8 μm, as shown in Figure 2b.

The SEM observations of the Ti-6Al-4V and Ti-4Al-2V-1Mo-1Fe alloys are shown in Figure 3. As seen in Figure 3, both alloys were composed of the dark α phase and the bright β phase. No other precipitates were observed either inside grains or at grain boundaries. As shown in Figure 3a, the β phase dispersed and scattered inside the equiaxed α phase. However, the distribution of the β phase in the Ti-4Al-2V-1Mo-1Fe alloy was more continuous. As shown in Figure 3b, clustered β phase distributed evenly inside the α phase in the Ti-4Al-2V-1Mo-1Fe alloy. The volume fraction of the β phase in the Ti-6Al-4V and Ti-4Al-2V-1Mo-1Fe alloys was about 7.4% and 47.3%, respectively. EDS analysis was conducted to investigate the composition of α and β phases of the tested alloys, and the results are shown in Table 2. It is reported that Fe, Mo, and V atoms have long been recognized as strong β-stabilizing elements [19,22,27], thus the atomic ratio of these elements in β phase is higher than that in α phase. This is mainly due to the higher element solid solubility and elemental diffusion rate in the β phase [27,28]. Based on the research [28], the volume fraction of the β phase in Ti-4Al-2V-1Mo-1Fe alloy may be higher than that of the Ti-6Al-4V alloy.

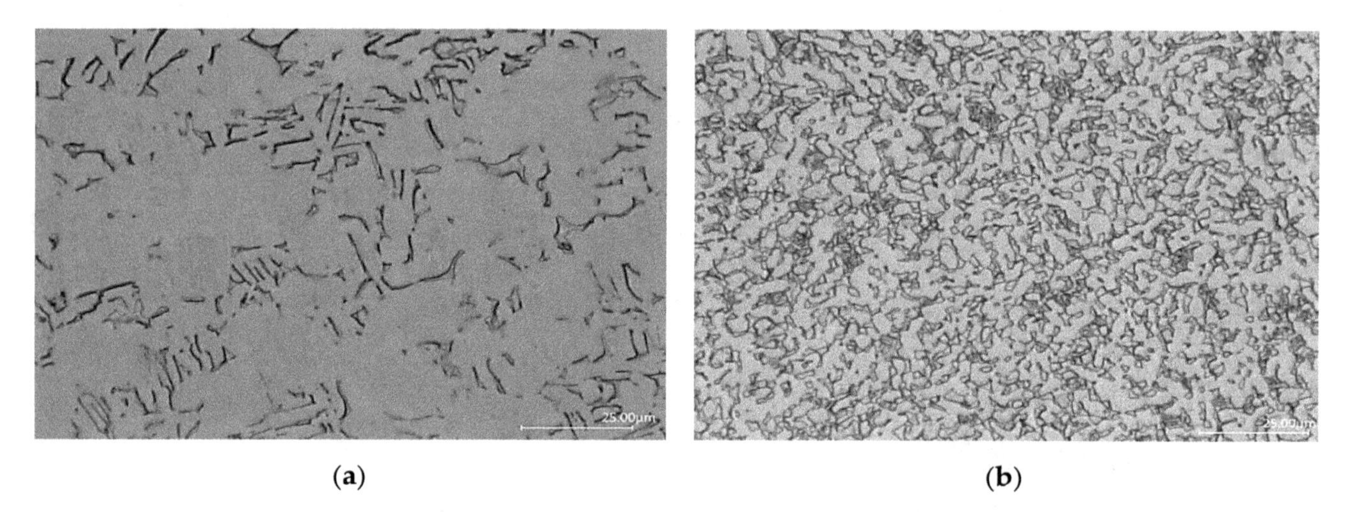

Figure 2. LM of the (**a**) Ti-6Al-4V and (**b**) Ti-4Al-2V-1Mo-1Fe alloys.

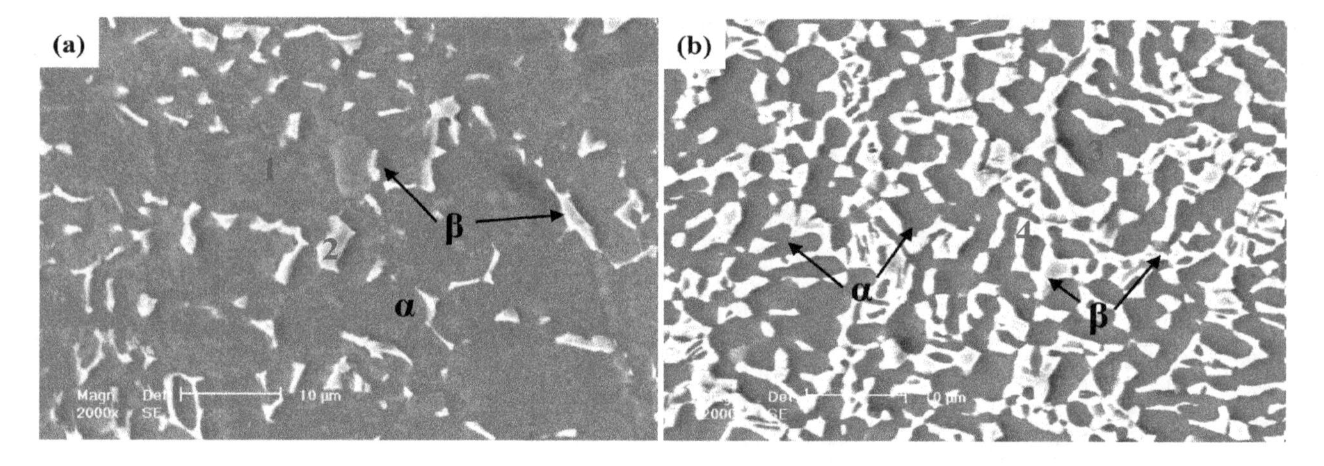

Figure 3. SEM microstructures of the (**a**) Ti-6Al-4V and (**b**) Ti-4Al-2V-1Mo-1Fe alloys.

Table 2. Chemical compositions (EDS) of α and β phases of the Ti-6Al-4V and Ti-4Al-2V-1Mo-1Fe alloys (wt.%).

Alloy		Ti	Al	V	Mo	Fe
Ti-6Al-4V	α (area 1)	88.6	6.9	4.5	-	-
	β (area 2)	80.7	2.4	15.9	-	1.0
Ti-4Al-2V-1Mo-1Fe	α (area 3)	93.5	5.3	1.2	-	-
	β (area 4)	85.4	2.6	4.7	4.4	2.9

3.2. Immersion Test

The microstructures of the Ti-6Al-4V and Ti-4Al-2V-1Mo-1Fe alloys after immersion in 3.5 wt.% NaCl solution for 180 days are shown in Figure 4. It can be seen that no traces of corrosion were observed, indicating the Ti-6Al-4V and Ti-4Al-2V-1Mo-1Fe alloys had excellent corrosion resistance in 3.5 wt.% NaCl solution.

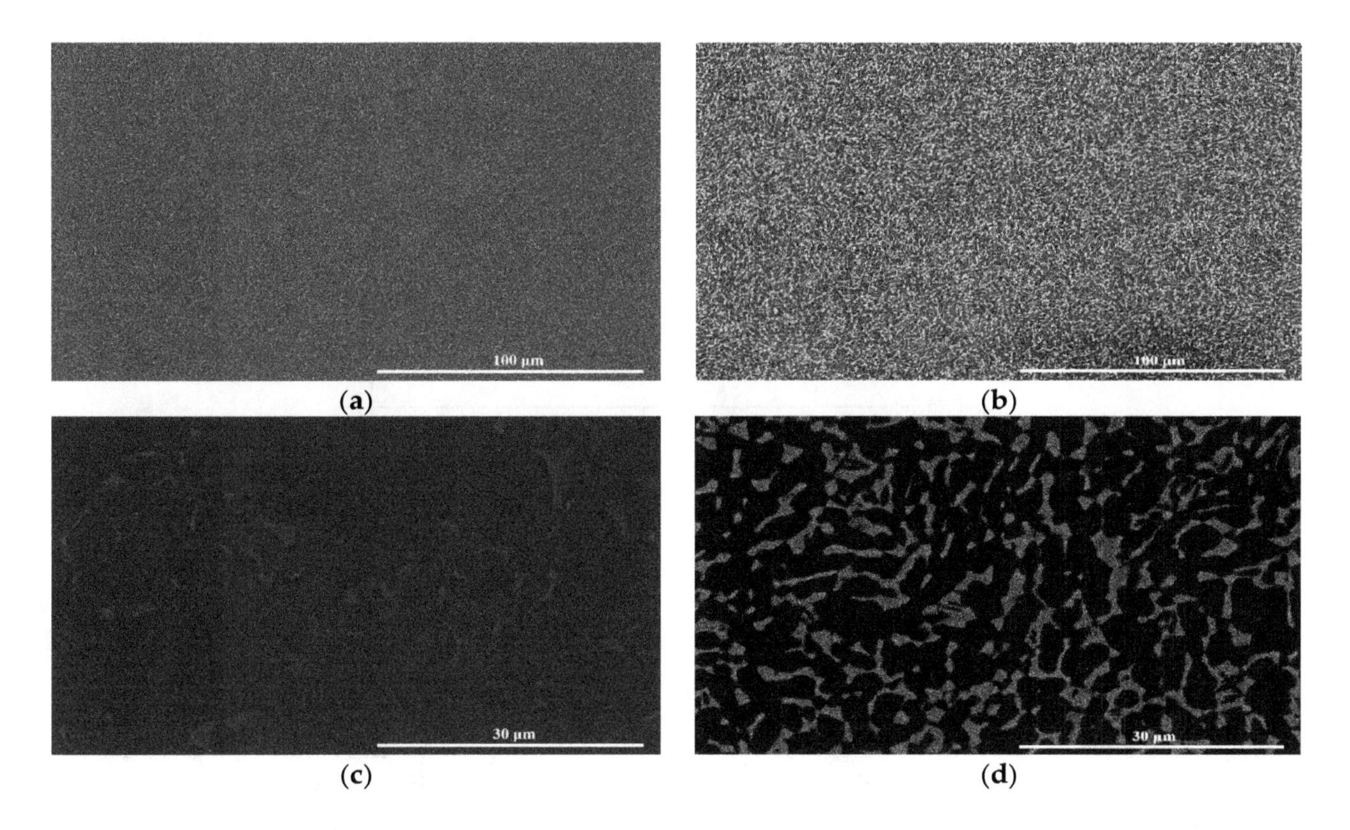

Figure 4. Microstructures of the (**a**,**c**) Ti-6Al-4V and (**b**,**d**) Ti-4Al-2V-1Mo-1Fe alloys after immersion in 3.5 wt.% NaCl solution for 180 days.

The mass loss rates of the Ti-6Al-4V and Ti-4Al-2V-1Mo-1Fe alloys after 180 days of immersion in 3.5% NaCl solution were calculated based on the mass loss, Δm, using Equation (1) shown as following:

$$\Delta m = \frac{m_0 - m_1}{S \times t} \tag{1}$$

where m_0 is the weight (mg) of the sample before the immersion test, m_1 is the weight (mg) of the sample after the immersion test, S is the surface area of the sample (cm^2), and t is the immersion time (180 days). The mass loss rate of the Ti-6Al-4V and Ti-4Al-2V-1Mo-1Fe alloys was 1.99×10^{-4} and 2.01×10^{-4} $mg{\cdot}cm^{-2}{\cdot}day^{-1}$, respectively. This indicates both behaved similarly, without any obvious occurrence of corrosion in the present tested solution.

3.3. Electrochemical Response

Once the Ti-6Al-4V and Ti-4Al-2V-1Mo-1Fe alloys were immersed in NaCl solution at OCP, the dissolution process of the naturally-formed oxide film (rutile TiO_2) in air began and the self-passivated film simultaneously formed [9]. The OCPs for the Ti-6Al-4V and Ti-4Al-2V-1Mo-1Fe alloys in 3.5 wt.% NaCl solution are shown in Figure 5. Since the specimens were exposed in ambient atmosphere for 1 h to allow the native growth of oxide film, the spontaneous OCP after the immersion indicated the stability of the naturally-formed oxide [5]. It is seen from Figure 5 that corrosion potential (E_{corr}) of the Ti-6Al-4V alloy shifted continuously to positive potential with the immersion time. As for the Ti-4Al-2V-1Mo-1Fe alloy, the E_{corr} gradually increased from -0.65 V_{SCE} to more noble potential, and finally achieved steady-state potential of -0.40 V_{SCE} when the immersion time was 1 h. The corrosion data in Figure 5 revealed that a shorter time was required for the steady-state potentials to be obtained in 3.5 wt.% NaCl solution for the Ti-4Al-2V-1Mo-1Fe alloy compared to the Ti-6Al-4V alloy.

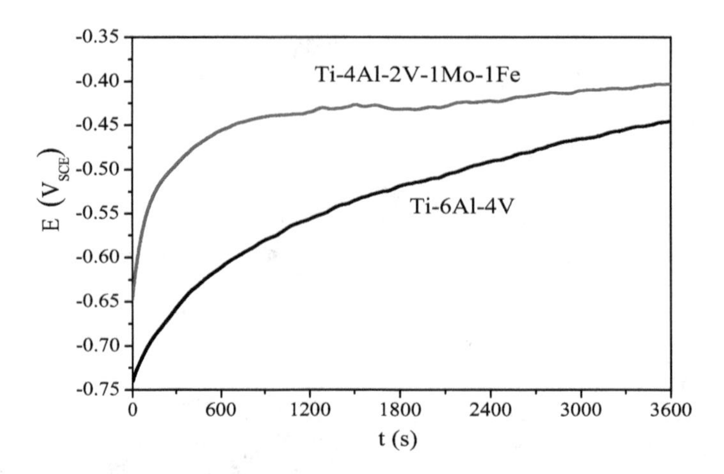

Figure 5. Evolution of open circuit potential (OCP) with immersion time for the Ti-6Al-4V and Ti-4Al-2V-1Mo-1Fe alloys in 3.5 wt.% NaCl solution.

To investigate the stability of the passive film formed on the Ti-6Al-4V and Ti-4Al-2V-1Mo-1Fe alloys, EIS measurements were carried out at OCP for both alloys in 3.5 wt.% NaCl solution. Figures 6 and 7 show the Nyquist and Bode plots for the Ti-6Al-4V and Ti-4Al-2V-1Mo-1Fe alloys for 1 h immersion in 3.5 wt.% NaCl solution. As seen from Figure 6, both the Ti-6Al-4V and Ti-4Al-2V-1Mo-1Fe alloys exhibited an unfinished single capacitive arc. It can be seen in Figure 7 that only one time constant was observed. The initial impedance (*Z*) recorded for the Ti-4Al-2V-1Mo-1Fe alloy was higher than that of the Ti-6Al-4V alloy, indicating that the Ti-4Al-2V-1Mo-1Fe alloy had a superior corrosion resistance [3]. The impedance data were analyzed using the equivalent circuit shown in Figure 8. The use of a constant phase element (CPE) was necessary [29–31] due to the distribution of relaxation times resulting from heterogeneities at the electrode surface. The CPE was used for the description of a frequency-independent phase shift between an applied AC potential and its current response [32], and has been extensively investigated [29,33]. The impedance of the CPE was given by:

$$Z_{CPE} = \frac{1}{Q}(j\omega)^{-n} \tag{2}$$

Therefore, the total impedance was [34]:

$$Z_{total} = R_s + (Q(j\omega)^n + \frac{1}{R_p})^{-1} \tag{3}$$

where n was the depression angle (in degrees) that evaluated the semicircle deformation, R_s was the electrolyte resistance, R_p represented the charge transfer resistance, and Q corresponded to the pseudo-capacitance of the film, expressed using the CPE. The reason may be as follows: the CPE accounted for two contributions, one arising from double-layer capacitance (C_H) and one arising from semiconductor capacitance relating to the passive film (Csc). The capacitance of the double layer seems to be neglected according to the result reported by Hirschorn et al. [35], so the capacitance behavior of Ti-6Al-4V and Ti-4Al-2V-1Mo-1Fe alloys was dominated by the passive film [36]. Table 3 shows the values of the electric parameters obtained using the equivalent electric circuit to fit the EIS data. It is seen in Table 3 that the film resistance for the Ti-6Al-4V and Ti-4Al-2V-1Mo-1Fe alloys was 5.69×10^5 $\Omega \cdot cm^{-2}$ and 6.50×10^5 $\Omega \cdot cm^{-2}$, respectively. It is consistent with the generally accepted sense that a larger capacitive arc indicates higher corrosion resistance. It can be inferred that the stability of the passive film formed on the Ti-4Al-2V-1Mo-1Fe alloy in the present solution was slightly better than that of the Ti-6Al-4V alloy.

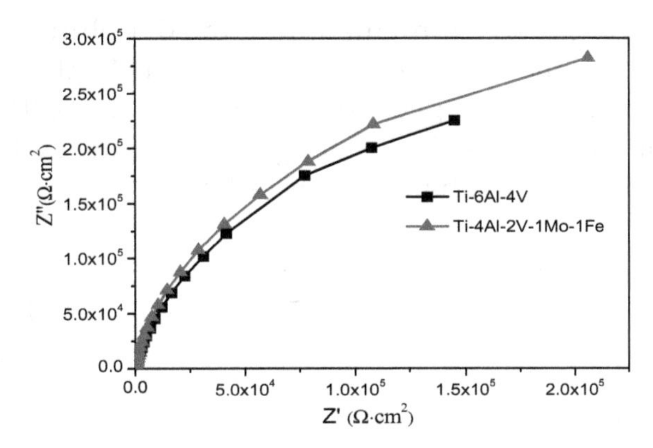

Figure 6. Nyquist plots for the Ti-6Al-4V and Ti-4Al-2V-1Mo-1Fe alloys for 1 h of immersion in 3.5 wt.% NaCl solution.

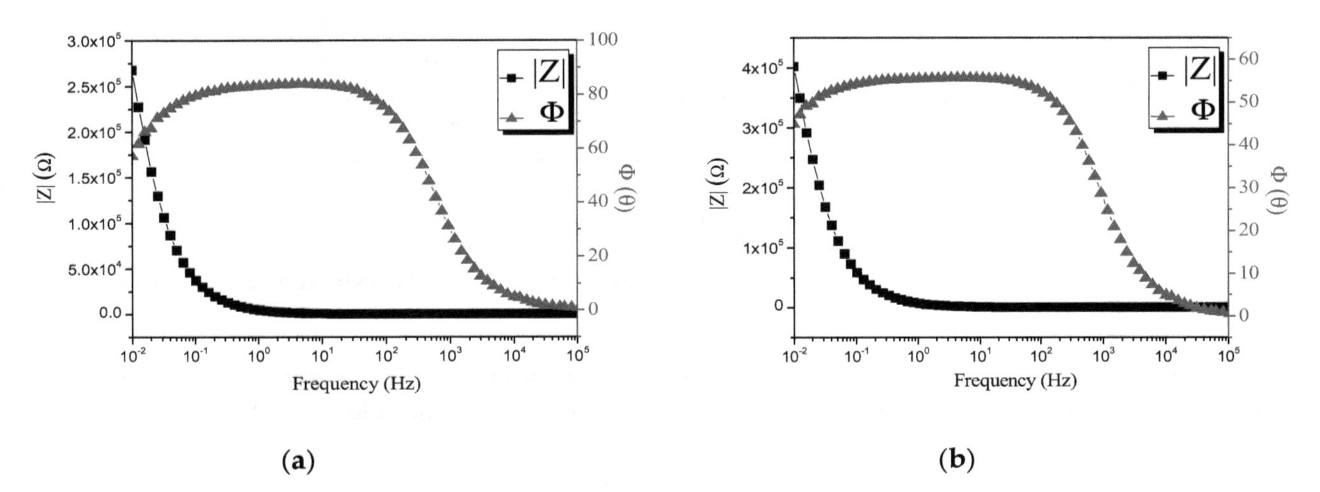

Figure 7. Bode plots for the (**a**) Ti-6Al-4V and (**b**) Ti-4Al-2V-1Mo-1Fe alloys for 1 h of immersion in 3.5 wt.% NaCl solution.

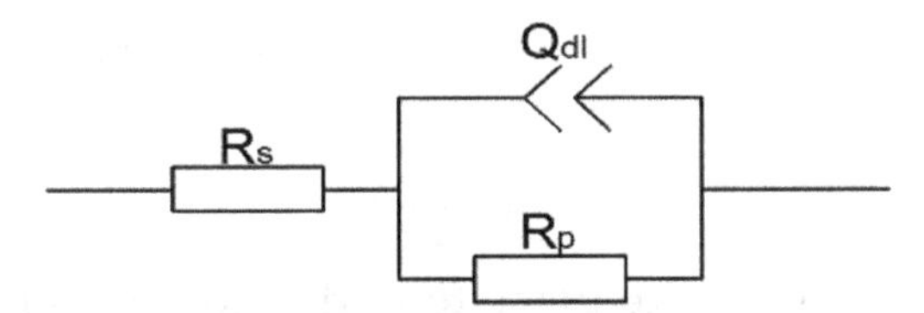

Figure 8. The equivalent circuit used for quantitative evaluation of electrochemical impedance spectroscopy (EIS).

Table 3. The EIS fitted data of Ti-6Al-4V and Ti-4Al-2V-1Mo-1Fe alloys.

Alloy	R_s ($\Omega \cdot cm^{-2}$)	R_p ($\Omega \cdot cm^{-2}$)	Q ($\Omega^{-1} \cdot S^n \cdot cm^{-2}$)	n
Ti-6Al-4V	10.89 ± 1.12	$5.69 \pm 0.13 \times 10^5$	$3.98 \pm 0.32 \times 10^{-5}$	0.93 ± 0.01
Ti-4Al-2V-1Mo-1Fe	8.71 ± 0.87	$6.50 \pm 0.35 \times 10^5$	$2.56 \pm 0.26 \times 10^{-5}$	0.95 ± 0.01

The potentiodynamic polarization curves for the Ti-6Al-4V and Ti-4Al-2V-1Mo-1Fe alloys in 3.5 wt.% NaCl solution are shown in Figure 9. It is seen from Figure 9 that the corrosion behaviors of the Ti-6Al-4V and Ti-4Al-2V-1Mo-1Fe alloys were similar. Both alloys were typically passive materials, displaying a wide passive region from 0.11 ± 0.03 V_{SCE} to 2.5 ± 0.05 V_{SCE}. After the potential was scanned to 2.5 V_{SCE} in 3.5 wt.% NaCl solution, no film breakdown was observed. This clearly indicates the passive film forming spontaneously on the Ti-6Al-4V and Ti-4Al-2V-1Mo-1Fe alloys surfaces was thermodynamically stable [37]. It is reported that Al or Fe could be oxidized and form a compact Al and Fe oxide layer on the top of passive film, inhibiting the dissolution of the oxide film [22]. The corrosion current density (i_{corr}) for the Ti-6Al-4V and Ti-4Al-2V-1Mo-1Fe alloys was 2.23 ± 0.41 × 10^{-7} A·cm^{-2} and 1.51 ± 0.22 × 10^{-7} A·cm^{-2}, respectively. Compared to the Ti-4Al-2V-1Mo-1Fe alloy, the lower E_{corr} and higher i_{corr} of the Ti-6Al-4V alloy suggested that the Ti-4Al-2V-1Mo-1Fe alloy had a higher corrosion resistance, consistent with the results of EIS test.

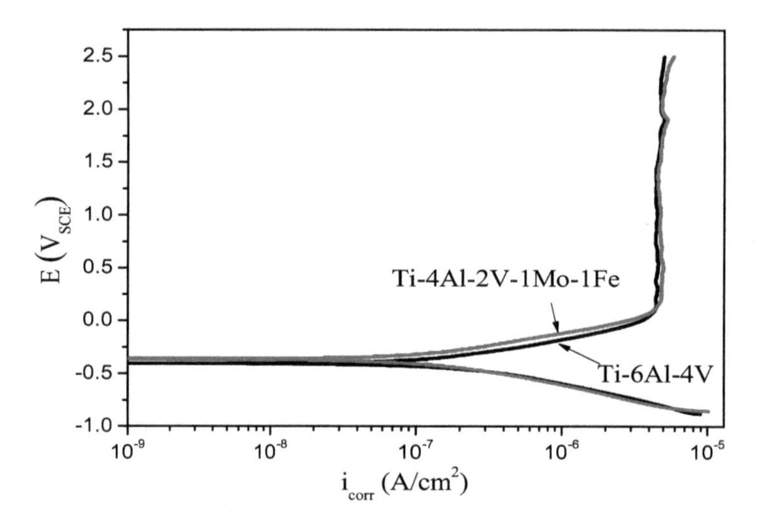

Figure 9. Potentiodynamic polarization curves of Ti-6Al-4V and Ti-4Al-2V-1Mo-1Fe alloys in 3.5 wt.% NaCl solution.

The corrosion resistance of titanium alloy relies on the stability of its passive film. The formation of passive layer requires the transfer of titanium and hydroxyl ions as follows [38]:

$$Ti \rightarrow Ti^{2+} + 2e^{-} \quad (4)$$

Since Ti^{2+} is unstable, it will react with H_2O and produce Ti^{3+} once it is formed;

$$2Ti^{2+} + 2H_2O \rightarrow 2Ti^{3+} + 2OH^{-} + H_2 \quad (5)$$

$$Ti^{2+} + 3OH^{-} \rightarrow Ti(OH)_3 \quad (6)$$

Transformation of $Ti(OH)_3$ might take place to hydrated TiO_2 layer in a dynamic equilibrium reaction as follows:

$$2Ti(OH)_3 \rightarrow TiO_2 \cdot H_2O + H_2 \quad (7)$$

In addition, the corrosion resistance of the Ti-4Al-2V-1Mo-1Fe alloy was slightly higher than the Ti-6Al-4V alloy, which was mainly due to the fact that the Ti-4Al-2V-1Mo-1Fe alloy contained Mo elements, making the passive film more stable [39].

3.4. Mechanical Properties

Figure 10 shows typical tensile curves of the Ti-6Al-4V and Ti-4Al-2V-1Mo-1Fe alloys. The specific mechanical property values including the tensile R_m, yield stress $R_{p0.2}$ strength, and elongation ε from the tensile curves, are summarized in Table 4. It shows that strength of the Ti-4Al-2V-1Mo-1Fe alloy was slightly lower but its elongation was higher compared to the Ti-6Al-4V alloy (yield strength of 838 MPa vs. 968 MPa, and the elongation of 15.8% vs. 13.8%). Since the volume fractions of the α and β phases were different in Ti-6Al-4V and Ti-4Al-2V-1Mo-1Fe alloys, these will have influences on their tensile properties. Moreover, the α phase had more slip systems and the β phases had limited slip systems according to their structure identified by XRD in Figure 1. Therefore, the stress concentration at α/β phase interfaces could be easily induced in the tensile tests due to the incompatible plastic deformation between two phases. Then, micro-cracks would preferentially initiate at the α/β phase interfaces. The fracture morphologies of the tensile test for the Ti-6Al-4V and Ti-4Al-2V-1Mo-1Fe alloys are displayed in Figure 11. Many dimples were observed in Figure 11, and the Ti-6Al-4V and Ti-4Al-2V-1Mo-1Fe alloys were typically transgranular with a dimple fracture [40].

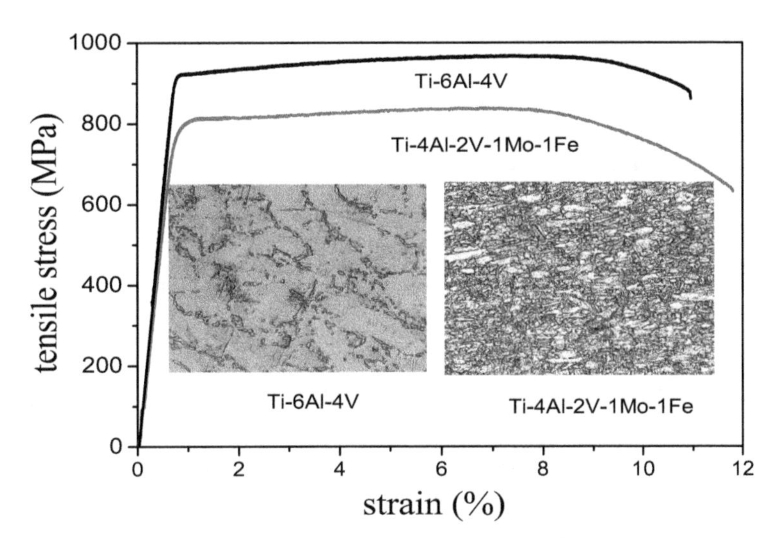

Figure 10. Tensile stress–strain curves of the Ti-6Al-4V and Ti-4Al-2V-1Mo-1Fe alloys.

Table 4. Mechanical parameters of the Ti-6Al-4V and Ti-4Al-2V-1Mo-1Fe alloys.

Alloy	R_m (MPa)	$R_{p0.2}$ (MPa)	ε (%)
Ti-6Al-4V	968 ± 22.1	921 ± 15.4	13.8 ± 1.1
Ti-4Al-2V-1Mo-1Fe	838 ± 16.3	796 ± 14.3	15.8 ± 1.3

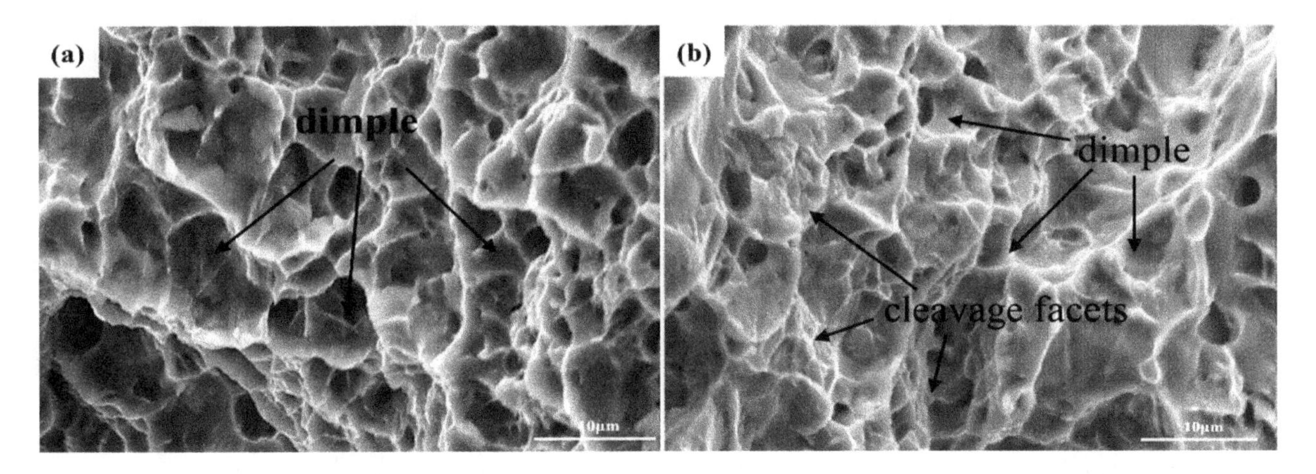

Figure 11. Fractography of the (**a**) Ti-6Al-4V and (**b**) Ti-4Al-2V-1Mo-1Fe alloys.

4. Conclusions

In this paper, the corrosion and tensile behaviors of the Ti-6Al-4V and Ti-4Al-2V-1Mo-1Fe alloys were investigated. The results were summarized as follows.

(1) Both Ti-6Al-4V and Ti-4Al-2V-1Mo-1Fe alloys were composed of the α and β phases. The volume fractions of the β phase in these two alloys were 7.4% and 47.3%.

(2) Both Ti-6Al-4V and Ti-4Al-2V-1Mo-1Fe alloys presented excellent corrosion resistance in 3.5 wt.% NaCl solution. No obvious corrosion was observed on the surface of the two alloys after immersion in 3.5 wt.% NaCl for 180 days. Compared to the Ti-6Al-4V alloy, the higher Mo content

in the Ti-4Al-2V-1Mo-1Fe alloy increased the stability of passivation film, and showed an increased corrosion resistance.

(3) Compared to the Ti-6Al-4V alloy, the Ti-4Al-2V-1Mo-1Fe alloy presented a slightly lower strength and higher ductility.

Author Contributions: Data curation, Y.W. and H.Z.; funding acquisition, Y.M.; methodology, D.X. and Y.M.; writing—original draft, Y.Q. and S.W.; writing—review and editing, J.C.

Acknowledgments: The authors acknowledge the financial support of the National Natural Science Foundation of China (No. 51871225), National Key Research and Development Program of China [2018YFC0310400], and Natural Science Foundation of the Higher Education Institutions of Jiangsu Province, China [18KJB460007].

References

1. Boyer, R.R. An overview on the use of titanium in the aerospace industry. *Mater. Sci. Eng. A* **1996**, *213*, 103–144. [CrossRef]
2. Zhang, H.; Li, J.L.; Ma, P.Y.; Xiong, J.T.; Zhang, F.S. Study on microstructure and impact toughness of TC4 titanium alloy diffusion bonding joint. *Vacuum* **2018**, *152*, 272–277. [CrossRef]
3. Nady, H.; El-Rabiei, M.M.; Samy, M. Corrosion behavior and electrochemical properties of carbon steel, commercial pure titanium, copper and copper–aluminum–nickel alloy in 3.5% sodium chloride containing sulfide ions. *Egypt. J. Pet.* **2017**, *26*, 79–94. [CrossRef]
4. Cui, Z.Y.; Wang, L.W.; Zhong, M.Y.; Ge, F.; Gao, H.; Man, C.; Liu, C.; Wang, X. Electrochemical behavior and surface characteristics of pure titanium during corrosion in simulated desulfurized flue gas Condensates. *J. Electrochem. Soc.* **2018**, *165*, C542. [CrossRef]
5. Wang, Z.B.; Hu, H.X.; Zheng, Y.G.; Ke, W.; Qiao, Y.X. Comparison of the corrosion behavior of pure titanium and its alloys in fluoride-containing sulfuric acid. *Corros. Sci.* **2016**, *103*, 50–65. [CrossRef]
6. Contu, F.; Elsener, B.; Böhni, H. Serum effect on the electrochemical behaviour of titanium, Ti6Al4V and Ti6Al7Nb alloys in sulphuric acid and sodium hydroxide. *Corros. Sci.* **2004**, *46*, 2241–2254. [CrossRef]
7. Narayanan, R.; Seshadri, S.K. Point defect model and corrosion of anodic oxide coatings on Ti-6Al-4V. *Corros. Sci.* **2008**, *50*, 1521–1529. [CrossRef]
8. Tamilselvi, S.; Raman, V.; Rajendran, N. Evaluation of corrosion behavior of surface modified Ti-6Al-4V ELI alloy in hanks solution. *J. Appl. Electrochem.* **2010**, *40*, 285–293. [CrossRef]
9. Rai, S.; Dihingia, P.J. *Optoelectronics of Cu^{2+}-Doped TiO_2 Films Prepared by Sol.–Gel Method*; Springer India: New Delhi, India, 2015; pp. 581–589.
10. Hugot-Le Goff, A. Structure of very thin TiO_2 films studied by Raman spectroscopy with interference enhancement. *Thin Solid Films* **1986**, *142*, 193–197. [CrossRef]
11. Diamanti, M.V.; Souier, T.; Stefancich, M.; Chiesa, M.; Pedeferri, M.P. Probing anodic oxidation kinetics and nanoscale heterogeneity within TiO_2 films by conductive Atomic Force Microscopy and combined techniques. *Electrochim. Acta* **2014**, *129*, 203–210. [CrossRef]
12. Yue, T.M.; Yu, J.K.; Mei, Z.; Man, H.C. Excimer laser surface treatment of Ti–6Al–4V alloy for corrosion resistance enhancement. *Mater. Lett.* **2002**, *52*, 206–212. [CrossRef]
13. Blanco-Pinzon, C.; Liu, Z.; Voisey, K.; Bonilla, F.A.; Skeldon, P.; Thompson, G.E.; Piekoszewski, J.; Chmielewski, A.G. Excimer laser surface alloying of titanium with nickel and palladium for increased corrosion resistance. *Corros. Sci.* **2005**, *47*, 1251–1269. [CrossRef]
14. Newman, R.C. The dissolution and passivation kinetics of stainless alloys containing molybdenum- I. Coulometric studies of Fe-Cr and Fe-Cr-Mo alloys. *Corros. Sci.* **1985**, *25*, 331–339. [CrossRef]
15. Newman, R.C. The dissolution and passivation kinetics of stainless alloys containing molybdenum-II. Dissolution kinetics in artificial pits. *Corros. Sci.* **1985**, *25*, 341–350. [CrossRef]
16. Marcus, P. On some fundamental factors in the effect of alloying elements on passivation of alloys. *Corros. Sci.* **1994**, *36*, 2155–2458. [CrossRef]
17. Ouchi, C.; Fukai, H.; Hasegawa, K. Microstructural characteristics and unique properties obtained by solution treating or aging in β-rich α+β titanium alloy. *Mater. Sci. Eng. A* **1999**, *263*, 132–136. [CrossRef]

18. Erween Abd, R.; Safian, S. Investigation on tool life and surface integrity when drilling Ti-6Al-4V and Ti-5Al-4V-Mo/Fe. *JSME Int. J. Ser. C* **2006**, *49*, 340–345.
19. Lin, D.J.; Lin, J.H.C.; Ju, C.P. Structure and properties of Ti-7.5Mo-xFe alloys. *Biomaterials* **2002**, *23*, 1723–1730. [CrossRef]
20. Prodana, M.; Bojin, D.; Ioniţă, D. Effect of hydroxyapatite on interface properties for alloy/biofluid. *UPB Sci. Bull. Ser. B Chem. Mater. Sci.* **2009**, *71*, 89–98.
21. Pimenova, N.V.; Starr, T.L. Electrochemical and corrosion behavior of Ti–xAl–yFe alloys prepared by direct metal deposition method. *Electrochim. Acta* **2006**, *51*, 2042–2049. [CrossRef]
22. Lu, J.W.; Zhao, Y.Q.; Niu, H.Z.; Zhang, Y.S.; Du, Y.Z.; Zhang, W.; Huo, W.T. Electrochemical corrosion behavior and elasticity properties of Ti-6Al-xFe alloys for biomedical applications. *Mater. Sci. Eng. C* **2016**, *62*, 36–44. [CrossRef] [PubMed]
23. Hsu, H.C.; Pan, C.H.; Wu, S.C.; Ho, W.F. Structure and grindability of cast Ti-5Cr-xFe alloys. *J. Alloy. Compd.* **2009**, *474*, 578–583. [CrossRef]
24. Hsu, H.C.; Hsu, S.K.; Wu, S.C.; Lee, C.J.; Ho, W.F. Structure and mechanical properties of as-cast Ti-5Nb-xFe alloys. *Mater. Charact.* **2010**, *61*, 851–858. [CrossRef]
25. Qiao, Y.X.; Cai, X.; Cui, J.; Li, H.B. Passivity and semiconducting behavior of a high nitrogen stainless steel in acidic NaCl solution. *Adv. Mater. Sci. Eng.* **2016**, *6065481*, 1–8. [CrossRef]
26. Qiao, Y.X.; Chen, J.; Zhou, H.L.; Wang, Y.X.; Song, Q.N.; Li, H.B.; Zheng, Z.B. Effect of solution treatment on cavitation erosion behavior of high-nitrogen austenitic stainless steel. *Wear* **2019**, *424–425*, 70–77. [CrossRef]
27. Huang, S.S.; Ma, Y.J.; Ping, Z.Y.; Zhang, S.L.; Yang, R. Influence of alloying elements partitioning behaviors on the microstructure and mechanical properties in $\alpha+\beta$ titanium alloy. *Acta Metall. Sin.* **2019**, *55*, 741–750.
28. Huang, S.S.; Zhang, J.H.; Ma, Y.J.; Zhang, S.L.; Youssef, S.S.; Qi, M.; Wang, H.; Qiu, J.K.; Xu, D.S.; Lei, J.F.; et al. Influence of thermal treatment on element partitioning in $\alpha+\beta$ titanium alloy. *J. Alloy. Compd.* **2019**, *791*, 575–585. [CrossRef]
29. Qiao, Y.X.; Tian, Z.H.; Cai, X.; Chen, J.; Wang, Y.X.; Song, Q.N.; Li, H.B. Cavitation erosion behaviors of a nickel-free high-nitrogen stainless steel. *Tribol. Lett.* **2019**, *67*, 1–9. [CrossRef]
30. Carnot, A.; Frateur, I.; Zanna, S.; Tribollet, B.; Dubois-Brugger, I.; Marcus, P. Corrosion mechanisms of steel concrete moulds in contact with a demoulding agent studied by EIS and XPS. *Corros. Sci.* **2003**, *45*, 2513–2524. [CrossRef]
31. Hitz, C.; Lasia, A. Experimental study and modeling of impedance of the her on porous Ni electrodes. *J. Electroanal. Chem.* **2001**, *500*, 213–222. [CrossRef]
32. Jeyaprabha, C.; Sathiyanarayanan, S.; Venkatachari, G. Influence of halide ions on the adsorption of diphenylamine on iron in 0.5 M H_2SO_4 solutions. *Electrochim. Acta* **2006**, *51*, 4080–4088. [CrossRef]
33. Qiao, Y.X.; Zheng, Y.G.; Ke, W.; Okafor, P.C. Electrochemical behaviour of high nitrogen stainless steel in acidic solutions. *Corros. Sci.* **2009**, *51*, 979–986. [CrossRef]
34. Grubač, Z.; Metikoš-Huković, M. EIS study of solid-state transformations in the passivation process of bismuth in sulfide solution. *J. Electroanal. Chem.* **2004**, *565*, 85–94. [CrossRef]
35. Hirschorn, B.; Orazem, M.E.; Tribollet, B.; Vivier, V.; Frateur, I.; Musiani, M. Determination of effective capacitance and film thickness from constant-phase-element parameters. *Electrochim. Acta* **2010**, *55*, 6218–6227. [CrossRef]
36. Wang, Z.B.; Hu, H.X.; Liu, C.B.; Zheng, Y.G. The effect of fluoride ions on the corrosion behavior of pure titanium in 0.05M sulfuric acid. *Electrochim. Acta* **2014**, *135*, 526–535. [CrossRef]
37. Cao, C.N. *Theory of Electrochemical*; Chemical Industry Press: Beijing, China, 2004; pp. 24–35.
38. Ibrahim, M.A.; Pongkao, D.; Yoshimura, M. The electrochemical behavior and characterization of the anodic oxide film formed on titanium in NaOH solutions. *J. Solid State Electrochem.* **2002**, *6*, 341–350. [CrossRef]
39. Wang, B.J.; Xu, D.K.; Wang, S.D.; Han, E.H. Recent progress in the research about fatigue crack initiation of Mg alloys under elastic stress amplitudes: A review. *Front. Mech. Eng.* **2019**, *14*, 113–127. [CrossRef]
40. Wang, B.J.; Xu, D.K.; Wang, S.D.; Sheng, L.Y.; Zeng, R.C.; Han, E.H. Influence of solution treatment on the corrosion fatigue behavior of an as forged Mg-Zn-Y.-Zr alloy. *Int. J. Fatigue* **2019**, *120*, 46–55. [CrossRef]

5

Microstructural Evolution during Pressureless Sintering of Blended Elemental Ti-Al-V-Fe Titanium Alloys from Fine Hydrogenated-Dehydrogenated Titanium Powder

Changzhou Yu, Peng Cao * and Mark Ian Jones *

Department of Chemical and Materials Engineering, The University of Auckland, Private Bag 92019, Auckland 1142, New Zealand; cyu060@aucklanduni.ac.nz

* Correspondences: p.cao@auckland.ac.nz (P.C.); mark.jones@auckland.ac.nz (M.I.J.)

Abstract: A comprehensive study was conducted on microstructural evolution of sintered Ti-Al-V-Fe titanium alloys utilizing very fine hydrogenation-dehydrogenation (HDH) titanium powder with a median particle size of 8.84 μm. Both micropores (5–15 μm) and macropores (50–200 μm) were identified in sintered titanium alloys. Spherical micropores were observed in Ti-6Al-4V sintered with fine Ti at the lowest temperature of 1150 °C. The addition of iron can help reduce microporosity and improve microstructural and compositional homogenization. A theoretical calculation of evaporation based on the Miedema model and Langmuir equation indicates that the evaporation of aluminum could be responsible for the formation of the macropores. Although reasonable densification was achieved at low sintering temperatures (93–96% relative density) the samples had poor mechanical properties due mainly to the presence of the macroporosity and the high inherent oxygen content in the as-received fine powders.

Keywords: titanium alloys; sintering; powder metallurgy; microstructural evolution

---

1. Introduction

Sintering is by far the most common consolidation method in titanium powder metallurgy. The initial stage of sintering can be empirically modeled in terms of isothermal neck growth as measured by the neck size ratio X/D [1]:

$$(X/D)^n = Bt/D^m \tag{1}$$

where D is the particle diameter, X = neck diameter, t = isothermal sintering time, and B is a collection of material and geometric constants. The values of n, m, B depend on the mechanism of mass transport. The above empirical equation indicates that sintering is highly sensitive to the particle size, with a smaller particle size giving rise to more rapid densification.

The sintering data compiled by Robertson et al. confirms that a finer particle size is beneficial for titanium powder densification [2]. However titanium powders with very fine particle size are not usually available, particularly if a low impurity level is required. A particle size of −100 mesh (<150 μm), −200 mesh (<75 μm), or −325 mesh (<45 μm), is most commonly used [3–5]. A recent study reports a novel technique to produce titanium powder with a mean particle size of <10 μm [6]. However, the powder reported has a high impurity content of oxygen (>0.8 wt %) [6]. It is known that high oxygen content can adversely affect mechanical properties, especially ductility [4,5,7], and therefore the aim of this research is primarily to investigate the feasibility of using a very fine titanium powder to sinter titanium products and investigate the effects of these fine powders on densification. Ti-6Al-4V is known as a "work-horse" $\alpha + \beta$ titanium alloy, which has high strength and good ductility. However, the β-stabilizing element vanadium is not only expensive but also toxic to human beings if used as an implant [8]. For blended elemental sintering of Ti-6Al-4V, better sinterability can be obtained using a master alloy powder (Al-V) than when using elemental aluminum and vanadium [9]. Our first objective of this work was to investigate the sinterability of Ti-6Al-4V by using fine titanium powder and Al-V master alloy powder (60 wt % Al, 40 wt % V).

Current titanium powder metallurgy research has been largely devoted to cost reduction in titanium components by both developing cost-effective powder manufacturing and developing low-cost titanium alloys [10–12]. In the development of low-cost titanium alloys, the introduction of iron (Fe) as a β-Ti stabilizer into the alloy compositions has been widely explored for powder metallurgical titanium alloys [13,14].

Iron is much cheaper than titanium, and the application of 5% Fe is equivalent to saving the use of 5% Ti [15,16]. The sinterability of titanium alloys can be enhanced by the addition of iron since the mobility of titanium atoms is accelerated by the rapid diffusion of iron [13,17,18]. The fast diffusion of iron in titanium alloy is evidenced by the fact that iron is essentially uniform in a blended elemental Ti-10V-2Fe-3Al alloy when heated at 5 °C/min to 1200 °C [19]. Ti-10V-2Fe-3Al ($O \leq 0.13$; $N \leq 0.05$; $C \leq 0.05$; $H \leq 0.015$; all in wt %), assigned formally as TIMETAL® 10-2-3, has been realized for aircraft under-carriage applications [14,19]. Further experiments indicate that the enhancement in sinterability observed through the addition of fine iron powder (mean particle size = 8 μm) is not seen when working with coarser iron powder (mean particle size = 97 μm) [20].

Iron is a strong β phase stabilizer and suppresses the formation of the α phase and the eutectoid transformation $\beta \rightarrow \alpha$ + TiFe when it cools below 595 °C, thus avoiding the formation of the brittle TiFe phase [21]. Such a phenomenon is also observed and validated by the recent research works conducted by Bolzoni et al. [11,12]. In addition to the fast diffusion mobility, iron additions also change the sequence of chemical homogenization for other alloying elements such as Al, V [14]. In another study, Yang et al. found a linear drop in ductility in Ti-xFe-0.5Si alloys with increasing Fe content from 3% to 6%. Such ductility drop is largely because the α-Ti phase becomes thinner and more acicular with increasing Fe [22].

The second objective of this work was to investigate the sintering behavior of low-cost Ti-Al-V-Fe titanium alloys with fine particle titanium powders through varying the amount of expensive master 60Al-40V powder and low-cost iron powder. The effect of the master alloy powder during sintering on sinterability is also discussed.

2. Experimental Procedures

2.1. Materials

The starting material for these experiments was a fine Ti powder produced from Ti sponge fines which were hydrogenated at 700 °C with a holding time of 2 h followed by 5 h ball milling. In this way, fine titanium hydride was produced with a median particle size <3 μm. An inhibitor (NaCl) was introduced for the coating of titanium hydride powder, which was further dehydrogenated at 630 °C for 2 h. The fine hydrogenation-dehydrogenation (HDH) titanium powders were water-leached by deionized water in order to remove the inhibitor. The median particle size was <10 μm. More detailed description of the fine Ti production process can be seen in the literature [6].

The fine titanium powder (O: 0.82 wt %, C: 0.058 wt %, H: 0.414 wt %, N: 0.575 wt %) was utilized to investigate the interaction effect of fine particle size and high impurity content on sintering and densification, with the understanding that the high impurity content of this powder may not result in outstanding mechanical properties. A commercially available 60Al-40V master alloy powder (−120 mesh; O: 0.18 wt %, C: 0.029 wt %, H: 0.0005 wt %, N: 0.16 wt %) and iron powder (−300 mesh; O: 0.67 wt %, C: 0.041 wt %, H: 0.0022 wt %, N: 0.262 wt %) supplied by Beijing Youxinglian Nonferrous

Metals Co. Ltd., Beijing, China, was employed to balance the designed alloy composition into Ti $(0.6Al\text{-}0.4V)_{10-x}Fe_x$, where $x = 0, 2, 4$, and 6. i.e., the Fe content ranged from 0 to 6 wt % as a replacement for the master alloy in a Ti-6Al-4V composition. The effect of beta stabilizing elements is generally quantified by an equivalent molybdenum (Mo), as given by:

$$
\begin{aligned}
(\mathrm{Mo})_{\mathrm{eq}} = \; & (\mathrm{Mo}) + 0.67\,(\mathrm{V}) + 0.44\,(\mathrm{W}) + 0.28\,(\mathrm{Nb}) + 0.22\,(\mathrm{Ta}) + \\
& 2.9\,(\mathrm{Fe}) + 1.6\,(\mathrm{Cr}) + 1.25\,(\mathrm{Ni}) + 1.7\,(\mathrm{Mn}) + 1.7\,(\mathrm{Co}) - 1.0\,(\mathrm{Al})
\end{aligned} \tag{2}
$$

The calculated equivalent Mo values for the Ti-4.8Al-3.2V-2Fe, Ti-3.6Al-2.4V-4Fe, and Ti-2.4Al-1.6V-6Fe are 3.1, 9.6, and 16 respectively. As such the Ti-4.8Al-3.2V-2Fe and Ti-3.6Al-2.4V-4Fe can be regarded α/β alloys while the Ti-2.4Al-1.6V-6Fe is a metastable β alloy.

2.2. Press-and-Sinter

Compacts were uniaxially pressed at a constant pressure of 300 MPa into two different cylindrical sizes: 16 mm diameter with 5–6 mm thickness, and 45 mm diameter with 3–4 mm thickness. No lubricants were added in the powder mixture, or applied on the die walls. The green density for all four alloys was 62.9 $\pm$ 0.5%. Compacts with dia = 16 mm were used for sintering densification calculations and the dia = 45 mm samples were sintered for tensile testing. Vacuum sintering (vacuum level: 2×10^{-3} Pa) was conducted in a high-temperature Mo-heating-element furnace (Dingli, Changsha, China) at 1150, 1250, and 1350 °C with a soaking time of 3 h. The racking material was molybdenum plate. The specific heating profile is demonstrated in Figure 1.

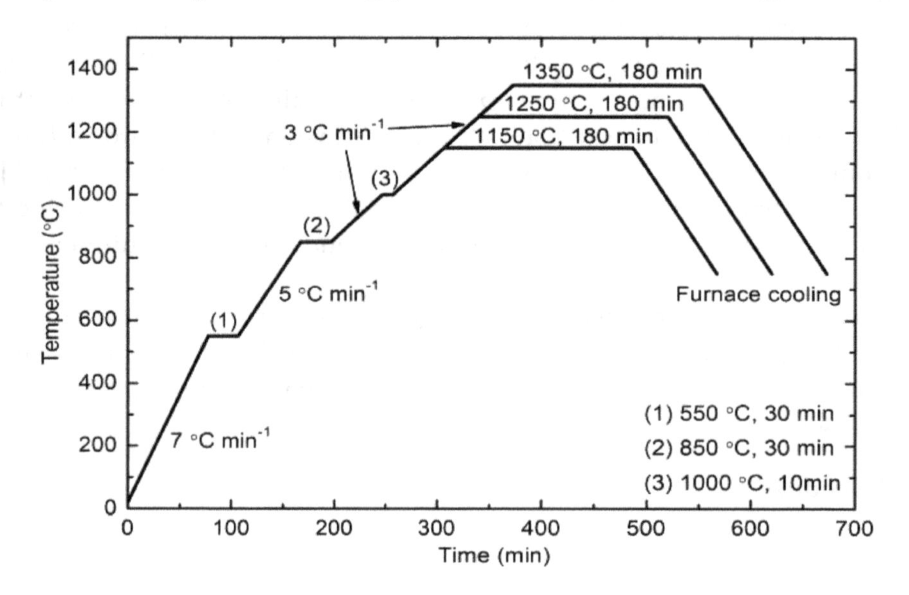

Figure 1. Heating schedule for blended elemental sintered titanium alloys.

The initial heating rate was set at 7 °C·min^{-1} from room temperature to 550 °C. A holding time of 30 min was introduced at both 550 and 850 °C and a 10-min holding time at 1000 °C were used to prevent the possible pore formation caused by the exothermal reaction between titanium and the master alloy or iron powder, and to allow sufficient time for homogeneous elemental diffusion [23]. A slower heating rate of 3–5 °C·min^{-1} was employed from 1000 °C to the final sintering temperature for the same reason.

2.3. Characterization and Mechanical Testing

The particle size distribution was analyzed by a particle size analyzer (Mastersizer 2000, Malvern Instruments, Malvern, UK). The sintered density was measured according to the Archimedes method described in ASTM B962-08. Each data point was averaged from 3–5 specimens. Fine polishing was conducted with OP-S, a colloidal silica suspension with 10% hydrogen peroxide (H_2O_2), where the size of the colloidal silica is ~40 nm. Microstructural morphology observations were conducted on an environmental scanning electron microscope (ESEM) (Quanta 200F, FEI, Hillsboro, OR, USA) using both back-scattered and secondary electron modes. Semi-quantitative compositional analysis was carried out by energy dispersive spectrometry (EDS, Pegasus detector, EDAX Inc., Mahwah, NJ, USA). Both elemental X-ray mapping and X-ray line scans were carried out using standardless quantification techniques. An accelerating voltage of 20 kV was employed. X-ray diffraction (XRD) analysis to determine phase constituents was carried out on a D2 PHASER (Bruker, Karlsruhe, Germany) equipped with Cu X-ray source. X-ray patterns were collected over a scan range from 2θ of 20–80° with a step size of 0.02°.

Tensile specimens were machined from the sintered specimens in compliance with ISO 2892-1:2009 (~2.65 mm × 2.65 mm cross-section, 15 mm gauge length and 20 mm parallel length), and tested on an Instron mechanical tester (Model: 3367, Instron Co., Norwood, MA, USA) with a crosshead speed of 0.1 mm·min^{-1} (initial strain rate: 1.1×10^{-4} s^{-1}). The precise displacement was determined using an extensometer with a gauge length of 8 mm (Model: 2630-120, Instron Co., Norwood, MA, USA). Three tensile specimens were prepared for each data point.

3. Results

3.1. Characteristics of As-Received Powders

The micrographs of as-received powders are shown in Figure 2 including fine titanium, iron and 60Al-40V master alloy powder as well as the particle size distribution of fine titanium and 60Al-40V powders.

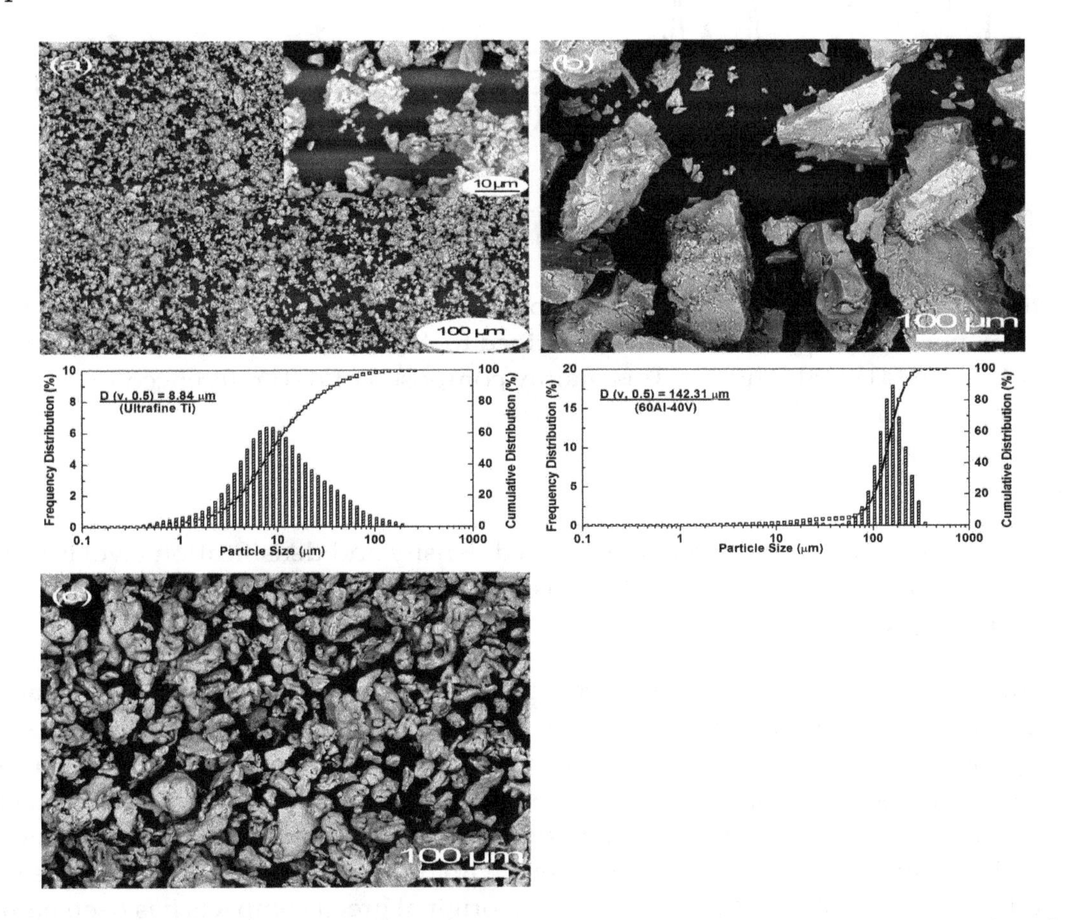

Figure 2. Scanning electron microscope (SEM) images and particle size distribution of (**a**) fine Ti, (**b**) 60Al-40V master alloy and (**c**) Fe powders.

All three powders show angular morphologies, with Figure 2a demonstrating that the fine Ti had a much finer particle size than the 60Al-40V (Figure 2b) and iron (Figure 2c) powders. This was validated by particle size distribution analysis, which illustrates the median particle sizes of fine Ti and master powder were 8.84 and 142.31 μm, respectively. The diameter of most iron particles was less than 50 μm estimated from Figure 2c, which is in accordance with its nominal particle size (−300 mesh). XRD patterns of each powder are given in Figure 3.

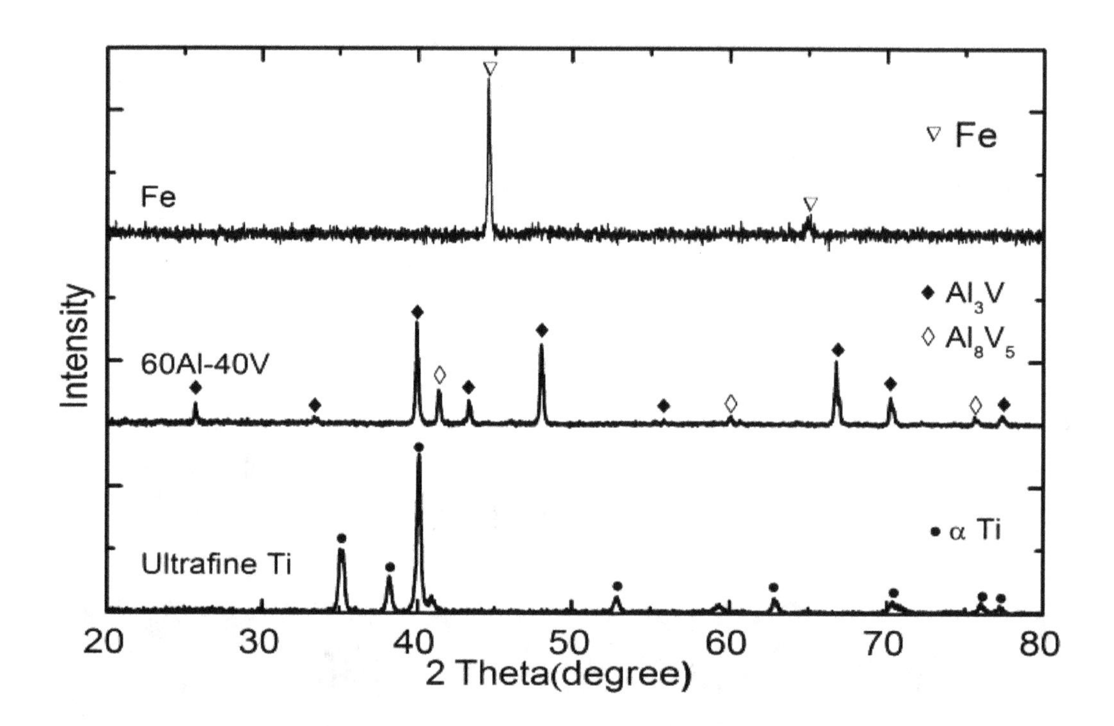

Figure 3. X-ray diffraction (XRD) patterns of as-received powders.

The pattern from the reduced elemental iron powder is indexed as pure iron by JCPDS 87-0721 [24]. The master alloy powder (60Al-40V) is mainly ascribed to Al_3V (JCPDS 07-0399) with a small amount of Al_8V_5 (JCPDS 71-0141) [25]. The fine Ti is mainly composed of α-Ti referenced by JCPDS 44-1294 and some minor titanium oxides.

3.2. Densification

The effect of iron content on the relative sintered density and densification level is illustrated in Figure 4 for specimens uniaxially pressed at a constant compaction pressure of 300 MPa and sintered at different temperatures.

The relative density increases with iron content at sintering temperatures of 1150 and 1250 °C (Figure 4a). However at a sintering temperature of 1350 °C the iron content seems to have little effect on the relative sintered density. As expected, a higher sintering temperature causes an increased sintered density and this effect was more pronounced for the Ti-6Al-4V than for the specimens with Fe. The densification level offers an evaluation method for porosity elimination level of green compacts by sintering. Figure 4b demonstrates that specimens with 6% Fe sintered at 1350 °C have the highest densification level, where ~90% of the porosity of the original green compacts has been eliminated by sintering. The lowest densification level is observed in the Ti-6Al-4V specimens sintered at 1150 °C, where ~20% of the original porosity of the green compacts remains after sintering.

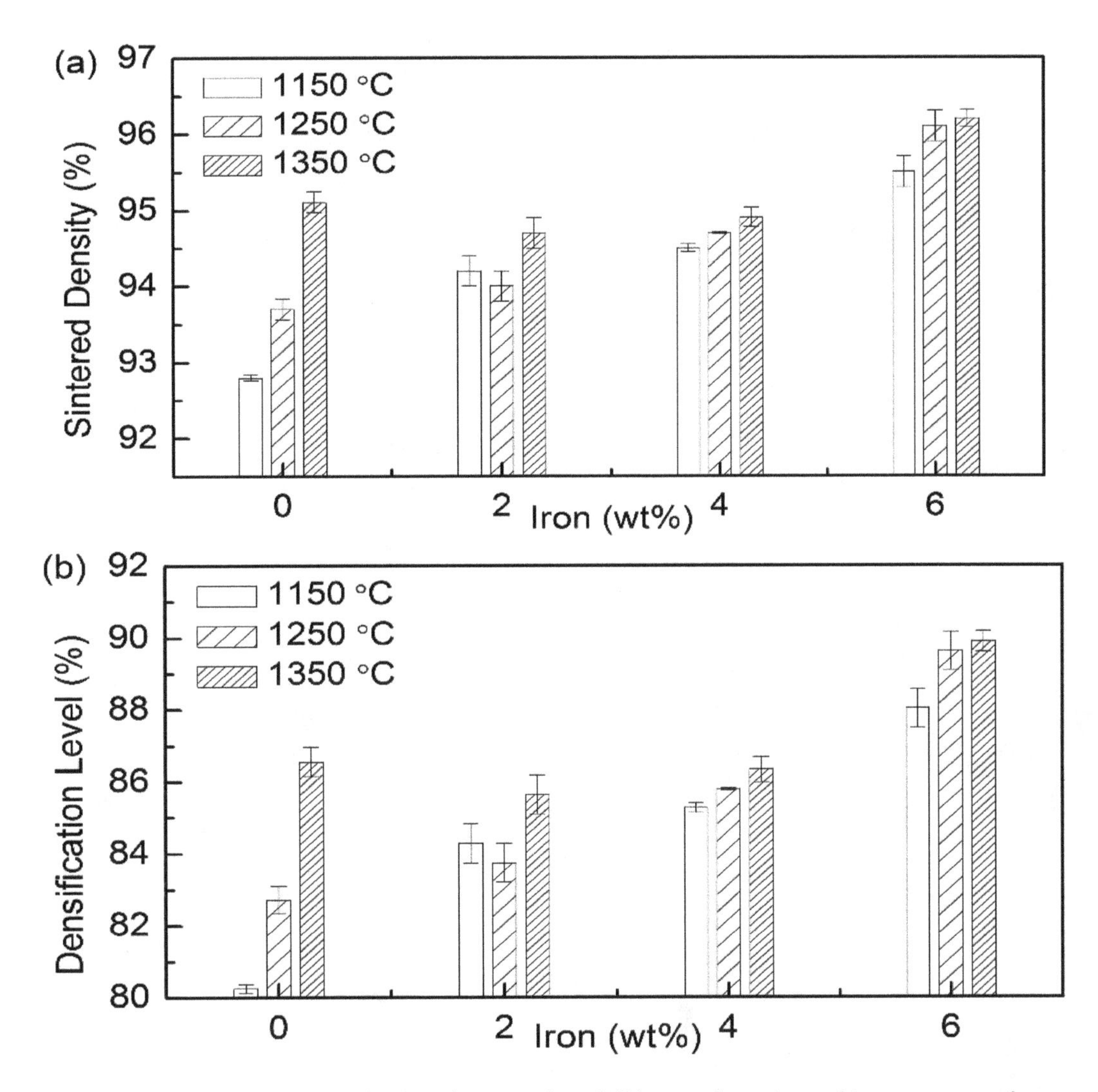

Figure 4. Relative density (**a**) and densification level (**b**) as a function of iron content for samples sintered at different temperatures.

3.3. Microstructure Observation and Compositional Analysis

Microstructural images taken from cross-sections of the specimens sintered at various temperatures and compositions are shown in Figure 5, and confirm the density measurements.

When sintered at temperatures of 1150 and 1250 °C, the porosity decreased with increasing Fe content. At the highest temperature (1350 °C) the relative sintered density of the sample with 2 wt % Fe has the lowest density and the microstructure shows the highest porosity seen in Figure 5(b3). In comparison, the least porosity is observed in the cross-section of the specimen with 6% Fe shown in Figure 5(d3) which has the highest relative sintered density.

Two different types of pores are also observed in Figure 5 in regard to the pore size and shape. Irregular-shaped macropores with an average size of 50–200 μm are noted in all specimens under different sintering temperatures and compositions. The amount of these macropores decreased with the addition of iron content when sintered at either 1150 or 1250 °C. At 1350 °C, a large quantity of irregular-shaped macropores is observed in specimens with 2 wt % Fe. The other type of pores observed is spherical with an average diameter of 5–15 μm. A large number of these micropores are distributed throughout the cross-section of the 1150 °C sintered Ti-6Al-4V, and decreases with increasing sintering temperature. Interestingly, there are only very few micropores observed on the cross-sections of sintered alloys containing iron although their diameter is slightly larger than those in Ti-6Al-4V.

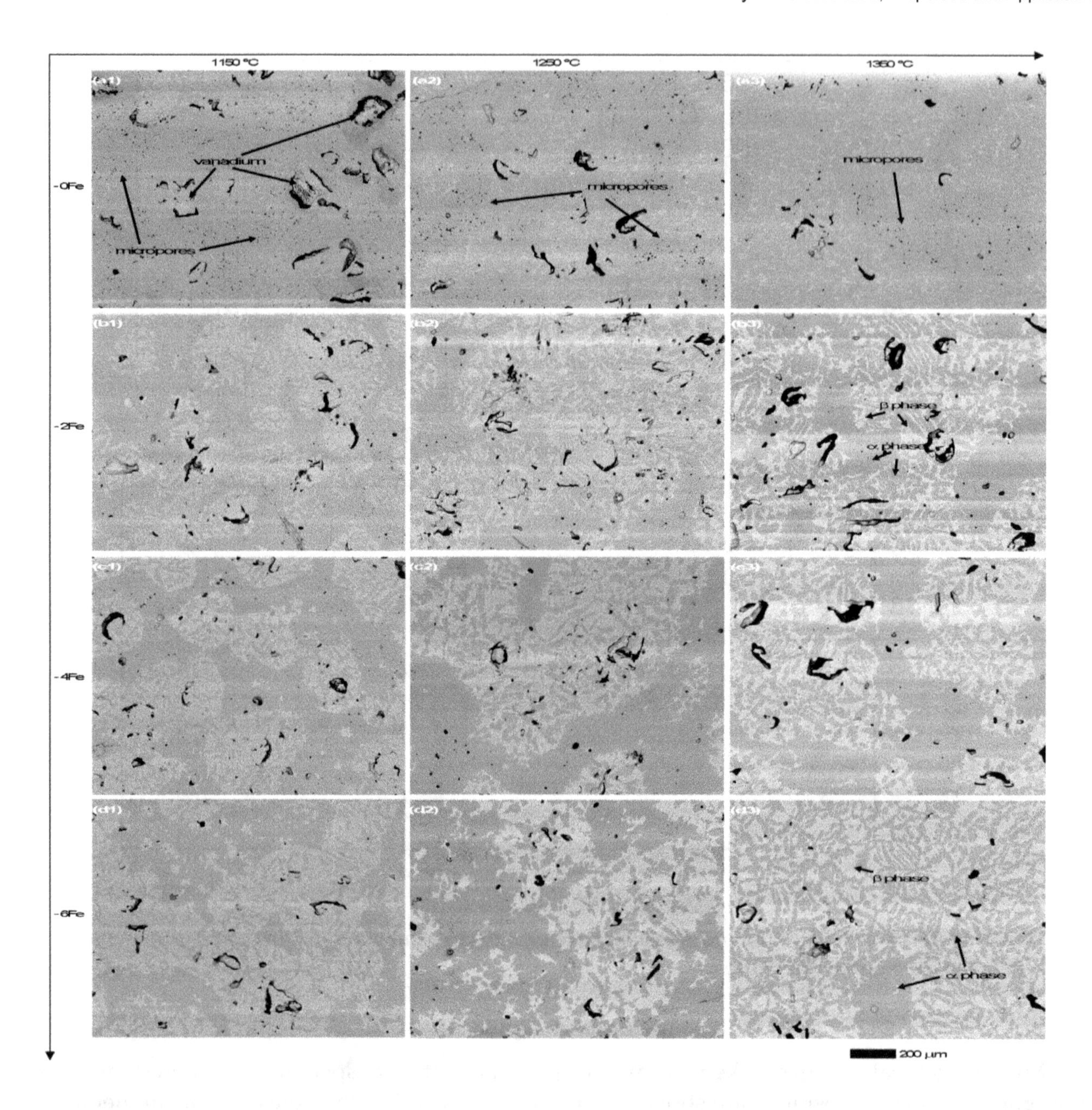

Figure 5. Cross-sectional SEM images of sintered titanium alloys at various temperatures and compositions. The horizontal axis represents the sintering temperature, while the vertical axis represents the iron content.

As also shown in Figure 5, the microstructures show two distinct regions or phases which appear as light and dark contrast in the SEM micrographs, with an increase in the light phase observed with increasing Fe content. It is also observed that the darker contrast regions appear to be in two different morphologies, with smaller needle-like or acicular grains and some much larger particles or grains. In order to further understand these microstructures, compositional analysis was conducted on Ti-6Al-4V sintered at 1150 and 1250 °C including EDS line scanning (Figure 6) and EDS mapping (Figure 7).

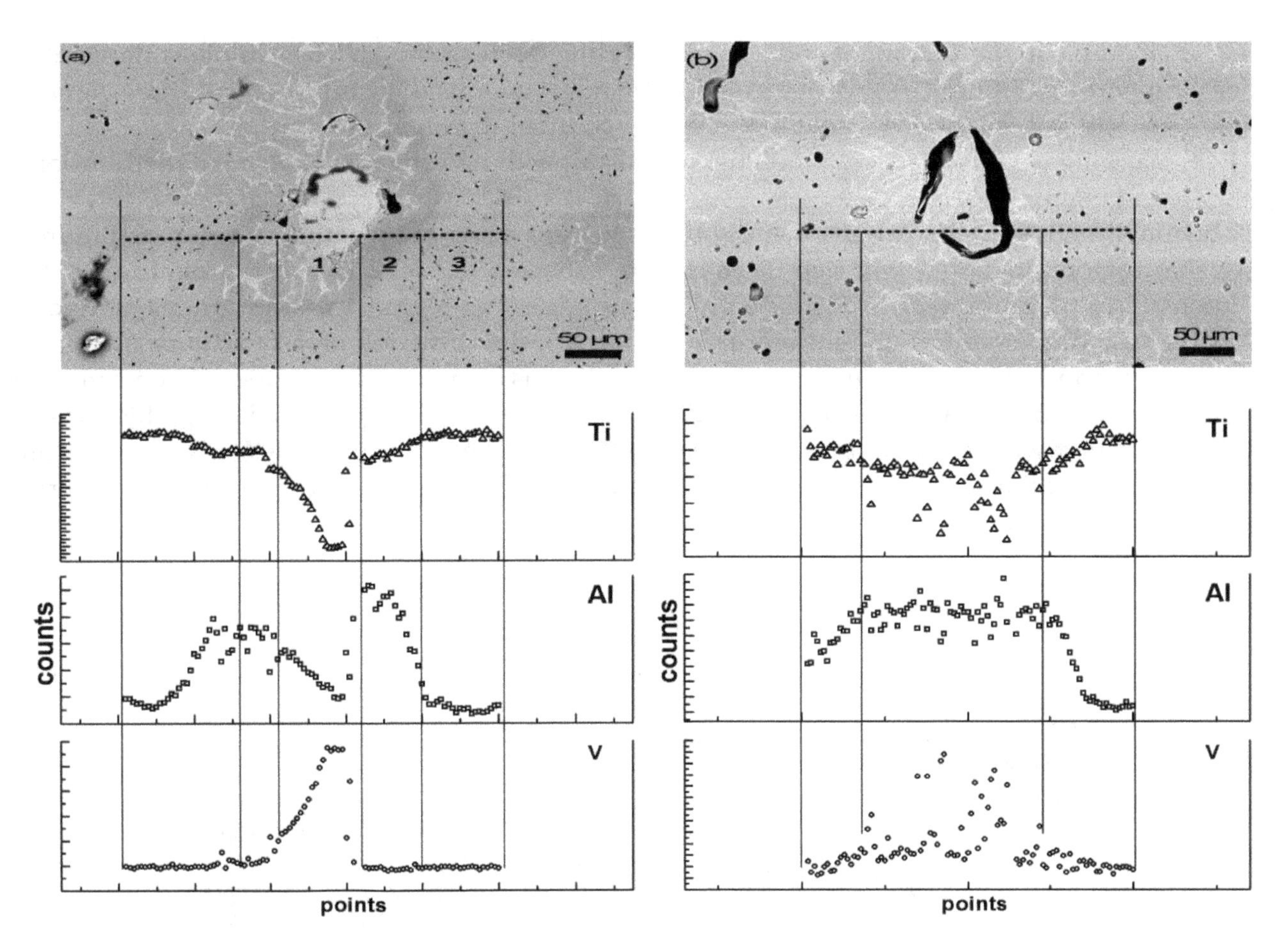

Figure 6. Energy dispersive spectrometry (EDS) line scanning of blended elemental Ti-6Al-4V specimens sintered at (**a**) 1150 °C and (**b**) 1250 °C.

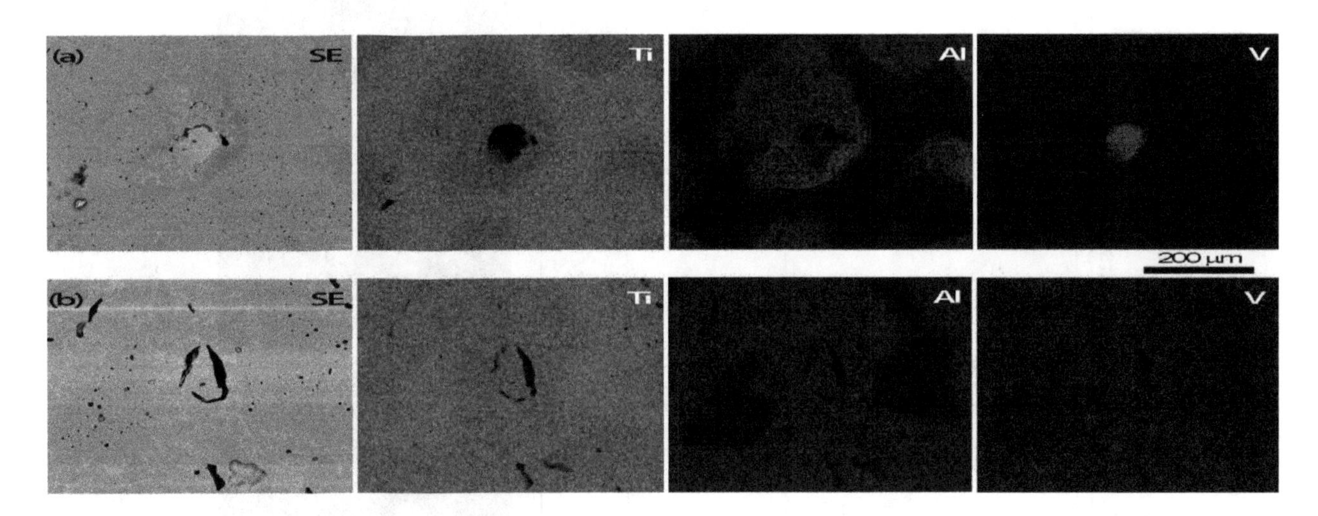

Figure 7. EDS mapping of blended elemental Ti-6Al-4V specimens sintered at (**a**) 1150 °C and (**b**) 1250 °C.

The image in Figure 6 shows a line scan across the area containing the macropores typically seen in the cross-section of Ti-6Al-4V (Figure 5).

For the sample sintered at 1150 °C, the line scan (Figure 6a) is classified into three different regions depending on the color contrast. The light contrast regions (particularly the particle at the center of the pore) are rich in vanadium and contain no Al or Ti. The darkest contrast region around the pore contains both Al and Ti but no V. Moving away from the pore into the mid-contrast, matrix region there is an increase in the Ti content and a corresponding decrease in Al. The sample sintered at the higher temperature (Figure 6b) shows a much more homogenous distribution of all three elements although vanadium is slightly elevated in the lighter contrast regions.

EDS mapping was also carried out on the same sample to investigate the element distribution as highlighted in Figure 7.

The mapping results are in agreement with the line scanning observation in Figure 6. Figure 7a shows that region 1 is noticeably rich in vanadium whereas region 2 is rich in aluminum. Also, the intensity of titanium decreases gradually from region 3 to region 1 for specimens sintered at 1150 °C. At the higher sintering temperature (Figure 7b), the diffusion area of aluminum became distinctly wider and titanium achieved a more homogeneous elemental distribution when compared with sintering at 1150 °C. A remarkable difference was also witnessed for vanadium, which was no longer isolated in discrete particles but was much more homogeneously distributed. EDS line scanning and mapping of specimens with 4 wt % iron are highlighted in Figure 8 when sintering at the highest temperature (1350 °C).

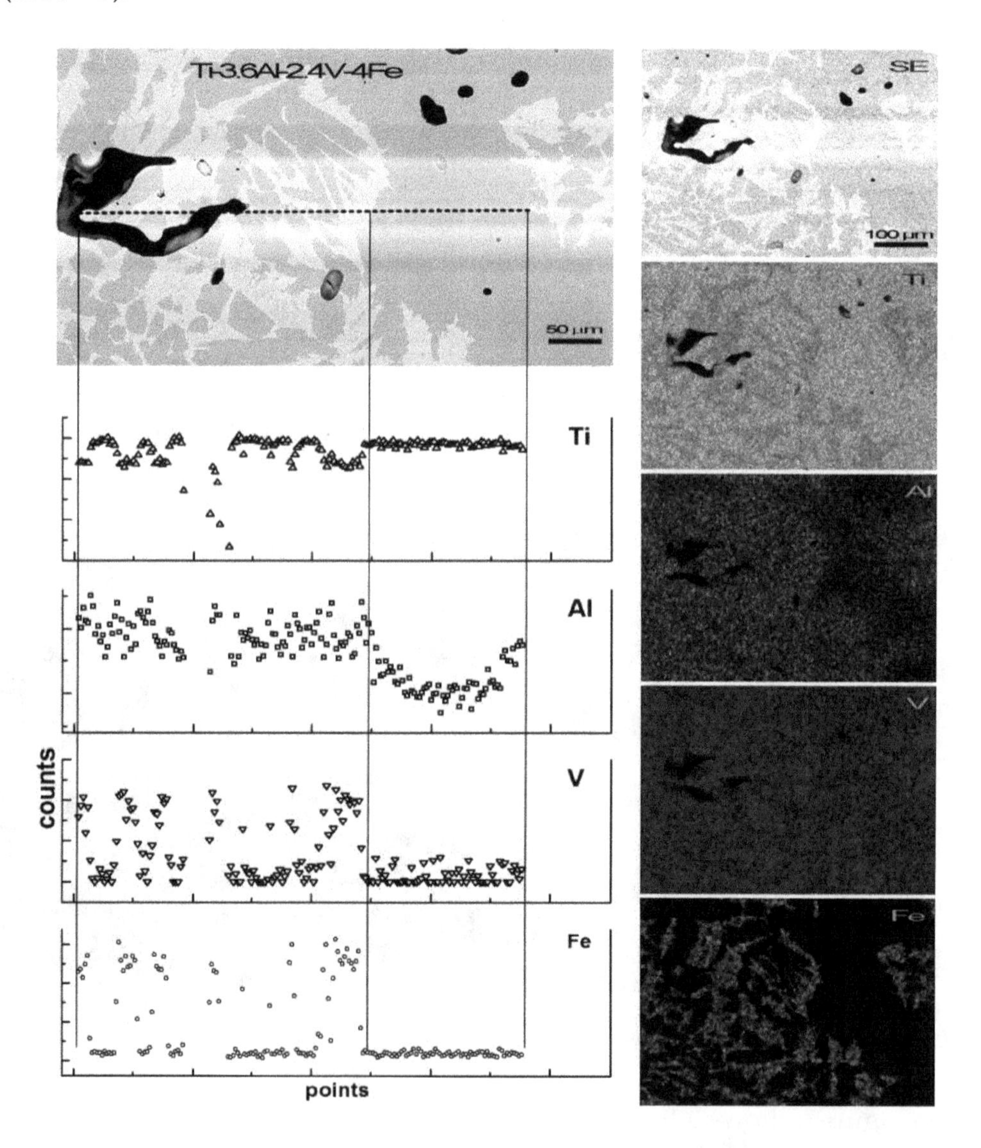

Figure 8. EDS mapping (right column) and EDS line scanning (left column) of Ti-3.6Al-2.4V-4Fe sintered at 1350 °C.

In Figure 8, inhomogeneous microstructural zones can also be distinguished by dark and light contrast regions. It is observed from the corresponding EDS line scanning and mapping that iron is rich in grains within the light region and is absent in the darker regions. The intensity of aluminum in the blocky darker regions is slightly weaker than those in the lighter contrast areas while the Ti concentration shows the opposite trend, being lower in the lighter regions than in the darker grains. Comparing sintering at the highest temperature (Figure 8) to lower temperatures shown in Figures 6 and 7, improved elemental distribution homogeneity is observed. For example, the contrast of aluminum shown by EDS mapping in Figure 8 is not as strong as shown in Figure 7.

3.4. Phase Determination

The phase analysis of sintered titanium alloys is demonstrated in Figure 9, including the XRD patterns of Ti-6Al-4V sintered under different sintering temperatures (Figure 9a) and the Ti-Al-V-Fe alloys sintered under a constant temperature of 1350 °C (Figure 9b).

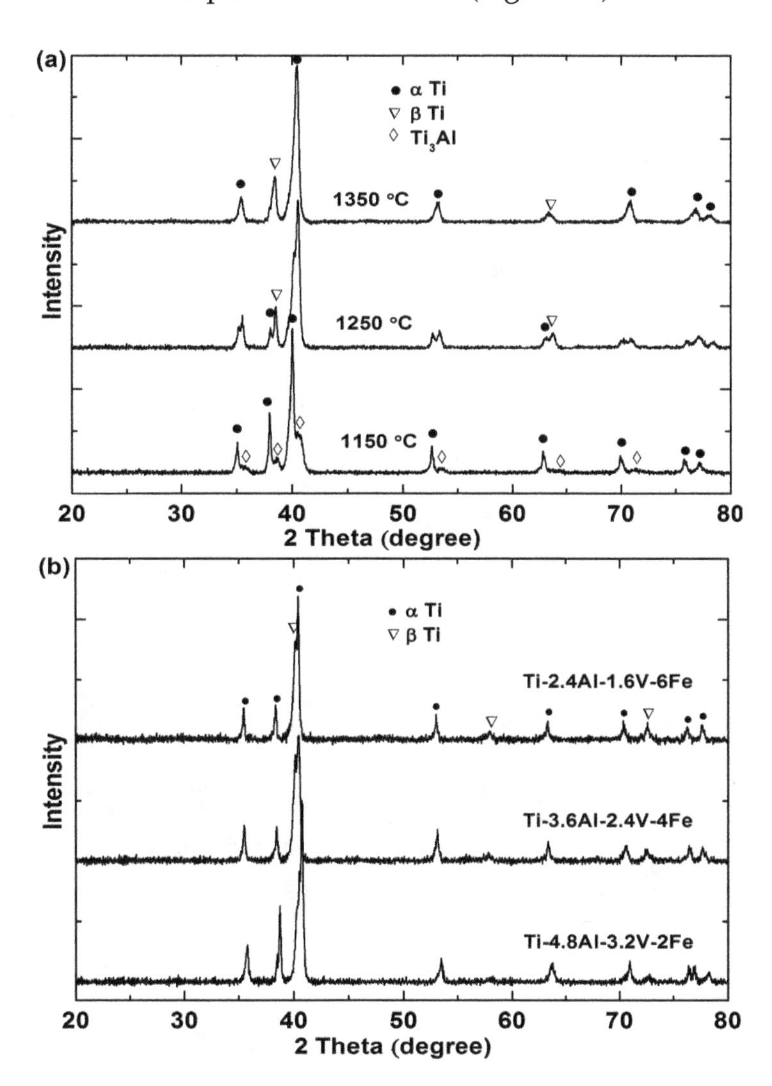

Figure 9. XRD patterns of (**a**) Ti-6Al-4V sintered at various temperatures and (**b**) Ti-Al-V-Fe of various Fe concentrations sintered at 1350 °C.

The XRD patterns of specimens sintered at 1150 °C (Figure 9a), show that the samples consisted of α-, β-Ti with some Ti_3Al peaks [26]. The presence of the Ti_3Al phase when sintered at lower temperatures is indicative of the alloying process being incomplete at these temperatures and supports the microstructural observations. In Figure 9b, peaks attributed to the β-Ti phase are observed for the sintered Ti-Al-V-Fe alloys, as suggested by the literature [19,21], and their intensity increases with increasing iron content in the alloy composition, This indicates that iron stabilizes the β phase in titanium alloys and more β can be retained after cooling from the sintering temperature if the iron content is increased.

4. Discussion

4.1. Microporosity Formation

Two different types of pores were formed during sintering, which are defined here as micropores (5–15 μm) and macropores (50–200 μm) as shown in Figure 5. The micropores observed in Figure 5 were spherical and isolated in sintered specimens especially in Ti-6Al-4V even at the lowest sintering temperature (1150 °C). Pore spheroidization indicates that sintering had entered into the final stage at this temperature (1150 °C). Although the green density was only about 63% for all of these samples, the sintered density at this lowest temperature was around 93% for the Ti-6Al-4V specimen and increased with increasing iron content up to around 96%. It is considered that the reasonably high levels of densification even at low temperatures and for samples with low green density, as evidenced by the densification results and the presence of isolated small pores, is due mainly to the high sintering driving force resulting from the high surface free energy when using the fine titanium powder, and improved diffusion with the additions of iron.

The evolution of the microporosity for samples with different iron contents as a function of temperature can be seen in the SEM fractographs shown in Figure 10.

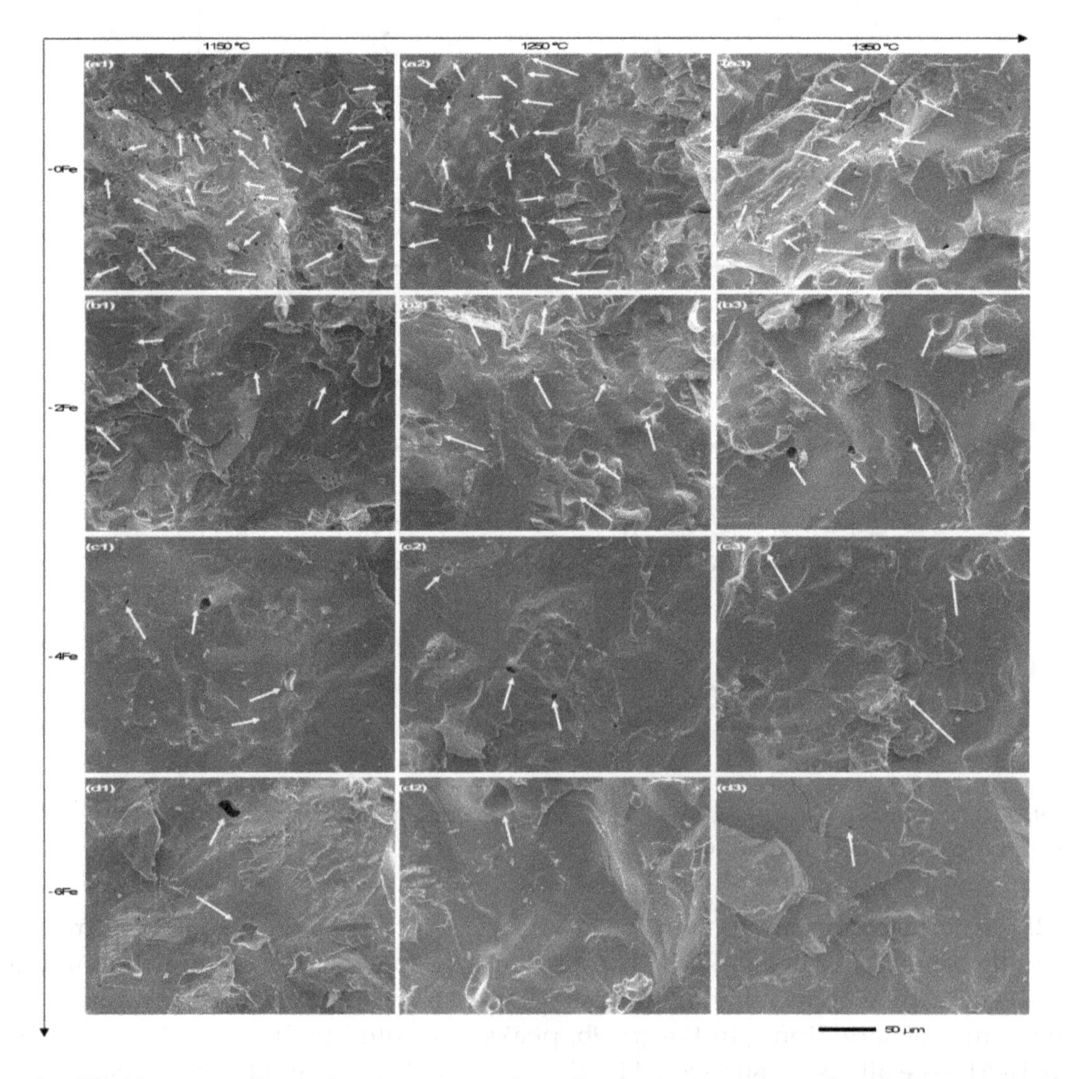

Figure 10. SEM fractographs with micropores of tensile specimens sintered at various temperatures and compositions.

The micropores shown in Figure 10 are in accordance with the observations made from Figure 5, whereby the Ti-6Al-4V specimens show the largest number of micropores (diameter < 10 μm). Although the amount of microporosity in these samples decreased with sintering temperature, the amount of micropores in the Ti-6Al-4V at any temperature was higher than the titanium alloy specimens containing iron. For the Fe containing samples, the microporisity decreased with increasing iron content. This indicates that the addition of iron is beneficial to elimination of micropores; in other words iron assists in densification during sintering.

The diffusion coefficient of elements follows an Arrhenius relationship:

$$D = D_0 \exp(-Q/RT) \tag{3}$$

where D_0 is the pre-exponential factor ($m^2 \cdot s^{-1}$), Q the activation energy ($kJ \cdot mol^{-1}$), R the molar gas constant, and T the absolute temperature (K). These parameters can be obtained for both the self-diffusion of β-Ti and the inter-diffusion of Al, V, and Fe in β-Ti [27] as shown in Table 1.

Table 1. Self-diffusion of β-Ti and inter-diffusion of Al, V, and Fe in β-Ti [27].

Specimen	Diffusion Type	Temperature (°C)	D_0 ($m^2 \cdot s^{-1}$)	Q ($kJ \cdot mol^{-1}$)
β-Ti	Self diffusion	899–1540	3.58×10^{-8}	130.6
Al in β-Ti	Inter-diffusion	920–1600	1.14×10^{-5}	213.1
V in β-Ti	Inter-diffusion	902–1543	3.1×10^{-8}	134.8
Fe in β-Ti	Inter-diffusion	969–1645	7.8×10^{-7}	132.3

Diffusion coefficients were calculated using the data presented in Table 1 and the ratio of inter-diffusion of the different elements to the self-diffusion in Ti are presented in Figure 11 over the temperature range of 1000–1400 °C.

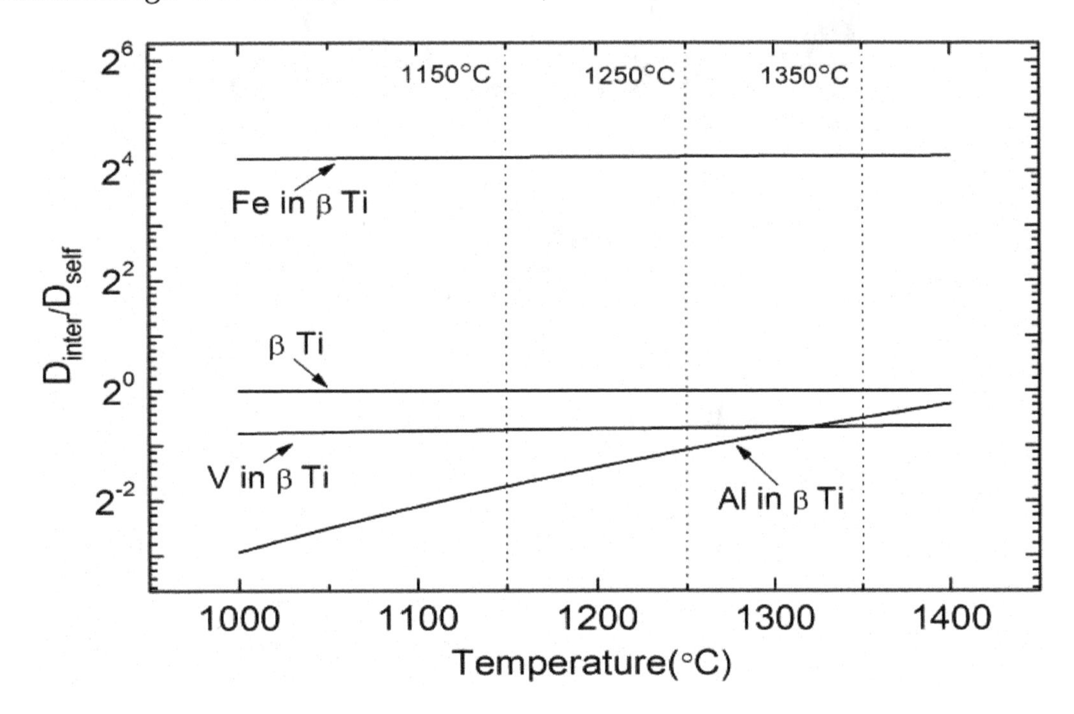

Figure 11. Temperature dependence of ratio of inter-diffusion coefficients to self-diffusion coefficient of β-Ti.

The inter-diffusion coefficient of Fe in β-Ti is over 16 times higher than the self-diffusion coefficient (β-Ti) and is also much higher than the inter-diffusion coefficients of Al and V in Ti (both <1). This indicates that iron is a fast diffuser in β-Ti, and empirical rules of diffusion suggest that the addition of fast diffusers can enhance self-diffusion rates of both the solute and solvent atoms [28]. Additions of a fast diffuser such as iron can therefore enhance the densification of titanium during sintering.

4.2. Macroporosity Formation

A fractograph of the Ti-6Al-4V sample sintered at 1150 °C is presented in Figure 12 as well as corresponding compositional analyses. Two macropores ~100 μm in diameter are observed in Figure 12, and material with a "coral-like" structure was observed inside the pores. Compositional analysis was conducted on the coral-like material (plot 1) and the more dense surrounding area (plot 2). The EDS spectra and analysis show that the coral-like material was predominantly vanadium (88.78 wt %) whilst the surrounding area was composed of titanium (93.82 wt %) and a small amount of aluminum (3.55 wt %) but no vanadium. The presence of vanadium-rich "coral-like" material with a smooth cellular wall is considered to be correlated with the evaporation of aluminum according to the following analysis.

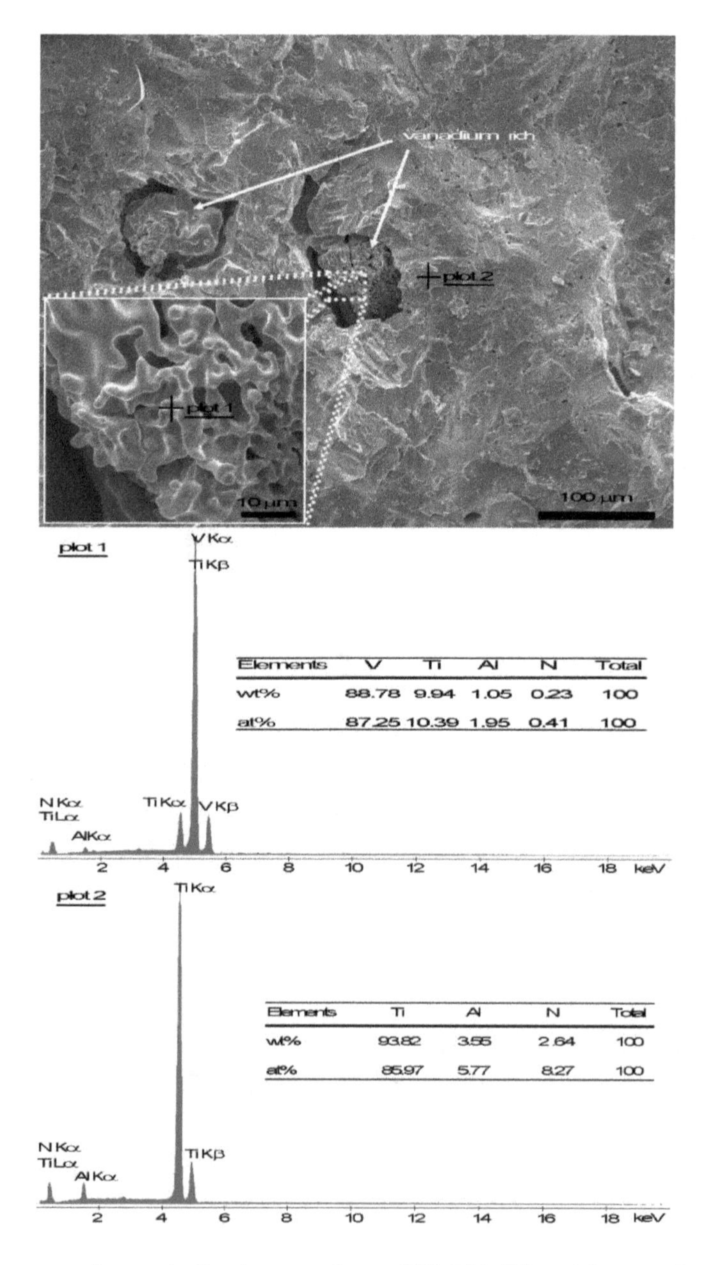

Elements	V	Ti	Al	N	Total
wt%	88.78	9.94	1.05	0.23	100
at%	87.25	10.39	1.95	0.41	100

Elements	Ti	Al	N	Total
wt%	93.82	3.55	2.64	100
at%	85.97	5.77	8.27	100

Figure 12. Macropores shown in fracture surface of Ti-6Al-4V specimens sintered at 1150 °C.

Theoretical calculation of the evaporation of binary alloys can be determined based on the Miedema model and Langmuir equation [23,29,30]. For instance, Guo and co-workers calculated the theoretical elemental evaporation of a Ti-Al melt during casting with this model, which had reasonable agreement with the experimental data [29,30]. Chen et al. [23] and Xu et al. [31] also utilized this model to calculate the evaporation of PM fabricated Fe-Al and Fe-Mn alloys in good accordance with experimental observations.

The evaporation loss rate ($N_{m,\ A}$) of the component A in a binary A-B alloy system is demonstrated by the Langmuir theory [29,32]:

$$N_{m,\ A} = \mathrm{K_L} \times \varepsilon \times (P^e{}_A - P^g{}_A) \times \sqrt{M_A/T} \tag{4}$$

The Langmuir constant ($\mathrm{K_L}$) is equal to 4.37×10^{-4} when the partial pressure is described in Pascal [29,30]. ε is the condensation constant (for metals = 1) [29,30]. $P^e{}_A$ indicates the saturated vapor partial pressure of element A in the system and $P^g{}_A$ is the partial pressure of the component. Since the specimens were sintered under high vacuum and the volatiles are reactive with titanium substrate, it is proposed that $P^g{}_A$ can be considered to be zero. M_A and T are the component's atomic mass and absolute temperature respectively.

The saturated vapor partial pressure $P^e{}_A$ can be defined by:

$$P^e{}_A = \chi_A \times \gamma_A \times P^0_A \tag{5}$$

where χ_A and γ_A indicate the molar fractions and activity coefficient of component A in the binary system. The equilibrium pressure of pure component A, P^0_A can be calculated for Al and V elements by the following formula [33]:

$$\log_{10} P^0_{\mathrm{Al}} = 14.465 - 17342T^{-1} - 0.7927 \log_{10} T \tag{6}$$

$$\log_{10} P^0_{\mathrm{V}} = 14.75 - 27132T^{-1} - 0.5501 \log_{10} T \tag{7}$$

The activity coefficient of component A (γ_A) in a binary system is given by [34]:

$$\ln \gamma_A = \frac{\Omega}{\mathrm{R}T}(1 - \chi_A)^2 \tag{8}$$

where R is the ideal gas constant and Ω is an interaction parameter which can be obtained by calculating the molar enthalpy of mixing (ΔH_{mix}) for the binary system [34]:

$$\Delta H_{mix} = \Omega \times \chi_A \times \chi_B \tag{9}$$

where χ_A and χ_B are the molar fractions of component A and B respectively.

Substituting Equation (9) into Equation (8), gives:

$$\ln \gamma_B = \frac{\chi_A}{\mathrm{R}T\chi_B} \Delta H_{mix} \tag{10}$$

$$\ln \gamma_A = \frac{\chi_B}{\mathrm{R}T\chi_A} \Delta H_{mix} \tag{11}$$

Thus, the theoretical evaporation loss rate is dependent on the molar enthalpy of mixing (ΔH_{mix}), which can be determined using the model of Miedema [35]:

$$\Delta H_{mix} = \frac{2\mathrm{P}f(c^s)(\chi_A \mathrm{V}_A^{2/3} + \chi_B \mathrm{V}_B^{2/3})}{(\mathrm{n}^{\mathrm{A}}_{\mathrm{ws}})^{-1/3} + (\mathrm{n}^{\mathrm{B}}_{\mathrm{ws}})^{-1/3}} \times \left[-(\varphi_A - \varphi_B)^2 + \frac{\mathrm{Q}}{\mathrm{P}}(\mathrm{n}^{A}_{\mathrm{ws}}{}^{1/3} - \mathrm{n}^{B}_{\mathrm{ws}}{}^{1/3})^2 - \mathrm{R/P}\right] \tag{12}$$

where $f(c^s)$ is given by:

$$f(c^s) = f(\chi_A{}^s \chi_B{}^s) = \chi_A \chi_B V_A^{2/3} V_B^{2/3} / (\chi_A V_A^{2/3} + \chi_B V_B^{2/3})^2 \quad (13)$$

$V_A^{2/3}$, $V_B^{2/3}$, ${n_{ws}^A}^{1/3}$, ${n_{ws}^B}^{1/3}$, Q, P, R, φ_A, and φ_B are all constants and are available in the literature [35]. Once the molar enthalpy of mixing is determined, the activity coefficients of Al and V can be calculated from Equations (10) and (11) and combined with the equilibrium pressure (Equations (6) and (7)) to give the saturated vapor pressure (Equation (5)) and ultimately the evaporation rate (Equation (4)).

The weight ratio of Al to V in the 60 wt % Al-40 wt % V master alloy results in an atomic ratio of Al to V of 17:6. Using these values and the model described above, the dependence of the saturated vapor pressure and evaporation rate on temperature is plotted in Figure 13.

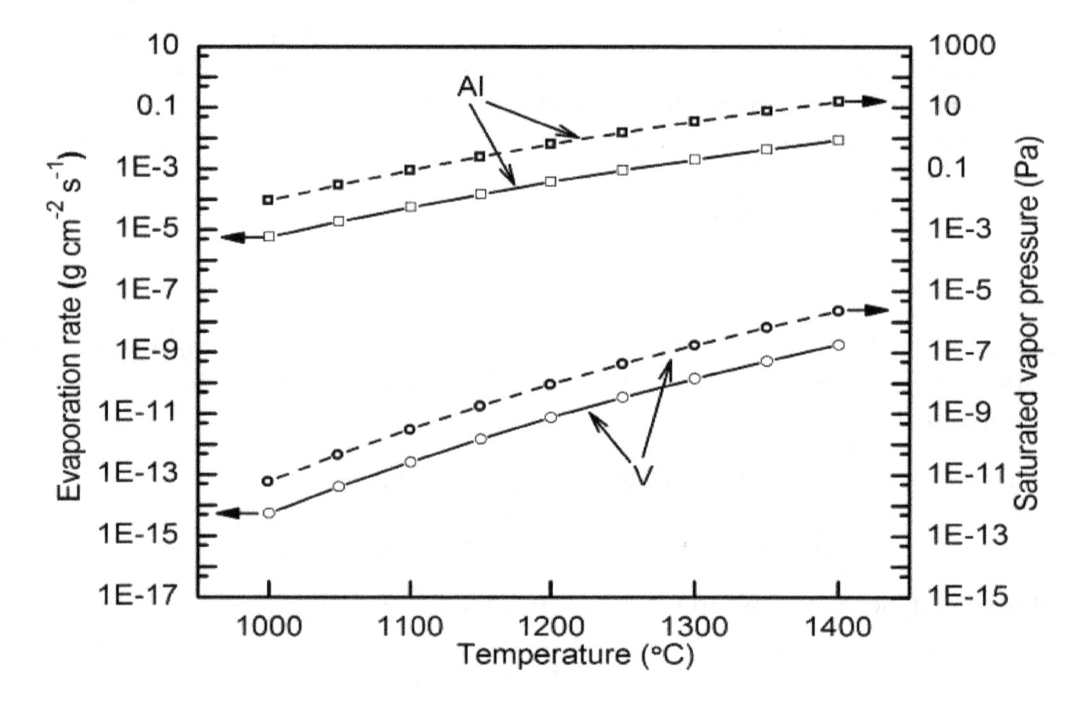

Figure 13. The dependence of evaporation rate and saturated vapor pressure on temperature.

From Figure 13 it can be seen that the saturation vapor pressure of aluminum is 9.2×10^{-3} Pa at a temperature of 1000 °C and increases with temperature. This is higher than the furnace vacuum level of 2×10^{-3} Pa thus suggesting that the Al is in a state of free evaporation [29,30]. The evaporation rate of aluminum and vanadium are 1.48×10^{-4} and 1.48×10^{-12} g·cm^{-2}·s^{-1} respectively at a temperature of 1150 °C seen from Figure 13. The median diameter of the master powder particles was 142.31 μm as shown in Figure 1. Assuming the particle is spherical and the compositional ratio is constant, the mass of aluminum and vanadium in this particle would be 3.15 and 2.1 μg, respectively. When sintering was conducted at the lowest temperature (1150 °C) with a holding time of 3 h, the total loss of aluminum and vanadium by evaporation would be 1.02×10^3 and 1.02×10^{-5} μg respectively. It can be seen that the evaporation loss of aluminum is three orders of magnitude higher than the actual content, whereas, the evaporation of vanadium can be neglected compared with its original mass. This indicates that aluminum had completed the evaporation process before the sintering ends, which is validated by the observation that aluminum is almost non-existent inside the pore shown in Figure 12. It is therefore suggested that the macropores are formed due to the high saturated vapor pressure resulting in the evaporation of aluminum and leaving a vanadium rich region inside the pores whose shape and size are taken from original master particles. These macropores would dramatically decrease both tensile strength and ductility.

4.3. Microstructural Evaluation and Phase Transformation

Haase et al. [25] sintered c.p. Ti powder (<150 μm) blended with a 60Al-40V master powder (<160 μm) prepared by equal-channel angular pressing and the investigation revealed retarded dissolution of master alloy particles due to the formation of Al-rich and V-rich layers, which were assumed to be comprised of intermetallic phases [25]. Therefore, for blended elemental Ti-6Al-4V sintered at 1150 °C shown in Figures 6a and 7a, the phase of region 2 (Figure 6a) can be suggested as α_2-Ti_3Al [36], which is in agreement with the phase determination illustrated in Figure 9a. Moreover, the diffusion of elements in α_2-Ti_3Al is much slower than in β-Ti, for example the diffusion coefficient of Al in β-Ti is almost 2–3 orders of magnitude higher than in α_2-Ti_3Al [36]. Therefore, such intermetallic phase may be considered to be a diffusion barrier for vanadium diffusion which was left by the evaporation of aluminum from the master alloy powders. As indicated by Equation (3) the diffusion coefficient increases with temperature and the intermetallic phases would become no longer stable at higher temperatures (1250 and 1350 °C) compared with 1150 °C. This is in accordance with the observation shown in the literature [25]. Therefore, only α-Ti phase was observed at higher temperature with no intermetallic phase as shown in Figure 9a.

The light region shown in Figure 5 is considered to be β phase because of the abundance of β-stabilizers such as iron and vanadium. The dark region is suggested to be α phase at high temperature (1250 and 1350 °C) including both the blocky regions and acicular grains. Interestingly, the two different "alpha" morphologies are not consistent in composition. In the larger, blocky regions the Al content is much lower and this is thought to be a result of incomplete diffusion of the alloying elements into the Ti matrix because of the distance from the master alloy particle which is the source of aluminum. This is compatible with the unusual microstructural observation.

4.4. Mechanical Properties

Although the microporosity was reduced with the addition of iron, the mechanical properties of these samples were poor: fracture strength in the range of 20 to 100 MPa, with no or little plastic deformation. This is thought to be due to the macroporosity in the sintered samples and high oxygen content in the as-received titanium powders. High oxygen levels in particular are detrimental to the mechanical properties in PM Ti components [4,5,7]. The dependence of oxygen content on median particle size for HDH titanium powder is summarized from the literature [4,37] in Figure 14.

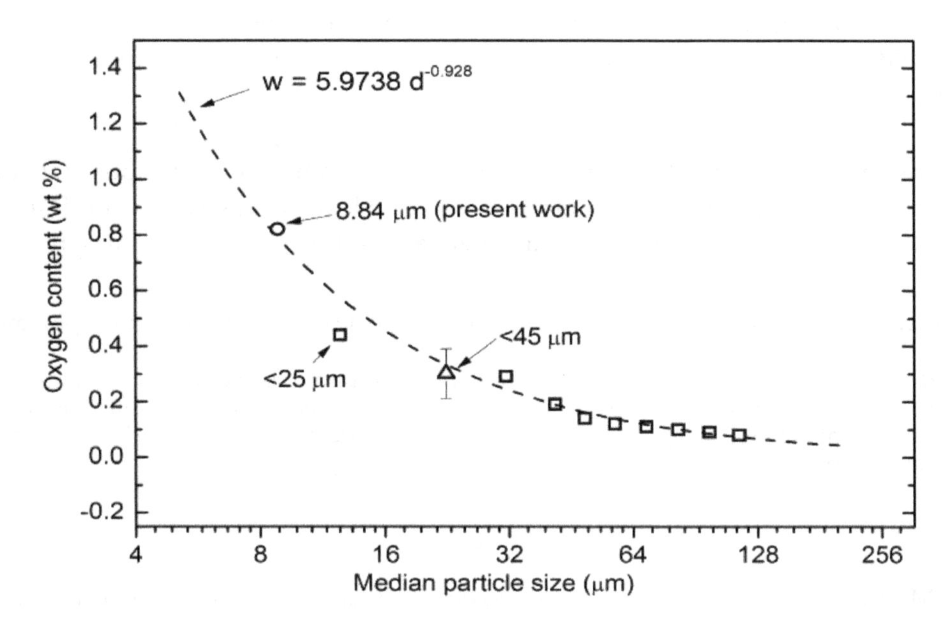

Figure 14. Dependence of oxygen content on median particle size of hydrogenation-dehydrogenation (HDH) titanium powder.

The correlation between oxygen content and median particle size shown in Figure 14 can be expressed by a power-law equation:

$$w = 5.9738\, d^{-0.928} \tag{14}$$

where w is the oxygen content (wt %) and d is the median particle size (μm). The relationship between oxygen content (0.82 wt %) and the median particle size (8.84 μm) of the fine powder used in this work fits well with this relationship. Since oxygen has high solubility in β-Ti [38], the oxygen element, in the form of oxides on titanium particle surfaces, will enter into the β-Ti lattice and therefore, the oxide layer on the titanium particle surface does not affect the densification of blended alloys [5]. However, the oxygen dissolved in β phase would precipitate as oxides at a temperature below the β transus, and these brittle oxides tend to precipitate at grain boundaries. This work has shown that although good densification can be obtained at low temperatures when using the fine titanium powder, the high levels of oxygen impurities results in poor mechanical properties. In order to remedy the poor mechanical properties, the amount of oxides should be reduced and the oxides should not be at the grain boundaries. One strategy is to use an oxygen scavenging element, which has significant solubility in titanium and has more affinity for oxygen than Ti, thus forming non-Ti oxides that are located in the titanium grain interior. Some rare earth elements might serve this purpose [39,40]. Another strategy is to use post-sinter thermomechanical treatments to mechanically move the oxides from grain boundaries to the grain interior.

5. Conclusions

Four different blended elemental titanium alloys were designed and sintered through varying the weight percentage of a fine Ti powder with (Al-V) master alloy and iron powders. Spherical micropores were formed during sintering with fine titanium powder even at the lowest sintering temperature for Ti-6Al-4V. With the addition of Fe, the microporosity was reduced and improved microstructural and compositional homogenization was observed. This is due to fast diffusion of iron in titanium thereby enhancing densification. However, the tensile mechanical properties of these blended elemental titanium alloys were poor with values of <100 MPa tensile strength and almost no elongation. This was mainly caused by high impurity content in the as-received powders and the formation of macropores around the original master alloy particles. Incomplete dissociation of the master alloy particles caused severe evaporation of aluminum resulting in the formation of macropores during sintering.

Although reasonable densification could be obtained with fine titanium powders and this could be improved with the addition of iron, the properties were low because of the high inherent oxygen and macroporosity and therefore special precautions have to be taken when using these powders.

Acknowledgments: This work was supported by Ministry of Business Innovation and Employment (MBIE).

Author Contributions: Changzhou Yu conducted the experimental work as part of his Ph.D. study. He also analyzed the data. Peng Cao and Mark Ian Jones conceived and designed the experiments and analyzed the data. Peng Cao revised the manuscript. Mark Ian Jones discussed the results. Peng Cao and Mark Ian Jones were the supervisors of Changzhou Yu.

References

1. German, R.M. *Powder Metallurgy and Particulate Materials Processing*; Metal Powder Industries Federation: Princeton, NJ, USA, 2005.
2. Robertson, I.M.; Schaffer, G.B. Some effects of particle size on the sintering of titanium and a master sintering curve model. *Metall. Mater. Trans. A* **2009**, *40*, 1968–1979. [CrossRef]

3. Luo, S.D.; Yan, M.; Schaffer, G.B.; Qian, M. Sintering of Titanium in Vacuum by Microwave Radiation. *Metall. Mater. Trans. A* **2011**, *42*, 2466–2474. [CrossRef]
4. Robertson, I.M.; Schaffer, G.B. Review of densification of titanium based powder systems in press and sinter processing. *Powder Metall.* **2010**, *53*, 146–162. [CrossRef]
5. Qian, M. Cold compaction and sintering of titanium and its alloys for near-net-shape or preform fabrication. *Int. J. Powder Metall.* **2010**, *46*, 29–44.
6. He, W.; Weng, Q.G.; He, Y.H.; Jiang, Y. Preparation of ultrafine Ti powder by inhibitor coated/HDH combined method. *Powder Metall.* **2013**, *56*, 239–244. [CrossRef]
7. Wang, H.T.; Fang, Z.Z.; Sun, P. A critical review of mechanical properties of powder metallurgy titanium. *Int. J. Powder Metall.* **2010**, *46*, 45–57.
8. Geetha, M.; Singh, A.K.; Asokamani, R.; Gogia, A.K. Ti based biomaterials, the ultimate choice for orthopaedic implants—A review. *Prog. Mater. Sci.* **2009**, *54*, 397–425. [CrossRef]
9. Ivasishin, O.M.; Savvakin, D.G.; Froes, F.; Mokson, V.C.; Bondareva, K.A. Synthesis of alloy Ti-6Al-4V with low residual porosity by a powder metallurgy method. *Powder Metall. Met. Ceram.* **2002**, *41*, 382–390. [CrossRef]
10. Peter, W.; Chen, W.; Yamamoto, Y.; Dehoff, R.; Muth, T.; Nunn, S.; Kiggans, J.; Clark, M.; Sabau, A.; Gorti, S.; et al. Current Status of Ti PM: Progress, Opportunities and Challenges. *Key Eng. Mater.* **2012**, *520*, 1–7. [CrossRef]
11. Bolzoni, L.; Ruiz-Navas, E.M.; Gordo, E. Understanding the properties of low-cost iron-containing powder metallurgy titanium alloys. *Mater. Des.* **2016**, *110*, 317–323. [CrossRef]
12. Bolzoni, L.; Ruiz-Navas, E.M.; Gordo, E. Quantifying the properties of low-cost powder metallurgy titanium alloys. *Mater. Sci. Eng. A* **2017**, *687*, 47–53. [CrossRef]
13. Wei, W.; Liu, Y.; Zhou, K.; Huang, B. Effect of Fe addition on sintering behaviour of titanium powder. *Powder Metall.* **2003**, *46*, 246–250. [CrossRef]
14. Savvakin, D.G.; Carman, A.; Ivasishin, O.M.; Matviychuk, M.V.; Gazder, A.A.; Pereloma, E.V. Effect of Iron Content on Sintering Behavior of Ti-V-Fe-Al Near-beta Titanium Alloy. *Metall. Mater. Trans. A* **2012**, *43*, 716–723. [CrossRef]
15. Qian, M.; Yang, Y.F.; Yan, M.; Luo, S.D. Design of low cost high performance powder metallurgy titanium alloys: Some basic considerations. *Key Eng. Mater.* **2012**, *520*, 24–29. [CrossRef]
16. Esteban, P.G.; Ruiz-Navas, E.M.; Bolzoni, L.; Gordo, E. Low-cost titanium alloys? Iron may hold the answers. *Met. Powder Rep.* **2008**, *63*, 24–27. [CrossRef]
17. Liu, Y.; Chen, L.F.; Tang, H.P.; Liu, C.T.; Liu, B.; Huang, B.Y. Design of powder metallurgy titanium alloys and composites. *Mater. Sci. Eng. A* **2006**, *418*, 25–35. [CrossRef]
18. Carman, A.; Zhang, L.C.; Ivasishin, O.M.; Savvakin, D.G.; Matviychuk, M.V.; Pereloma, E.V. Role of alloying elements in microstructure evolution and alloying elements behaviour during sintering of a near-beta titanium alloy. *Mater. Sci. Eng. A* **2011**, *528*, 1686–1693. [CrossRef]
19. Yang, Y.F.; Luo, S.D.; Schaffer, G.B.; Qian, M. Sintering of Ti-10V-2Fe-3Al and mechanical properties. *Mater. Sci. Eng. A* **2011**, *528*, 6719–6726. [CrossRef]
20. Esteban, P.G.; Ruiz-Navas, E.M.; Gordo, E. Influence of Fe content and particle size the on the processing and mechanical properties of low-cost Ti-xFe alloys. *Mater. Sci. Eng. A* **2010**, *527*, 5664–5669. [CrossRef]
21. Chen, B.Y.; Hwang, K.S.; Ng, K.L. Effect of cooling process on the alpha phase formation and mechanical properties of sintered Ti-Fe alloys. *Mater. Sci. Eng. A* **2011**, *528*, 4556–4563. [CrossRef]
22. Yang, Y.F.; Luo, S.D.; Schaffer, G.B.; Qian, M. The Sintering, Sintered Microstructure and Mechanical Properties of Ti-Fe-Si Alloys. *Metall. Mater. Trans. A* **2012**, *43A*, 4896–4906. [CrossRef]
23. Chen, G.; Cao, P.; He, Y.H.; Shen, P.Z.; Gao, H.Y. Effect of aluminium evaporation loss on pore characteristics of porous FeAl alloys produced by vacuum sintering. *J. Mater. Sci.* **2012**, *47*, 1244–1250. [CrossRef]
24. Kang, K.S.; Kim, C.H.; Cho, W.C.; Bae, K.K.; Woo, S.W.; Park, C.S. Reduction characteristics of $CuFe_2O_4$ and Fe_3O_4 by methane; $CuFe_2O_4$ as an oxidant for two-step thermochemical methane reforming. *Int. J. Hydrogen Energy* **2008**, *33*, 4560–4568. [CrossRef]
25. Haase, C.; Lapovok, R.; Ng, H.P.; Estrin, Y. Production of Ti-6Al-4V billet through compaction of blended elemental powders by equal-channel angular pressing. *Mater. Sci. Eng. A* **2012**, *550*, 263–272. [CrossRef]

26. Bolzoni, L.; Esteban, P.G.; Ruiz-Navas, E.M.; Gordo, E. Mechanical behaviour of pressed and sintered titanium alloys obtained from prealloyed and blended elemental powders. *J. Mech. Behav. Biomed. Mater.* **2012**, *14*, 29–38. [CrossRef] [PubMed]
27. Neumann, G.; Tuijn, C. *Self-Diffusion and Impurity Diffusion in Pure Metals: Handbook of Experimental Data*; Elservier: Amsterdam, The Netherlands, 2008; Volume 14, pp. 1–349.
28. Yang, Y.F.; Luo, S.D.; Bettles, C.J.; Schaffer, G.B.; Qian, M. The effect of Si additions on the sintering and sintered microstructure and mechanical properties of Ti-3Ni alloy. *Mater. Sci. Eng. A* **2011**, *528*, 7381–7387. [CrossRef]
29. Su, Y.Q.; Guo, J.J.; Jia, J.; Liu, G.Z.; Liu, Y.A. Composition control of a TiAl melt during the induction skull melting (ISM) process. *J. Alloys Compd.* **2002**, *334*, 261–266.
30. Guo, J.J.; Liu, Y.; Su, Y.Q.; Ding, H.S.; Liu, G.Z.; Jia, J. Evaporation behavior of aluminum during the cold crucible induction skull melting of titanium aluminum alloys. *Metall. Mater. Trans. B* **2000**, *31B*, 837–844. [CrossRef]
31. Xu, Z.; Hodgson, M.A.; Chang, K.; Chen, G.; Yuan, X.; Cao, P. Effect of Sintering Time on the Densification, Microstructure, Weight Loss and Tensile Properties of a Powder Metallurgical Fe-Mn-Si Alloy. *Metals* **2017**, *7*, 81. [CrossRef]
32. Langmuir, I. The Vapor Pressure of Metallic Tungsten. *Phys. Rev.* **1913**, *2*, 329–342. [CrossRef]
33. Alcock, C.B. Vapor pressure of the metallic elements—Equations. In *CRC Handbook of Chemistry and Physics (Internet Version 2013)*, 93rd ed.; Haynes, W.M., Ed.; CRC Press: Boca Raton, FL, USA, 2013; pp. 125–126.
34. Porter, D.A.; Easterling, K.E.; Sherif, M.Y. *Phase Transformations in Metals and Alloys*, 3rd ed.; CRC Press: Boca Raton, FL, USA, 2009.
35. Miedema, A.R.; de Chatel, P.F.; de Boer, F.R. Cohesion in alloys—Fundamentals of a semi-empirical model. *Physica B* **1980**, *100*, 1–28. [CrossRef]
36. Mishin, Y.; Herzig, C. Diffusion in the Ti-Al system. *Acta Mater.* **2000**, *48*, 589–623. [CrossRef]
37. McCracken, C. Production of Fine Titanium Powders via the Hydrid-Dehydride (HDH) Process. *PIM Int.* **2008**, *2*, 55–57.
38. Massalski, T.D. *Binary Alloys Phase Diagams*; Okamoto, H., Subramanian, P.R., Kasprzak, L., Eds.; ASM International: Geauga County, OH, USA, 1990.
39. Yan, M.; Liu, Y.; Liu, Y.B.; Kong, C.; Schaffer, G.B.; Qian, M. Simultaneous gettering of oxygen and chlorine and homogenization of the beta phase by rare earth hydride additions to a powder metallurgy Ti-2.25Mo-1.5Fe alloy. *Scr. Mater.* **2012**, *67*, 491–494. [CrossRef]
40. Luo, S.D.; Yang, Y.F.; Schaffer, G.B.; Qian, M. The effect of a small addition of boron on the sintering densification, microstructure and mechanical properties of powder metallurgy Ti-7Ni alloy. *J. Alloys Compd.* **2013**, *555*, 339–346. [CrossRef]

6

High Temperature Oxidation and Wear Behaviors of Ti–V–Cr Fireproof Titanium Alloy

Guangbao Mi [1], Kai Yao [2], Pengfei Bai [2], Congqian Cheng [2] and Xiaohua Min [2,*]

[1] Key Laboratory of Science and Technology on Advanced Titanium Alloys, AECC Beijing Institute of Aeronautical Materials, Beijing 100095, China; miguangbao@163.com
[2] School of Materials Science and Engineering, Dalian University of Technology, Dalian 116024, China; kaiyao@mail.dlut.edu.cn (K.Y.); bpf2014@mail.dlut.edu.cn (P.B.); cqcheng@dlut.edu.cn (C.C.)
* Correspondence: minxiaohua@dlut.edu.cn

Abstract: The high temperature oxidation and wear behaviors of Ti–35V–15Cr–0.3Si–0.1C fireproof titanium alloy were examined at 873 and 1073 K. The oxidation weight gain after oxidation at 1073 K for 100 h was significantly larger than that at 873 K. Based on the analyses of the oxidation reaction index and oxide layer, the oxidation process at 1073 K was mainly controlled by oxidation reaction at the interface between the substrate and oxide layer. Dry sliding wear tests were performed on a pin-on-disk tester in air conditions. The friction coefficient was smaller at 1073 K than that at 873 K, while the volume wear rate at 1073 K was larger due to formation of amount of oxides on the worn surface. When the wearing temperature increased from 873 to 1073 K, the wear mechanism underwent a transition from a combination of abrasive wear and oxidative wear to only oxidative wear.

Keywords: titanium alloy; fireproof; oxidation; high temperature wear; mechanism

---

1. Introduction

Titanium and its alloys are widely used in aerospace, chemical, and biomedical industries because of high specific strength and excellent corrosion resistance, especially for their applications in aero-engines [1–3]. However, conventional titanium alloys can be ignited and burned, known as "titanium fire," under specific conditions of high temperature, high pressure, and high-speed airflow, which limits their application for advanced aero-engines [4–6]. With the improvement of the thrust-weight ratio, the work conditions of components such as the case, blade, and disk becomes more complex and severe, resulting in an increase in the occurrence of titanium fire. Compared with other techniques including structural optimization design, surface alloying, and surface coating, the development of advanced fireproof titanium alloys is more desirable to completely prevent the occurrence of titanium fire in aero-engines, as well as to meet its high thrust-weight ratio [4,7].

Titanium fire is known to a typical accident of igniting, which occurs usually through high-energy friction and load impact as an ignition source, for example, between the blade and case. The whole combustion behavior goes through high temperature oxidation, super high temperature oxidation, burning, self-sustained combustion, and burning out [4]. If titanium alloys are ignited, the components in the compressor burn continually only for 4–20 s under gas flow with high temperature and pressure. The burning time is too short to take extinguishing measures. There are two types of fireproof titanium alloys: Ti–V–Cr base alloys, such as Ti–35V–15Cr (Alloy C), Ti–35V–15Cr–0.6Si–0.05C (Alloy C^+), Ti–25V–15Cr–2Al–0.2C, Ti–25V–15Cr–0.2Si (Ti40), and Ti–35V–15Cr–0.3Si–0.1C (TF550) alloys [8–11]; and Ti–Cu–Al base alloys, such as Ti–13Cu–4Al–4Mo–2Zr (BTT–1), Ti–18Cu–2Al–2Mo (BTT–3), and Ti–13Cu–1Al–0.2Si (Ti14) [12–14]. Note that abovementioned and other compositions in this study are expressed in mass%. Ti–V–Cr base alloys are highly stabilized β-type titanium alloys with better performance than that of Ti–Cu–Al base alloys.

In case of TF550 fireproof titanium alloy, we previously investigated the fireproof behavior by the frictional ignition method, and found that the presence of oxides on the worn surface such as TiO_2, V_2O_5, and Cr_2O_3 affected the lubrication behavior in the local area of friction [15–17]. However, the high temperature oxidation and wear behaviors remain unclear for this alloy, which are considered to be closely associated with their fireproof properties. The shortage of studies results in uncertainty of the fireproof level of titanium alloys used in aero-engines and restricts the material selection of fireproof titanium alloys for engine designers. Thus, the purpose of present study is to examine the oxidation and wear behaviors of TF550 alloy at high temperatures of 873 and 1073 K, and to discuss the mechanisms of high temperature oxidation and wear.

2. Experimental Procedures

2.1. Materials Preparation

An ingot of approximately 150 kg of TF550 alloy was fabricated by consumable vacuum arc melting (ZHT-650, Baoji institute of rare metal research, Baoji, China) with the raw materials of titanium sponge, pure vanadium rods, and chromium sponge. Subsequently, it was subjected to sheathed extrusion, forging and rolling, and its actual composition was Ti–35.5V–14.6Cr–0.32Si–0.11C. For oxidation testing, samples with a dimension of 12 mm (l) × 10 mm (w) × 6 mm (t) were cut from the rolled plate and their surfaces were mechanically polished with SiC abrasive paper in order to evaluate their high temperature oxidation behavior. As-received Ti–6Al–4V alloy was used to evaluate the oxidation behavior as a comparison. For wear testing, pins and disks were also cut from the rolled plate (HQ1UP, Hanqi Company, Suzhou, China). The rod as pins was designed in the form of a cylinder with a diameter of 5 mm and a height of 9 mm, and the contact surface was a half-ball with a diameter of 5 mm. Disks with the dimensions of 25 mm diameter and 10 mm thickness were used. The contact surfaces of pins and disks were mechanically polished (UNIPOL-1200M, Kejing Company, Shenyang, China) with SiC abrasive paper and then cleaned in alcohol.

2.2. Oxidation Testing

Before oxidation testing, the samples and crucibles were ultrasonically cleaned (XY-CS-S, Xinyi Company, Shanghai, China) in alcohol. In addition, the crucibles were baked at 1373 K for 20 h to ensure their constant weight during the whole oxidation process. Oxidation tests were carried out at 873 and 1073 K for a total oxidation time of 100 h. During the oxidation period, they were cooled down to room temperature at an interval of 10 h in order to measure their weight and observe their surface morphology. Oxidation kinetics were examined based on the weight gain per unit area versus oxidation time.

2.3. Wear Testing

Dry sliding wear tests were performed on a pin-on-disk tester (THT-04015, CSM Company, Carouge, Switzerland) in air conditions at temperatures of 873 and 1073 K. A pin was loaded against a rotating disk through the mechanical loading system with a normal load of 4.9 N for 30 min. The sliding velocity was 0.2 m/s along a diameter of 10 mm circular path with total sliding distance of 3.6×10^2 m, which was proved to be sufficient to attain a steady-state condition [18–21]. The friction coefficient was recorded by the tester automatically, and the mean friction coefficient was calculated based on the whole data, which was recorded during the sliding wear tests. Volume wear rate was calculated as follows:

$$Ws = \Delta V/(PL) \tag{1}$$

where ΔV, P, and L are the volume loss of disk, applied loading and sliding distance, respectively. For the disk, four different regions of the wearing indent were measured by using nanomap-3D microscopy (AEP Company, Columbus, OH, USA) for 3D micrograph and the corresponding cross-section area of the wearing indent, and then the volume loss was obtained by the mean cross-section area multiplied by the circumference of the wearing indent. For the pin, the volume loss was obtained by the wearing volume of the half-ball of pin.

2.4. *Microstructural Characterization*

Phase identification was made by X-ray diffraction analysis (XRD) using an EMPYREN diffractometer with Cu-Kα radiation operated at 40 kV and 300 mA (Panalytical, Almelo, The Netherlands). Morphologies including backscattered electron image (BEI) and secondary electron image (SEI), and chemical compositions were examined by a SUPRA55 type scanning electron microscope (SEM, Carl Zeiss Jena Company, Oberkochen, Germany) equipped with an energy dispersive spectrometer (EDS).

3. Results and Discussion

Figure 1 shows the oxidation kinetics curves of TF550 alloy along with Ti–6Al–4V alloy after oxidation at 873 and 1073 K for 100 h. As shown in Figure 1a, with increasing oxidation time, the oxidation weight gain of each alloy showed an increase tendency, and it was much larger at 1073 K than that at 873 K. The oxidation weight gain of TF550 alloy was 2 mg/cm^2 after oxidation at 873 K for 100 h, which was similar with that (1.4 mg/cm^2) of Ti–6Al–4V alloy, while the oxidation weight gain of TF550 alloy (90 mg/cm^2) was larger than that (40 mg/cm^2) of Ti–6Al–4V alloy at 1073 K. The smaller oxidation weight gain of Ti–6Al–4V alloy was attributed to the formation of the compact Al_2O_3 layer, which could prevent the further oxidation of the substrate [22,23]. Although the oxidation weight gain of Ti–6Al–4V alloy was smaller than that of TF550 alloy, the oxide layer of Ti–6Al–4V alloy flaked seriously at 1073 K, as shown in Figure 2a. While V_2O_5 which formed in the oxide layer of TF550 alloy, it melted and flowed into the crucible (yellow matters in Figure 2b) due to its low melting point (about 948 K) [24,25]. The formation of V_2O_5 in TF550 alloy at high temperatures improved the wear properties resulting from their lubrication, which could enhance its burn-resistant behavior. The detailed discussion is shown in Section 3.

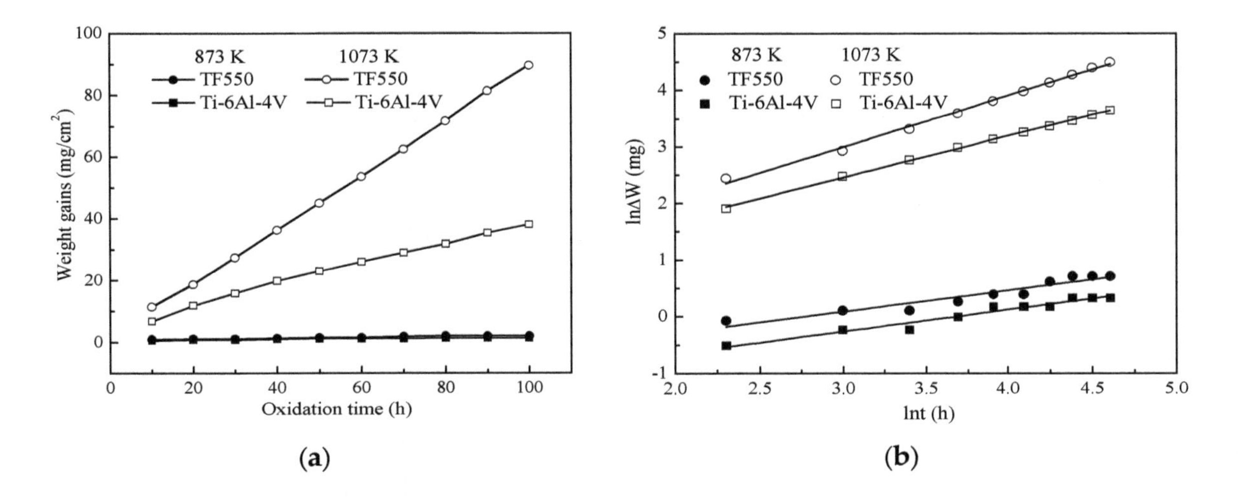

Figure 1. Oxidation kinetics curves of TF550 alloy along with Ti–6Al–4V alloy at temperatures of 873 and 1073 K. (**a**) Oxidation weight gain curves and (**b**) double logarithmic curves.

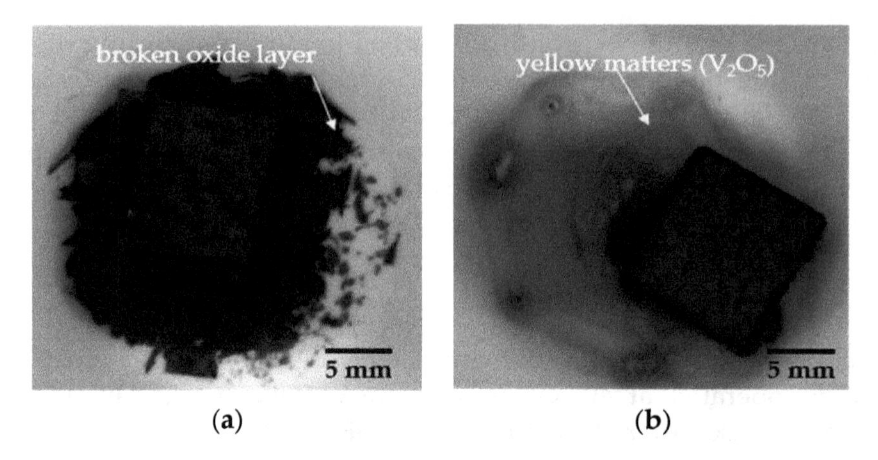

Figure 2. Macroscopic morphologies of TF550 alloy along with Ti–6Al–4V alloy after oxidation at 1073 K for 100 h. (**a**) Ti–6Al–4V and (**b**) TF550.

Figure 3 shows the X-ray diffraction profiles of TF550 alloy at 873 and 1073 K for 100 h. The surface oxide product was identified as TiO_2 and V_2O_5 at 873 K, while it was detected to be only TiO_2 at 1073 K. In addition, the peak intensity of V_2O_5 was weaker than that of TiO_2 at 873 K. Based on the oxidation kinetic relationship [26],

$$\Delta W = kt^n \tag{2}$$

where ΔW is the oxidation weight gain, n is the oxidation reaction index and t is the oxidation time, respectively. The oxidation reaction index (n) could be obtained to discuss the oxidation mechanism [26–29]. For example, when n is equal to 1, the oxidation reaction rate follows the linear law and the oxidation process is controlled by the reaction rate of oxygen and the substrate. When n is equal to 0.5, the oxidation reaction rate follows the parabolic law and the oxidation process is controlled by the diffusion of reactants in the oxide film. Based on the oxidation weight gain curves (Figure 1a), oxidation reaction indices were obtained through double logarithmic transformation, as shown in Figure 1b. However, considering the influencing factors of a lower oxidation weight gain and measurement deviation at 873 K, it was difficult to discuss the oxidation mechanism of TF550 alloy based only on the oxidation reaction index.

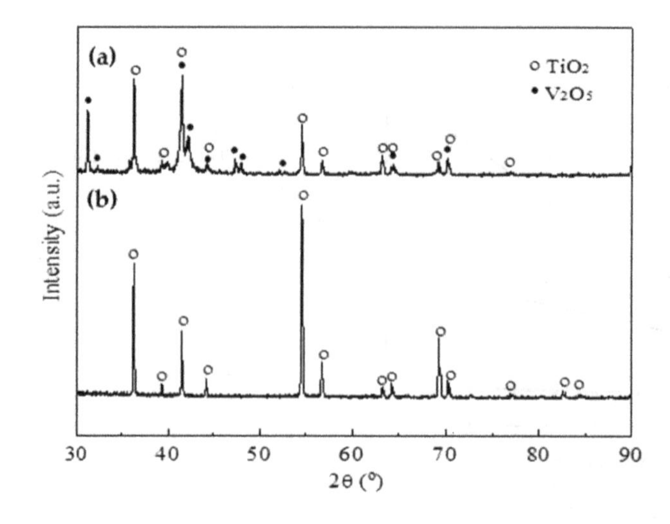

Figure 3. X-ray diffraction (XRD) profiles of TF550 alloy after oxidation at different temperatures for 100 h. (**a**) 873 K and (**b**) 1073 K.

Figure 4a shows the morphology of the oxide layer in TF550 alloy after oxidation at 873 K for 100 h, by SEM observation. The average thickness of the oxide layer was measured to be 7 μm. Figure 4b–f show the EDS mapping of main elements of Ti, V, Cr, Si, and O. The concentration of O element in the oxide layer was higher than that of the substrate, while those of Ti and V elements in the oxide layer were lower. Combined with XRD results (Figure 3), the surface oxide layer consisted of TiO_2 and V_2O_5. Consequently, the oxidation process of TF550 alloy at 873 K was deduced to be controlled by the oxidation reaction at the interface between the substrate and oxide layer.

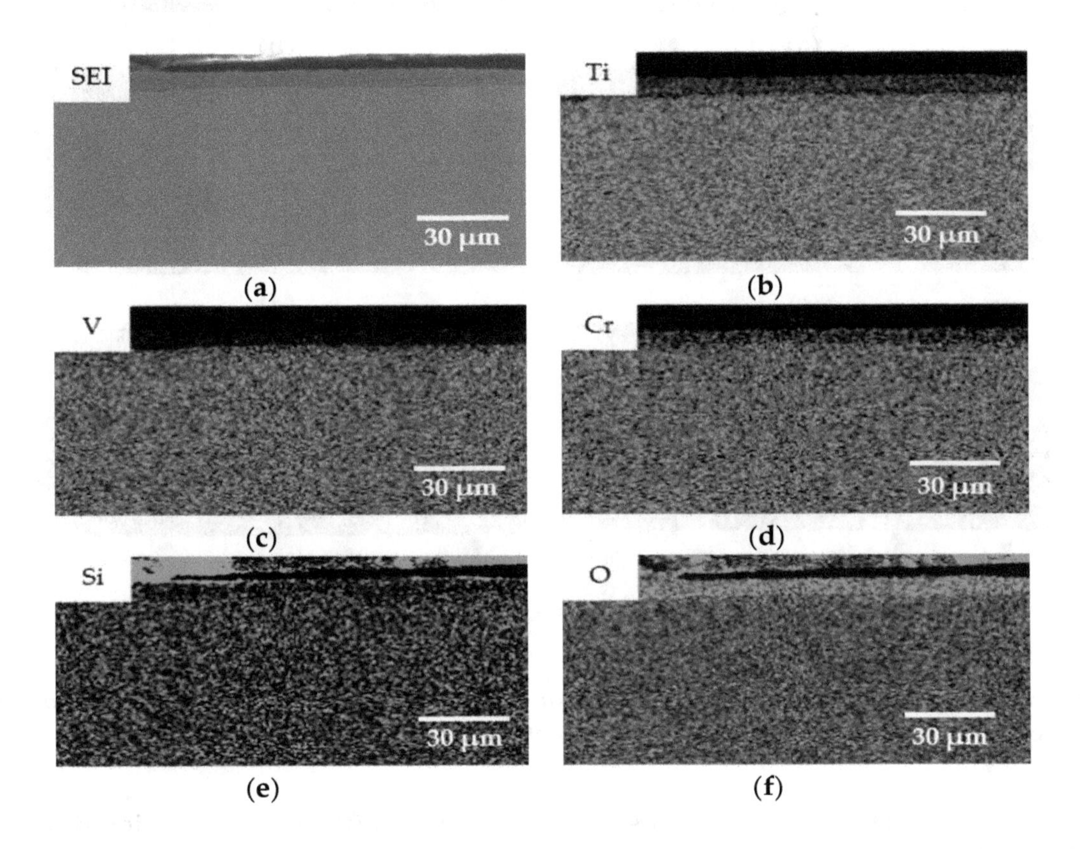

Figure 4. Scanning electron microscope (SEM) morphology and energy dispersive spectrometer (EDS) mapping of the oxide layer in TF550 alloy after oxidation at 873 K for 100 h. (**a**) Secondary electron image (SEI); (**b**) Ti element; (**c**) V element; (**d**) Cr element; (**e**) Si element; and (**f**) O element.

After oxidation at 1073 K for 100 h, the oxidation reaction index of TF550 alloy was analyzed to be 0.91, which was slightly larger than that (0.71) of Ti–6Al–4V alloy. The average thickness of the oxide layer as shown in Figure 5a was 460 μm after oxidation at 1073 K, which was much thicker than that of the oxide layer at 873 K. Based on the EDS mapping from Figure 5b–f, the Ti and O elements were clearly detected in the oxide layer, indicating that the oxide layer mainly consisted of TiO_2 combined with XRD results (Figure 3). The detected Si element on the top of Figures 4e and 5e was a residual polishing solution (SiO_2), which was considered to have no effect on the oxidation mechanism. Thus, the oxidation reaction rate at 1073 K followed the linear law and its oxidation process was controlled by the oxidation reaction at the interface between the substrate and oxide layer [26]. As shown in Figure 5c, the V element was clearly identified in the oxide layer, while the V_2O_5 was not detected by XRD as shown in Figure 3. Due to the low melting point of V_2O_5, they flowed into the crucible rather than adhered to the specimen, resulting in an absence of V_2O_5, which was in accordance with aforementioned occurrence of yellow matters in the crucible (Figure 2b). Furthermore, a positive concentration gradient of V occurred from the substrate to the surface due to the continuous consumption of V_2O_5 in the oxide layer, which accelerated the diffuse rate of V [30]. Consequently, V element was clearly detected in the oxide layer in terms of the EDS analysis.

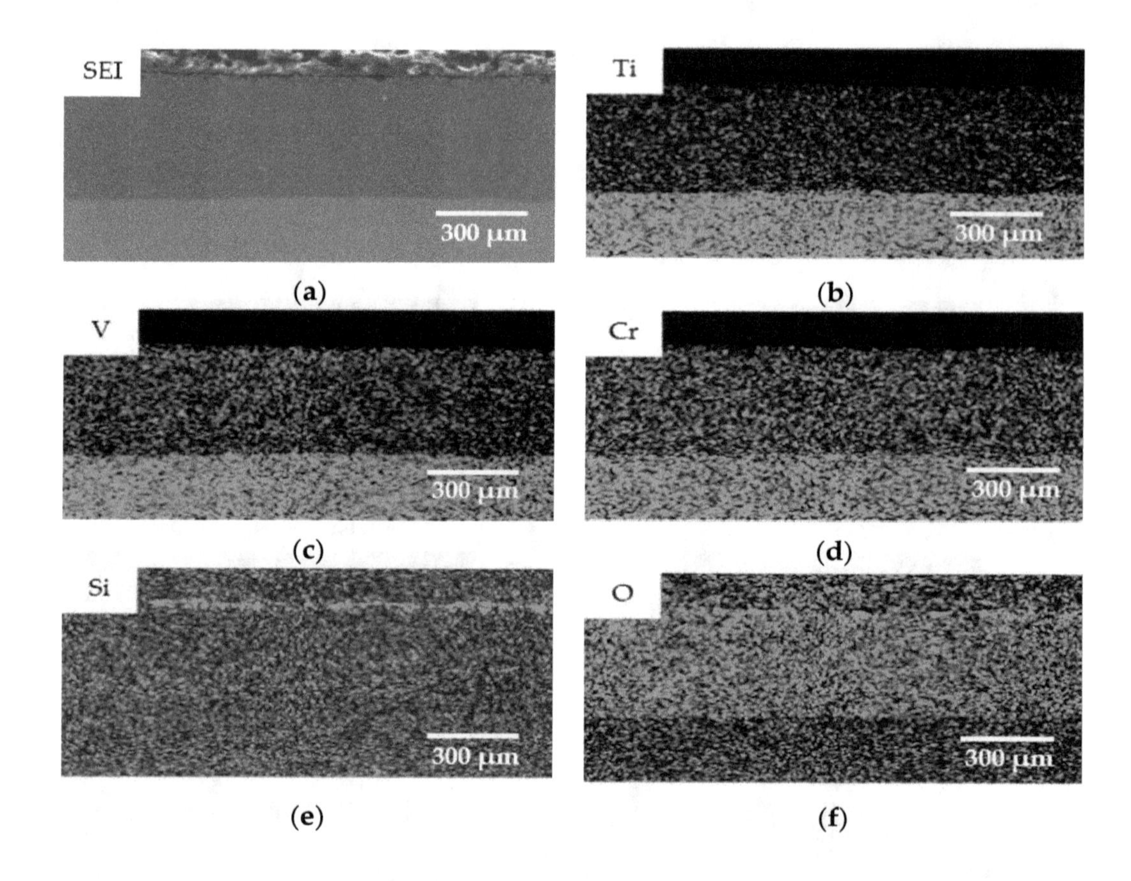

Figure 5. SEM morphology and EDS mapping of the oxide layer in TF550 alloy after oxidation at 1073 K for 100 h. (**a**) SEI; (**b**) Ti element; (**c**) V element; (**d**) Cr element; (**e**) Si element; and (**f**) O element.

Figure 6 shows the friction coefficients of TF550 alloy as a function of wearing time at 873 and 1073 K. The friction coefficient at 873 K was largely fluctuated with wearing time up to 10 min, and then reached a steady-state condition. Compared with 873 K, the friction coefficient at 1073 K became smaller, and its fluctuation became much weaker. The mean friction coefficient (f_m) at 873 and 1073 K, as listed in Table 1, were calculated to be 0.394 and 0.286, respectively. On the other hand, the volume wear rate of the disk (W_{sd}) at 873 K was obtained to be 2.21 $\times$ 10^{-13} m^3/Nm (Table 1), while it could not be measured at 1073 K because of the amount of oxides formed on the worn surface. For a comparison, the volume wear rate of the pin (W_{sp}) at 873 K was 0.197 $\times$ 10^{-13} m^3/Nm, which was smaller than that (0.566 $\times$ 10^{-13} m^3/Nm) of the pin at 1073 K (Table 1). Figure 7 shows the XRD profiles of worn surfaces for disks at different testing temperatures. Besides the β-phase, weak peaks of TiO_2 were detected on the worm surface at 873 K. Only peaks for both oxides, i.e., TiO_2 and V_2O_5, were present at 1073 K, while the peaks of the β-phase disappeared.

Table 1. Mean friction coefficient (f_m), volume wear rate of the disk (W_{sd}), and the pin (W_{sp}) in TF550 alloy after wearing at high temperatures of 873 and 1073 K.

Temperature (K)	f_m	W_{sd} (10^{-13}·m^3/Nm)	W_{sp} (10^{-13}·m^3/Nm)
873	0.394 ± 0.095	2.21 ± 0.35	0.197 ± 0.072
1073	0.286 ± 0.050	-	0.566 ± 0.039

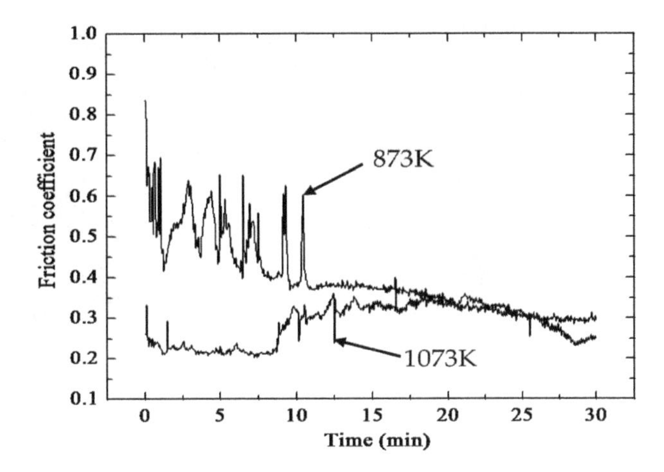

Figure 6. Friction coefficient curves of TF550 alloy as a function of wearing time at high temperatures of 873 and 1073 K.

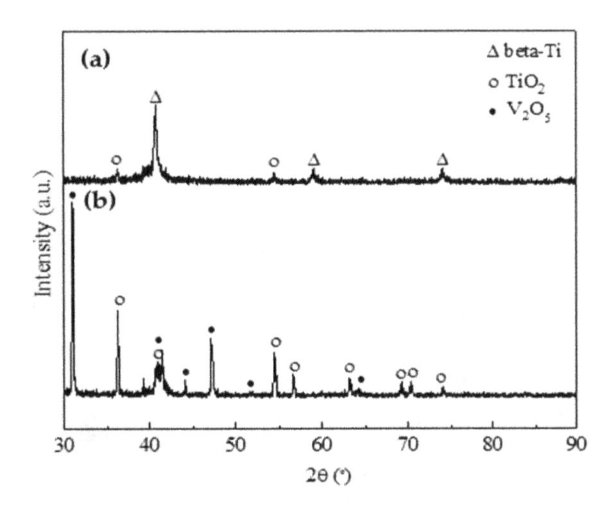

Figure 7. X-ray diffraction profiles of worn surfaces in TF550 alloy after wearing at high temperatures. (**a**) 873 K and (**b**) 1073 K.

Figure 8a shows the morphology of worn surfaces with a width of the wearing indent equal to about 1.2 mm at 873 K. Furrows indicated by arrows in Figure 8b were readily identified on the worn surface. Black smooth regions were also observed, which were regarded as tribo-layers containing the oxide of TiO_2 in terms of the EDS analysis (Figure 8c). In addition, some white particles consisting of not only metal debris but also the oxide of TiO_2 were confirmed by the EDS analysis as well (Figure 8d). Figure 9 shows the morphologies and compositions of worn surfaces at 1073 K. An average width of the wearing indent was 1.6 mm, and snowflake-like features were present on the worn surface (Figure 9a). The furrows were not observed and few black smooth regions appeared, as shown in Figure 9b; instead, a large amount of particles were present on the worn surface. The particles were further observed in Figure 9c and they were confirmed to be a mixture of the oxides of TiO_2 and V_2O_5, as shown in Figure 9d. Figure 10 shows the morphologies of the cross-section near the worn surfaces at 873 and 1073 K. The tribo-layer with an average thickness of 4 μm formed on the worn surface at 873 K (Figure 10a). However, this tribo-layer was not compact and it contained trace oxide of $TiO_{2.}$, identified by EDS in Figure 10b. At 1073 K, the tribo-layer became thicker (7 μm) and more continuous, and it contacted with the matrix compactly in Figure 10c. This tribo-layer was identified as both TiO_2 and V_2O_5 oxides by EDS in Figure 10d.

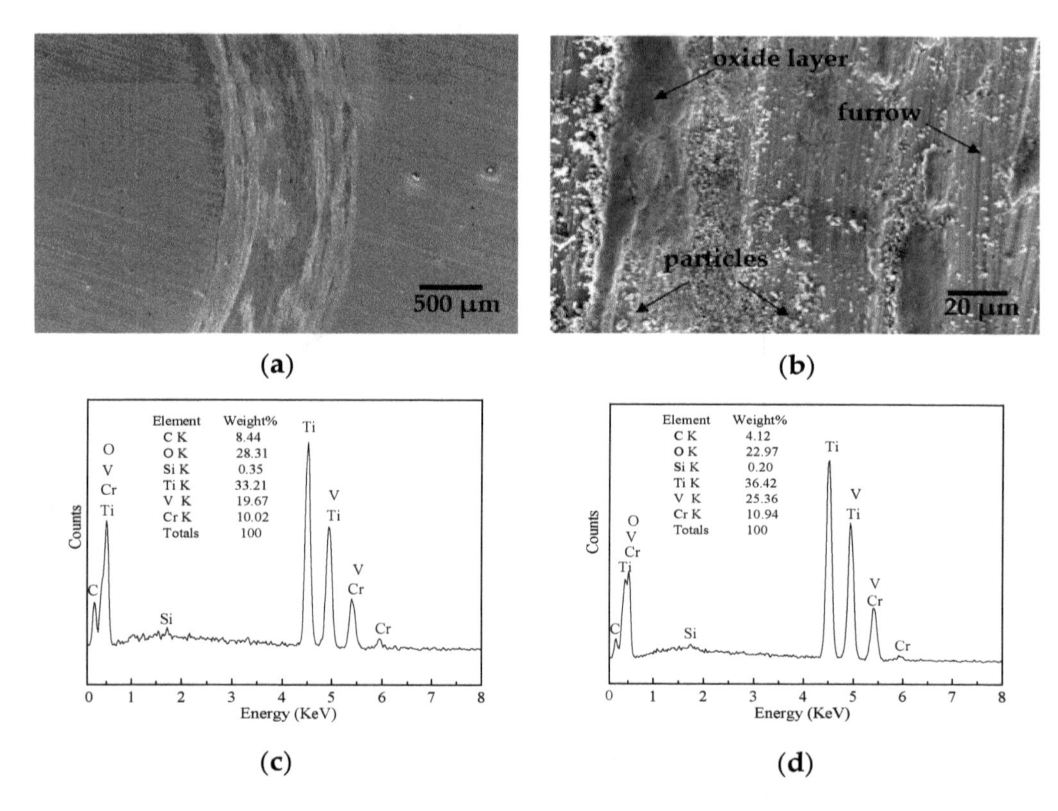

Figure 8. SEM morphologies and EDS spot-analyses of the worn surface in TF550 alloy after wearing at 873 K. (**a**) Low magnification SEI; (**b**) high magnification SEI; (**c**) compositions of the oxide layer region; and (**d**) compositions of the particles region.

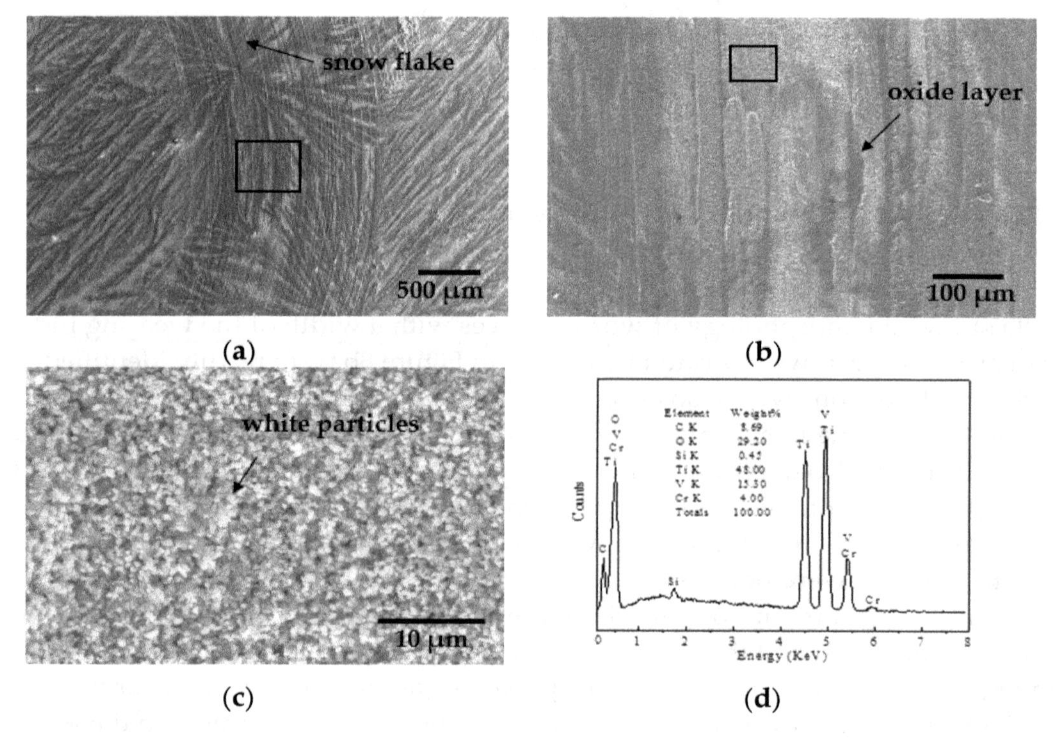

Figure 9. SEM morphologies and EDS spot-analysis of the worn surface in TF550 alloy after wearing at 1073 K. (**a**) Low magnification SEI; (**b**) high magnification SEI of area in black square in (**a**); (**c**) SEI of area in black square in (**b**); (**d**) compositions of the white particles region.

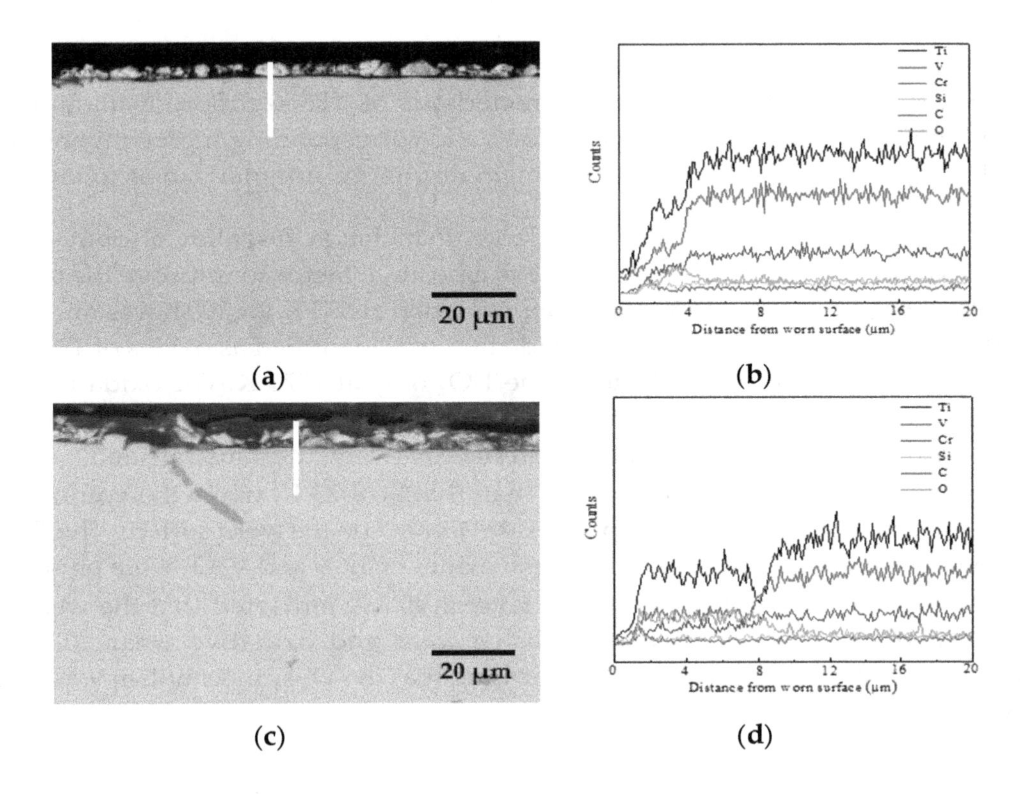

Figure 10. SEM cross-section morphologies and EDS line-analyses of the worn surface in TF550 alloy after wearing at high temperatures. (**a**) backscattered electron image (BEI) at 873 K; (**b**) compositions along the white line at 873 K; (**c**) BEI at 1073 K; and (**d**) compositions along the white line at 1073 K.

Based on the results of the friction coefficients, volume wear rates, and morphologies of worn surfaces, TF550 alloy presented different dry sliding wear behaviors at different testing temperatures. At 873 K, the tribo-oxide layer was not formed in the initial stage of dry sliding wear, and direct contact was not avoided completely, which resulted in a large friction coefficient. With an extension of wear time, a tribo-layer containing the oxide of TiO_2, i.e., black smooth regions formed, as shown in Figure 8b [31]. This tribo-oxide layer presented a partial protection from wearing for a decrease in the friction coefficient and its fluctuation. Some furrows appeared on the worn surface, while they were not covered with the tribo-oxide layer. Thus, the wear mechanism at 873 K was dominated by a combination of abrasive wear and oxidative wear. Amounts of oxides formed on the worn surface before the sliding test wear at 1073 K, because the temperature rose to the testing value in advance. The oxides formed before the sliding wear were loose, and thus were easily peeled off under the sliding in the initial period. Furthermore, V_2O_5 oxides could be melted during the sliding wear according to the oxidation morphology of TF550 alloy at 1073 K (Figure 2b). The molten V_2O_5 as a soft phase provided a lubricating effect, which led to a decrease in the friction coefficient, leading to a reduction of calorific value. Also, they filled defects between the hard phases, such as TiO_2 and metal debris, which could release the internal force between them, as shown in Figure 10c. Not only the loose oxides but also the molten V_2O_5 acted as a lubricant, leading to the absence of the running-in period and a relatively small friction coefficient during the wear process. The molten V_2O_5 flowed into the wear scratch and was subjected to solidification after wear testing, to exhibit the aforementioned snowflake-like features (Figure 9a), which was not in agreement with that of the single oxidation at 1073 K. In addition, the presence of amounts of oxides at relatively high temperatures made it difficult to measure the volume wear rate of the disk and resulted in an overestimated volume wear rate of the pin (Table 1). The furrows were not observed on the worn surface and amounts of oxide particles were present at 1073 K, indicating a typical feature of the oxidative wear mechanism.

4. Conclusions

In this study, the oxidation and wear characteristics of TF550 fireproof titanium alloy were investigated at high temperatures of 873 and 1073 K, and corresponding high temperature oxidation and wear mechanisms were also discussed. The main results are summarized as follows.

(1) The oxidation gain of TF550 alloy was much larger than that of Ti64 alloy, attributed the abundant addition of vanadium, which was regarded as a positive factor to improve the burn resistance. The oxidation weight gain of TF550 alloy after oxidation at 873 K for 100 h was much smaller than that at 1073 K for 100 h. The thin surface oxide film was identified as oxides of TiO_2 and V_2O_5 at 873 K, while the thick one was detected to be TiO_2 only at 1073 K. The oxidation reaction rate at 1073 K followed the linear law, indicating that the oxidation process of this alloy was mainly controlled by the oxidation reaction at the interface between the substrate and oxide layer.

(2) The friction coefficient at 873 K was larger than that at 1073 K, while the volume wear rate at 1073 K was larger because a large amount of oxides formed on the worn surface. The TiO_2 oxide was detected on the worn surface at 873 K, and both oxides of TiO_2 and V_2O_5 were present at 1073 K.

(3) The presence of furrows and a tribo-oxide layer at 873 K indicated that the wear mechanism was dominated by a combination of abrasive wear and oxidative wear. The continuous tribo-oxide layer was present at 1073 K, indicating that the wear mechanism was dominated by oxidative wear.

Acknowledgments: This work was financially supported by the National Natural Science Foundation of China, China (Grants Nos. 51471155 and 51471040). The authors would like to thank Jie Zhao and Wei Zhang of Dalian University of Technology for their assistance.

Author Contributions: Xiaohua Min and Guangbao Mi conceived and designed the experiments; Kai Yao and Pengfei Bai performed the experiments, analyzed the data and wrote the paper; Congqian Cheng contributed some analysis tools.

References

1. Banerjee, D.; Williams, J.C. Perspectives on Titanium Science and Technology. *Acta Mater.* **2013**, *61*, 844–879. [CrossRef]
2. Min, X.H.; Tsuzaki, K.; Emura, S.; Tsuchiya, K. Heterogeneous twin formation and its effect on tensile properties in Ti–Mo based β titanium alloys. *Mater. Sci. Eng. A* **2012**, *554*, 53–60. [CrossRef]
3. Min, X.H.; Emura, S.; Zhang, L.; Tsuzaki, K.; Tsuchiya, K. Improvement of strength-ductility tradeoff in β titanium alloy through pre-strain induced twins combined with brittle ω phase. *Mater. Sci. Eng. A* **2015**, *646*, 279–287. [CrossRef]
4. Luo, Q.S.; Li, S.F.; Pei, H.P. Progress in titanium fire resistant technology for aero-engine. *J. Aerosp. Power* **2012**, *27*, 2763–2767.
5. Zhang, P.Z.; Xu, Z.; Zhang, G.H.; He, Z.Y. Surface plasma chromized burn-resistant titanium alloy. *Surf. Coat. Technol.* **2007**, *201*, 4884–4887. [CrossRef]
6. Lv, D.S.; Xu, J.H.; Ding, W.F.; Fu, Y.C.; Yang, C.Y.; Su, H.H. Tool wear in milling Ti40 burn-resistant titanium alloy using pneumatic mist jet impinging cooling. *J. Mater. Process. Technol.* **2016**, *229*, 641–650. [CrossRef]
7. Zhang, X.J.; Cao, Y.H.; Ren, B.Y.; Tsubaki, N. Improvement of high-temperature oxidation resistance of titanium-based alloy by sol-gel method. *J. Mater. Sci.* **2010**, *45*, 1622–1628. [CrossRef]
8. Wang, M.M.; Zhao, Y.Q.; Zhou, L.; Zhang, D. Study on creep behavior of Ti–V–Cr burn resistant alloys. *Mater. Lett.* **2004**, *58*, 3248–3252. [CrossRef]
9. Xin, S.W.; Zhao, Y.Q.; Zeng, W.D.; Wu, H. Research on thermal stability of Ti40 alloy at 550 °C. *Mater. Sci. Eng. A* **2008**, *477*, 372–378. [CrossRef]
10. Li, Y.G.; Blenkinsop, P.A.; Loretto, M.H.; Walkern, N.A. Structure and stability of precipitates in 500 °C exposed in Ti–25V–15Cr–xAl. *Acta Mater.* **1998**, *46*, 5777–5794. [CrossRef]
11. Seagle, S.R. The state of the USA titanium industry in 1995. *Mater. Sci. Eng. A* **1996**, *213*, 1–7. [CrossRef]

12. Chen, Y.N.; Huo, Y.Z.; Song, X.D.; Bi, Z.Z.; Gao, Y.; Zhao, Y.Q. Burn-resistant behavior and mechanism of Ti14 alloy. *Int. J. Miner. Metall. Mater.* **2016**, *23*, 215–221. [CrossRef]
13. Campo, K.N.; Lima, D.D.; Lopes, E.S.N.; Caram, R. On the selection of Ti–Cu alloys for thixoforming processes: Phase diagram and microstructural evaluation. *J. Mater. Sci.* **2015**, *50*, 8007–8017. [CrossRef]
14. Chen, Y.N.; Yang, W.Q.; Zhan, H.F.; Zhang, F.Y.; Huo, Y.Z.; Zhao, Y.Q.; Song, X.D.; Gu, Y.T. Tailorable burning behavior of Ti14 alloy by controlling semi-solid forging temperature. *Materials* **2016**, *9*, 697. [CrossRef]
15. Mi, G.B.; Huang, X.; Cao, J.X.; Wang, B.; Cao, C.X. Microstructure characteristics of burning products of Ti–V–Cr fireproof titanium alloy by frictional ignition. *Acta Phys. Sin.* **2016**, *65*, 1–10.
16. Cao, J.X.; Huang, X.; Mi, G.B.; Sha, A.X.; Wang, B. Research progress on application technique of Ti–V–Cr burn resistant titanium alloys. *J. Aeronaut. Mater.* **2014**, *34*, 92–97.
17. Mi, G.B.; Cao, C.X.; Huang, X.; Cao, J.X.; Wang, B.; Sui, N. Non-isothermal oxidation characteristic and fireproof property prediction of Ti–V–Cr type fireproof titanium alloy. *J. Mater. Eng.* **2016**, *44*, 1–10.
18. Molinari, A.; Straffelini, G.; Tesi, B.; Bacci, T. Dry sliding wear mechanisms of the Ti6Al4V alloy. *Wear* **1997**, *208*, 105–112. [CrossRef]
19. Cui, X.H.; Mao, Y.S.; Wei, M.X.; Wang, S.Q. Wear Characteristics of Ti–6Al–4V Alloy at 20–400 °C. *Tribol. Trans.* **2012**, *5*, 185–190. [CrossRef]
20. Sun, Q.C.; Hu, T.C.; Fan, H.Z.; Zhang, Y.S.; Hu, L.T. Dry sliding wear behavior of TC11 alloy at 500 °C: Influence oflaser surface texturing. *Tribol. Int.* **2015**, *92*, 136–145. [CrossRef]
21. Mao, Y.S.; Wang, L.; Chen, K.M.; Wang, S.Q.; Cui, X.H. Tribo-layer and its role in dry sliding wear of Ti–6Al–4V alloy. *Wear* **2013**, *297*, 1032–1039. [CrossRef]
22. Zeng, S.W.; Jiang, H.T.; Zhao, A.M. High Temperature Oxidation Behavior of TC4 Alloy. *Rare Met. Mater. Eng.* **2015**, *44*, 2812–2816.
23. Zeng, S.W.; Zhao, A.M.; Jiang, H.T.; Fan, X.; Duan, X.G.; Yan, X.Q. Cyclic Oxidation Behavior of the Ti–6Al–4V Alloy. *Oxid. Met.* **2014**, *81*, 467–476. [CrossRef]
24. Mendez, S.F.O.; Rodriguez, C.R.S.; Venegas, K.C.; Aquino, J.A.M.; Magana, F.E. Magnetism and decarburization-like diffusion process on V_2O_5-doped ZnO ceramics. *Ceram. Int.* **2015**, *41*, 6802–6806. [CrossRef]
25. Li, Y.; Bai, C.Y.; Deng, X.Y.; Li, J.B.; Jing, Y.N.; Liu, Z.M. Effect of V_2O_5 addition on the properties of reaction-bonded porous SiC ceramics. *Ceram. Int.* **2014**, *40*, 16581–16587. [CrossRef]
26. Birks, N.; Meier, G.H.; Pettit, F.S. *Introduction of Metal Oxidation at High Temperature*; Higher Education Press: Beijing, China, 2010; p. 43.
27. Tomasi, A.; Gialanella, S. Oxidation phenomena in a Ti_3Al base-alloy. *Therm. Acta* **1995**, *269*, 133–143. [CrossRef]
28. Frangini, S.; Mignone, A.; De Riccardis, F. Various aspects of the air oxidation behavior of a Ti6Al4V alloy at temperatures in the range 600–700 °C. *J. Mater. Sci.* **1994**, *29*, 714–720. [CrossRef]
29. Jia, W.J.; Zeng, W.D.; Zhang, X.M.; Zhou, Y.G.; Liu, J.R.; Wang, Q.J. Oxidation behavior and effect of oxidation on tensile properties of Ti60 alloy. *J. Mater. Sci.* **2011**, *46*, 1351–1358. [CrossRef]
30. Huang, X.; Cao, C.X.; Ma, J.M.; Wang, B.; Gao, Y. High temperature oxidation behavior of a fire-resistant titanium alloy. *Rare Met. Mater. Eng.* **1997**, *26*, 27–30.
31. Wang, L.; Zhang, Q.Y.; Li, X.X.; Cui, X.H.; Wang, S.Q. Dry Sliding Wear Behavior of Ti–6.5Al–3.5Mo–1.5Zr–0.3Si Alloy. *Metall. Mater. Trans. A* **2014**, *45A*, 2284–2296. [CrossRef]

7

Evaluation of Fatigue Behavior in Dental Implants from In Vitro Clinical Tests

Rosa Rojo [1,*,†], María Prados-Privado [2,3,†], Antonio José Reinoso [4] and Juan Carlos Prados-Frutos [1]

1 Department of Medicine and Surgery, Faculty of Health Sciences, Rey Juan Carlos University, 28922 Alcorcon, Spain; juancarlos.prados@urjc.es
2 Department Continuum Mechanics and Structural Analysis Higher Polytechnic School, Carlos III University, 28911 Leganes, Spain; mprados@ing.uc3m.es
3 Asisa Dental (Engineering Researcher), José Abascal 32, 28003 Madrid, Spain
4 Department of ICT Engineering, Alfonso X El Sabio University, 28691 Madrid, Spain; areinpei@myuax.com
* Correspondence: rosa.rojo@urjc.es
† These authors contributed equally to this work.

Abstract: In the area of dentistry, there is a wide variety of designs of dental implant and materials, especially titanium, which aims to avoid failures and increase their clinical durability. The purpose of this review was to evaluate fatigue behavior in different connections and implant materials, as well as their loading conditions and response to failure. In vitro tests under normal and dynamic loading conditions evaluating fatigue at implant and abutment connection were included. A search was conducted in PubMed, Scopus, and Science Direct. Data extraction was performed independently by two reviewers. The quality of selected studies was assessed using the Cochrane Handbook proposed by the tool for clinical trials. Nineteen studies were included. Fourteen studies had an unclear risk and five had high risk of bias. Due to the heterogeneity of the data and the evaluation of the quality of the studies, meta-analysis could not be performed. Evidence from this study suggests that both internal and morse taper connections presented a better behavior to failure. However, it is necessary to unify criteria in the methodological design of in vitro studies, following methodological guidelines and establishing conditions that allow the homogenization of designs in ISO (International Organization for Standardization) standards.

Keywords: biomechanics; dental implant(s); in vitro; systematic reviews; evidence-based medicine

---

1. Introduction

The use of dental implants has become a common practice for replacing missing teeth in different clinical situations [1,2]. The used materials are chosen according to both their mechanical and chemical properties, as well as to their biocompatibility [3]. Commercially, pure titanium and its alloys are widely used for manufacturing dental implants because of excellent mechanical and physical properties, and favorable rates for long-term clinical survival [4]. In addition, titanium-based implants have a good resistance to corrosion with an excellent biocompatibility and high modulus of elasticity [5,6]. The use of Ti–6Al–4V [7] alloy is employed for biomedical application and, also, in dental implants due to its high mechanical resistance, which ensures load transmission to bone tissues over a long time, which is necessary when damaged hard tissues are replaced by prostheses [3]. This alloy presents a drawback due to the use of vanadium and aluminum that can cause some toxic effects. Other titanium grades can be employed in dental implants but they also have disadvantages like the Young modulus, relatively low mechanical strength, poor wear resistance and difficulty to improve the mechanical properties without reducing biocompatibility. New β-type titanium alloys for dental implants have been developed. These have good properties and less toxicity, good ductility, high resistance. They also have elastic modules closer to those of human bone compared to other alloys [7].

Although the mechanical strength of dental implants is important, they must also present adequate stiffness to avoid shielding the bones from stress. This stress shielding induces loss of bone density, leading to bone atrophy. Moreover, the interaction between titanium and tissues is a key factor in the success of dental implants and, for this reason, surfaces of used alloys are conveniently treated. If the implant is manufactured with titanium grade 5, those implants must have a surface treatment to improve the corrosion [3]. This interaction between titanium and tissues is affected by the implant surface composition, as well as by its hydrophilicity, morphology, and roughness [8]. Different surface treatments have been tried and developed with the aim of obtaining titanium surfaces with better biological properties. This surface treatment yields a good osseointegration and obtain an improvement on the success of dental implants [9] with a change in the chemical composition. Nano roughness, texture, and porosity are some of the most important factors in the surface of an implant because they affect the ability of cells to adhere to a solid substrate [10].

The manufacturing process also influences the alloy's characteristics. The tensile strength of titanium alloy can range from 369 to 3267 MPa depending on the process employed. Fatigue behavior is also affected by the manufacturing process and it can be also improved by combining the material properties, surface properties, and design optimization of implants.

Despite all the advantages of titanium, considered as the "gold standard" material for the manufacture of dental implants, its biggest drawback are aesthetic considerations. Therefore, manufacturers began to use other types of materials such as ceramics [4,11] or polymers [4].

Osseointegrated dental implants are described in detail in a great number of studies [12], although implant-supported connection have been less disclosed in the literature [13].

Nowadays, a vast number of implant designs are available. The first dental implants had an external connection, where the hexagonal anti-rotational component is the most common design. Figure 1a shows an example of this connection. Due to the high rate of rejected implants, a new connection was designed. In this case, internal connection (Figure 1b) allowed a better union between the implant and the abutment. Finally, morse taper connection (Figure 1c), which is another option for an internal connection, was introduced because of its improvement on screw loosening [14–16].

These designs are different in terms of the connection of the implant-abutment (external or internal connections) [17,18], which from a mechanical point of view, is the weakest area of the implant system [19].

Applying static load tests to evaluate the strength of the implants and their components is a common practice. However, these tests do not simulate real situations for implants [2]. Considering that masticatory forces are cyclic, a fatigue testing should be carried out to predict how long an implant system is going to function properly [20]. In vitro tests should better simulate the clinical situations [21–23] and allow clinicians to understand the probability of survival of prosthetic components and implants [13]. Before implant components are launched to the market, they should satisfy the ISO 14801 specifications [24]. This ISO recommendation was planned for single, endosteal, transmucosal dental implants tested under worst case applications. Nevertheless, several testing protocols for evaluating the mechanical reliability of dental implants are available in the literature [25]. Different loading angles, frequency of loads, and application load levels have been employed in several published cyclic testing protocols [26].

The fracture of an implant or of any of its components is an important complication which limits the lifetime of the reconstruction. Although most of the studies available are limited to 5–7 years of follow-up [26], Snauwaert et al. found in a 15-year study, an early implant fracture (up to one year after abutment connection) in 3.4% and late implant failures of 7.4% [27]. Considering that implants should serve for decades, these type of studies are inadequate to analyze implant failure or fracture and make essential to examine dental implants under fatigue testing approaches. Once the number of cycles an implant can support until failure are known, its expected life can be predicted accurately [28].

The aim of this systematic review was to evaluate fatigue behavior in different connections and types of implants, their loading conditions, and their response to failure between implant and abutment.

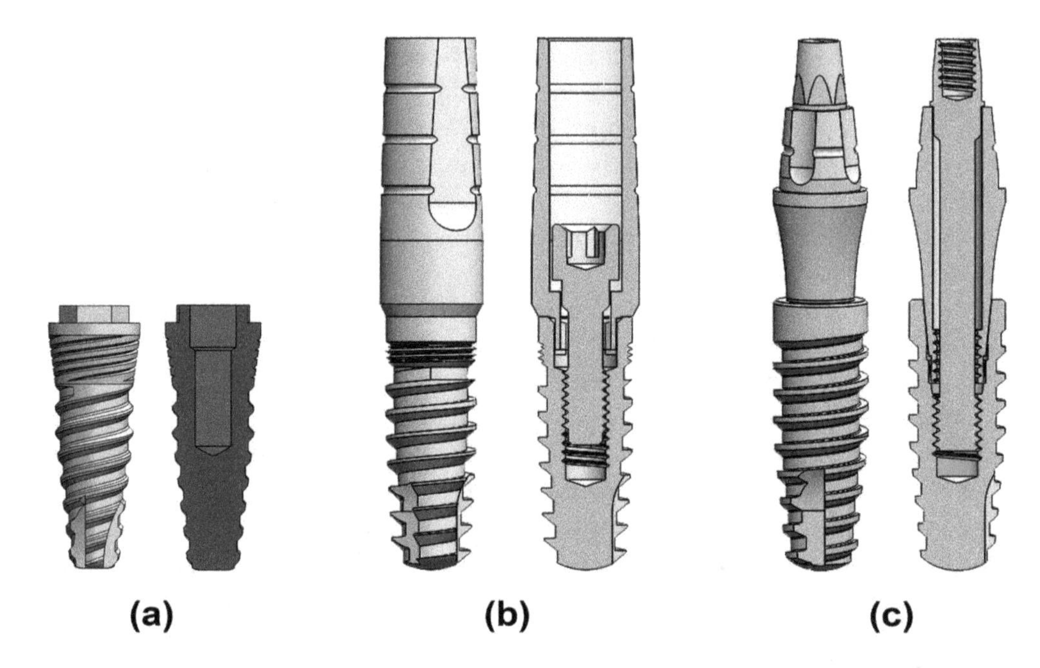

Figure 1. Schematic views of dental implant connections: (**a**) External connection (**b**) Internal connection (**c**) Morse taper connection.

2. Materials and Methods

This systematic review follows the Preferred Reporting Items for Systematic Reviews and Meta-Analyses (PRISMA) guidelines [29].

2.1. Focused Questions

First question (A): Does a certain number of cycles and a certain force (Newtons) exist between a defined implant connection and abutment failure?

Second question (B): How many cycles does a certain implant connection and abutment fail?

2.2. Inclusion and Exclusion Criteria

Inclusion criteria were in vitro clinical studies on dental implants in which fatigue at the implant and abutment connection were evaluated by subjecting them to dynamic cyclic loads. Studies were been carried out under normal environmental conditions. Dental implants are included disregarding the type of connection. There were no restrictions on the language or date of publication.

Exclusion criteria were all designs of observational studies, reviews, thermal fatigue assessment, static tests, or using incorrect units of measure.

2.3. Search Strategy

In this paper the research questions were elaborated considering each of the components of the PICO(S) [30] strategy research questions which is explained as follows: (P) dental implants and abutments; (I) cyclic loads; (C) studies with or without a comparison group where external, internal or morse taper connections were evaluated with implant materials and/or abutment of titanium, zirconium or others; (O) the evaluation of fatigue in terms of failure; (S) in vitro study.

An electronic search was performed in MEDLINE/PubMed, Scopus and Science Direct, database until the 15 March 2018. The search strategy used is detailed in Table 1.

Table 1. Search strategies carried out in databases.

Database	Search Strategy	Search Data
MEDLINE/PubMed	(dental AND (implant OR abutment) OR tooth implant) AND (cyclic loading) AND ((internal OR external) connection) AND (fatigue OR moment OR stress) AND ("in vitro" OR "experimental study") NOT (review)	15 March 2018
Scopus	(dental AND (implant OR abutment) OR tooth implant) AND (cyclic loading) AND ((internal OR external) connection) AND (fatigue OR moment OR stress) AND ("in vitro" OR "experimental study") AND NOT (review)	15 March 2018
Science Direct	(dental AND (implant OR abutment) OR tooth implant) AND (cyclic loading) AND ((internal OR external) connection) AND (fatigue OR moment OR stress) AND ("in vitro" OR "experimental study") AND NOT (review)	15 March 2018

2.4. Study Selection

Two authors (Rosa Rojo and María Prados-Privado) performed all the search operations and selected articles fulfilling the inclusion criteria independently and in duplicate. Additionally, the references of the articles included in this work were manually reviewed. Disagreements between the two authors were reviewed in a complete text by a third author (Juan Carlos Prados-Frutos) to make the final decision. The level of agreement between the reviewers regarding study inclusion was calculated using Cohen's kappa statistic.

2.5. Data Extraction

Two of the authors (María Prados-Privado and Rosa Rojo) collected all the data from the selected articles in duplicate and independently.

2.6. Study Quality Assessment

The assessment of risk of bias from clinical in vitro studies was evaluated by two of the authors (María Prados-Privado and Antonio José Reinoso), who were previously trained by an expert in evaluation of systematic reviews. For the assessment of risk of bias the Cochrane Handbook [31] was followed which incorporates seven domains: random sequence generation (selection bias); allocation concealment (selection bias); masking of participants and personnel (performance bias); masking of outcome assessment (detection bias); incomplete outcome data (attrition bias); selective reporting (reporting bias); and other bias.

The articles that did not achieve consensus between the two authors were reviewed by a third author (Rosa Rojo) to make the final decision.

The studies were classified into the following categories: low risk of bias—low risk of bias for all key domains; unclear risk of bias—unclear risk of bias for one or more key domains; high risk of bias—high risk of bias for one or more key domains.

2.7. Statistical Analysis

To evaluate the agreement between the inter-examiner, the statistic Cohen's kappa and the interpretation proposed by Landis & Koch [32] was used. Statistical calculations were performed with R software version 3.4.1 (R Core Development Team, R Foundation, Vienna, Austria) with the interrater reliability (irr) package.

3. Results

3.1. Study Selection

Figure 2 shows a flowchart of the study selection. All electronic search strategies provided 161 potential articles. Two of the authors (Rosa Rojo and María Prados-Privado) independently identified 48 eligible documents. The general agreement of eligibility of the studies between the authors was high ($k = 0.87$; $p = 0.049$.) Specifically, agreement between authors on the selection of articles in each considered database was high for all of them: Medline/PubMed ($k = 0.93$; $p = 0.046$), Scopus ($k = 0.80$; $p = 0.043$), and Science Direct ($k = 0.82$; $p = 0.05$). A total of 38 studies were excluded because they did not meet the defined inclusion criteria. Additionally, a manual search has been carried out to analyze the references cited in 10 of the articles that were included in this work. We reviewed 342 references. After removing duplicates, we analyzed the titles, abstracts and, when required, the full-text from 272 citations. A total of 263 studies were excluded as they did not match the inclusion criteria. As a result, nine additional articles were incorporated from the manual search. Finally, a total of nineteen in vitro studies were analyzed.

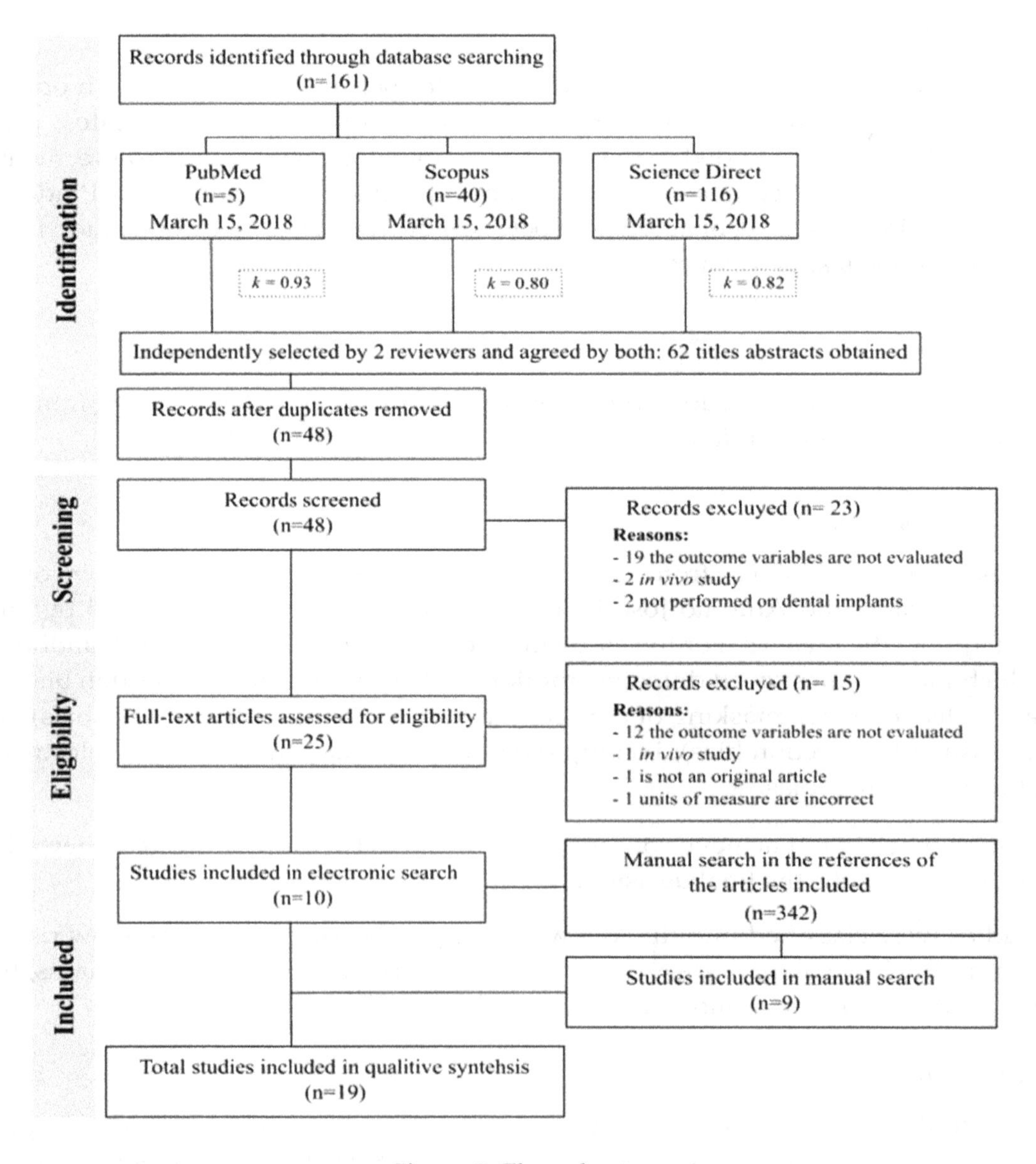

Figure 2. Flow chart.

3.2. *Relevant Data from Studies*

Two of the authors (María Prados-Privado and Rosa Rojo) extracted all the data from the selected works whose characteristics are shown in Table 2. The 19 in vitro studies analyzed [20,33–50] were carried out in several countries: Japan, Italy, Germany, Australia, Brazil, Switzerland, Turkey, Republic of Singapore, Greece, and the United States.

For all the studies, we collected the most important variables that could affect the results, such as sample size, the existence of funding, main characteristic of dental implants (connection, material, diameter and length), main properties of abutments (material and length), and applied load (magnitude, angulation, frequency and cycles).

Table 3 shows the 14 studies that answered the first question (A). The best behavior analyzed corresponded to a study where the zirconium-based implant hardened with alumina-doped yttrium-stabilized yttrium zirconium polyurethane and abutment of the same material was applied with a force of 98 N to 10,000,000 cycles [48]. No fault displayed.

Moreover, were are studies where the zirconium-based implant and abutment with an applied force of 50 N at 100 N, 45° of the axial axis of the dental implant and from 1,200,000 [46] to 3,600,000 cycles [50], respectively, showed failure under similar conditions.

It was observed that the titanium had a better behavior if the implant and the abutment were made of the same material [42,44,46,50]. The results also suggest that the behavior of the titanium worsens when the materials of the implant and the abutment are different [20,36,38].

Table 4 shows the six studies that answer the second question (B). The maximum number of reported cycles where the connection between the implant and the pillar fails is 5,000,000 cycles [33,41,45]. From the studies presenting better behavior fatigues, it has been found that the implant and the abutment are both made from titanium [33,45], or the abutment is combated with zirconium [41].

For both questions, Tables 3 and 4 show that the implant and the abutment behave better in the internal connections [33,41–46,48,50] and morse taper [41,47,49]. The study conducted by Mitsias et al. [47] answers the two research questions addressed in this work.

3.3. *Study Quality Assessment*

Evaluation of selection bias: They were only included in two of the analyzed studies of the method of randomization used [43,45]. However, it does not indicate whether there was concealment of this allocation.

Evaluation of performance bias: In all the studies analyzed there was no blinding of staff or assessors. Moreover, we found that in [34] the evaluator who prepared the specimens and who performed the tests were the same person. This fact may lead to a high risk of bias.

Assessment of detection bias: The results were not blinded in any of the studies.

Evaluation of attrition bias: All studies reported the complete results of the specimens defined in the clinical trial, although some reported inaccurately without indicating in the results the variable descriptions in their methodology [39,42].

Evaluation of notification bias: All studies provide detailed results with the exception of one, this study did not describe correctly whether the variables are quantitative or qualitative [39].

Evaluation of other bias: Funding was considered as another possible risk of bias in study designs. Eleven trials were funded by commercial firms [20,35,36,40–43,46,48–50], five were not reported [33,34,37,38,44], and three reported that no funding existed [39,45,47].

Using the evaluation of the seven domains for risk of bias it was determined that five had a high risk of bias [20,33–36], 14 an unclear risk [37–50], and none had a low risk of bias. Figure 3 shows a detailed description of the risk assessment of bias in the included studies.

Table 2. Main characteristics of the included studies.

Author/Year	Country	Journal	n	G	Financing
Balfour et al. [33] 1995	United States	Journal of Prosthetic Dentistry	21	3	U
Khraisat et al. [34] 2002	Japan	Journal of Prosthetic Dentistry	14	2	U
Çehreli et al. [35] 2004	Turkey	Clinical Oral Implants Research	8	1	Y
Butz et al. [36] 2005	United States	Journal of Oral Rehabilitation	48	3	Y
Gehrke et al. [37] 2006	Germany	Quintessence International	7	1	U
Kohal et al. [38] 2009	Germany	Clinical Implant Dentistry and Related Research,	48	3	U
Scarano et al. [39] 2010	Italy	Italian Oral Surgery	20	1	N
Magne et al. [40] 2011	Switzerland	Clinical Oral Implants Research	28	2	Y
Seetoh et al. [41] 2011	Republic of Singapore	The International Journal of Oral & Maxillofacial Implants	30	6	Y
Dittmer et al. [20] 2012	Germany	Journal of Prosthodontic Research	60	2	Y
Stimmelmayr et al. [42] 2012	Germany	Dental Materials	6	2	Y
Foong et al. [43] 2013	Australia	Journal of Prosthetic Dentistry	22	2	Y
Pintinha et al. [44] 2013	Brazil	Journal of Prosthetic Dentistry	48	2	U
Marchetti et al. [45] 2014	Italy	Implant Dentistry	15	2	N
Rosentritt et al. [46] 2014	Germany	Journal of Dentistry	64	8	Y
Mitsias et al. [47] 2015	Greece	The International Journal of Prosthodontics	36	2	N
Spies et al. [48] 2016	Germany	Journal of the Mechanical Behavior of Biomedical Materials	48	3	Y
Guilherme et al. [49] 2016	United States	Journal of Prosthetic Dentistry	57	3	Y
Preis et al. [50] 2016	Germany	Dental Materials	60	6	Y

n: sample size; G: Number of groups; Y: Yes; N: No; U: Unclear.

Table 3. Variables analyzed from the answer to question A.

Author/Year	Implant				Abutment		Applied Load			Cycles	Failure
	Connection	Material	Diameter	Length	Material	Length	Magnitude (N)	Angulation (°)	Frequency (Hz)		
Çehreli et al. [35] 2004	-	-	10	-	-	-	75 ± 5	20	0.5	500,000	N
Butz et al. [36] 2005	E	-	4	13	Ti	-	30	130	1.3	1,200,000	Y
	-	-	4	13	Zr	-	30	130	1.3	1,200,000	Y
	-	-	4	13	Ti	-	30	130	1.3	1,200,000	Y
Gehrke et al. [37] 2006	I	-	4.5	18	Zi	-	100–450	-	15	5,000,000	Y
	I	-	4.5	18	Zi	-	100–450	-	15	5,000,000	Y
	I	-	4.5	18	Zi	-	100–450	-	15	5,000,000	Y
Kohal et al. [38] 2009	M	Zr	-	-	Zi	-	45	-	-	1,200,000	Y
	M	Zr	-	-	Zi	-	45	-	-	1,200,000	Y
	M	Ti	-	-	P	-	45	-	-	1,200,000	Y
Scarano et al. [39] 2010	M	Ti	4	13	-	-	5–230	30	4	1,000,000	N

Table 3. *Cont.*

Author/Year	Implant				Abutment		Applied Load			Cycles	Failure
	Connection	Material	Diameter	Length	Material	Length	Magnitude (N)	Angulation (°)	Frequency (Hz)		
Magne et al. [40] 2011	I	-	4.1	12	Metal	12	80–280	30	5	20,000	Y
Dittmer et al. [20] 2012	I	-	4.5	13	Ti	1.5	100	30	2	1,000,000	Y
	I	-	4.5	13	Ti	4.1	100	30	2	1,000,000	Y
	E	-	4.3	13	-	11	100	30	2	1,000,000	Y
	I	-	4.5	14	-	-	100	30	2	1,000,000	Y
	E	-	4	13	-	1	100	30	2	1,000,000	Y
	I	-	4.1	14	Ti	5.5	100	30	2	1,000,000	Y
Stimmelmayr et al. [42] 2012	I	Ti	3.8	13	Ti	10	100	-	1.2	1,200,000	N
	I	Ti	3.8	13	Zr	10	100	-	1.2	1,200,000	N
Pintinha et al. [44] 2013	I	Ti	4	10	Ti	8.7	100 ± 5	20	2	500	N
	I	Ti	4	10	Ti	9	100 ± 5	20	2	500	N
Rosentritt et al. [46] 2014	I	Zr	4.1	10	Zr	-	50	45	1.6	1,200,000	Y
	I	Zr	4	10	Zr	-	50	45	1.6	1,200,000	N
	I	Zr	4.1	11	Zr	-	50	45	1.6	1,200,000	Y
	I	Zr	4.1	14	Zr	-	50	45	1.6	1,200,000	Y
	I	Ti	4	10	Ti	-	50	45	1.6	1,200,000	Y
	I	Ti	4.1	15	Ti	-	50	45	1.6	1,200,000	N
	I	Zr	4.5	12	Zr	-	50	45	1.6	1,200,000	N
	I	Zr	4	10	Zr	-	50	45	1.6	1,200,000	N
Mitsias et al. [47] 2015	M	-	-	-	Y-TZP	-	400	30	-	100,000	Y
	M	-	-	-	Y-TPZ	-	400	30	-	100,000	Y
Spies et al. [48] 2015	E	ATZ	4.4/4.1/4.2	12	-	-	98	-	2	10,000,000	N
	I	Y-TZP-A	4.1	12	Y-TZP-A	6	98	-	2	10,000,000	N
	E	Y-TZP-A	4.2	12	ATZ	6	98	-	2	10,000,000	N
Guilherme et al. [49] 2016	M	-	4.3	10	Zr	-	150–200	-	2	100	N
	M	-	4.3	10	LD	-	150–200	-	2	100	N
	M	-	4.3	10	R-BC	-	150–200	-	2	100	N
Preis et al. [50] 2016	I	Zr	4.1	10	Zr	-	100	45	1.6	3,600,000	N
	I	Zr	4.1	10	Zr	-	100	45	1.6	3,600,000	Y
	I	Zr	3.8	11	Zr	-	100	45	1.6	3,600,000	Y
	I	Zr	4.6	11	Zr	-	100	45	1.6	3,600,000	Y
	I	Zr	4.1	10	Zr	-	100	45	1.6	3,600,000	Y
	I	Zr	4.1	10	Zr	-	100	45	1.6	3,600,000	Y
	I	Ti	4.1	12	Ti	-	100	45	1.6	3,600,000	N

I: Internal connection; E: external connection; M; morse taper connection; Ti: Titanium; Zr: Zirconia; ATZ: alumina-toughened zirconia; Y-TZP-A: yttrium stabilized tetragonal zirconium dioxide polycrystal doped with alumina; LD: lithium disilicate; R-BC: resin-based composite; Y: Yes; N: No; -: No data.

Table 4. Variables analyzed from the answer to question B.

Author/Year	Implant				Abutment		Applied Load			Cycles	Failure
	Connection	Material	Diameter	Length	Material	Length	Magnitude	Angulation	Frequency		
Balfour et al. [33] 1995	E	Ti	-	-	Ti	-	242	30	14	5,000,000	Y
	I	Ti	-	-	Ti	-	400	30	14	5,000,000	Y
	I	Ti	-	-	Ti	-	367	30	14	5,000,000	Y
Khraisat et al. [34] 2002	E	Ti	4	10	Ti	3	100	90	1,25	1,800,000	Y (1,178,023 and 1,733,526)
	M	Ti	4.1	10	Ti	10	100	90	1,25	1,800,000	Y (more 1,800,000)
Seetoh et al. [41] 2011	M	-	4.5	15	Zr/Ti	-	21	45	10	5,000,000	Y
	I	-	4	15	Zr/Ti	-	21	45	10	5,000,000	Y
	M	-	4.1	14	Zr/Ti	-	21	45	10	5,000,000	Y
Foong et al. [43] 2013	I	Ti	4	9	Ti	1.5	50–400	30	2 to 5	5,000–20,000	Y (mean of 81,935)
	I	Ti	4	9	Zr	1.5	50–400	30	2 to 5	5,000–20,000	Y (mean of 26,926)
Marchetti et al. [45] 2014	I	Ti	3.8	13	Ti	-	400	30 ± 2	15	5,000,000	Y (12,678 and 15,387)
	I	Ti	3.8	13	Ti	-	300	30 ± 2	15	5,000,000	Y (more 27,732)
Mitsias et al. [47] 2015	M	-	-	-	Y-TZP	-	400	30	-	100,000	Y (less than 50,000)
	M	-	-	-	Y-TPZ	-	400	30	-	100,000	Y (less than 50,000)

I: Internal connection; E: external connection; M; morse taper connection; Ti: Titanium; Zr: Zirconia; Y: Yes; N: No; -: No data.

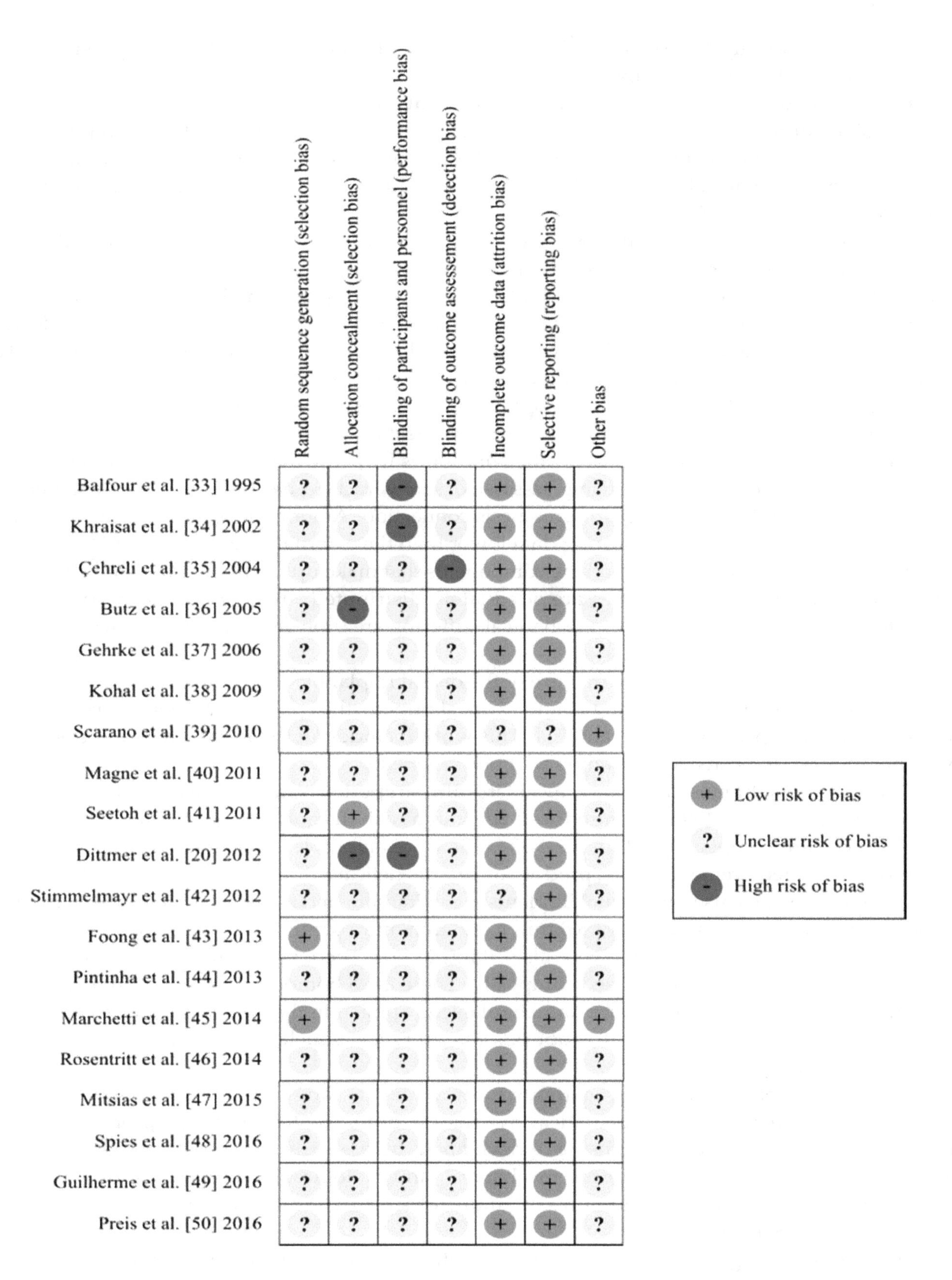

Figure 3. Assessment of risk of bias of included studies.

4. Discussion

There are several factors that influence the behavior of dental implants such as the biological effects of the location and magnitude of applied force [51], occlusal forces following implant treatment [52], immediate or early implant loading [53], the influence of bone quality, effects of prosthesis type, prosthesis material, or implant support [51].

The clinical long-term success of restorations in oral implantology depends, among other factors on the stable connection between the dental implant and the abutment [54]. Therefore, a good knowledge of the biomechanical behavior of dental implants is essential for clinical decision making and thus, avoid mechanical failures. These are mainly due to fatigue caused by overload or loss of bone around the implant [55–57].

The most common connections between the implant and the prosthesis are the external, internal hexagon and morse taper connections [58]. The main advantages of the external connection are the compatibility with a wide variety of implants, its economical price, the long-term follow-up data available, and the literature provides solutions to the main drawbacks associated with its use. However, among the main disadvantages of this connection are the loosening of the screws, the possibility of screw fracture, worse aesthetic results, and an inadequate microbial seal [59–61].

The appearance of the internal connections was due to the interest in trying to reduce the aesthetic and microbial filtration problems of the external connection, while improving the behavior of the implant and peri-implant bone against masticatory forces [60]. The main advantages of these internal connections are less or no screw loosening, less risk of screw fracture, aesthetic improvement, microbial sealing, and the stability of the implant-prosthesis connection. Among the disadvantages are a higher economic cost and long-term monitoring data lower than the external connection [59,62].

In the morse taper connections, all the components that make up the implantoprosthesis assembly behave as a single whole, the forces are adequately distributed, the stability of the prosthesis is guaranteed, and the areas of greatest mechanical suffering are protected, such as the crestal region of the implant [17,63].

Other factors that influence the fracture of the dental implant are the materials, the diameter and length of the implant, the material and length of the abutment, the applied force, its frequency, and the angulation with respect to the implant [51]. This work includes all the parameters described above even though there is a considerable amount of heterogeneity.

Titanium is still the most used material and it is also supported by long-term clinical studies [4], but also in vitro studies since 1995 [33] introduced zirconium with the gold standard, titanium, to evaluate the fatigue of dental implants. Although the results were favorable, its biggest drawback is its high degree of quality requirement during production and its delicate clinical management [4].

The materials most used for dental implants were titanium and zirconia, with ranges of measures in relation to diameter and height, of 3.8 mm to 4.6 mm and 10 mm to 15 mm, respectively. For the abutment the most used materials were also titanium and zirconia but in the one study [49] also used lithium disilicate and resin-based composite. There is no homogeneous criterion regarding the magnitude, angulation, frequency, and number of cycles applied in the dynamic loads to the implants. Therefore, with the data reported it is not possible to report conclusive results from the different in vitro studies analyzed.

There are several engineering methods to evaluate the fatigue behavior of dental implants such as finite elements [64], mathematical models for probabilistic fatigue [65], or other in vitro studies. These type of study designs have developed, before their use in patients, new materials to understand their physical, chemical, mechanical, and biological properties. These designs can be developed under normal conditions or under other conditions such as in water [66,67] or hydrothermal [68,69] environments.

There is not a large number of studies evaluating fatigue in dental implants in the scientific literature. However, a greater number of in vitro studies which evaluated the same objectives under normal conditions have been addressed. It is therefore desirable to conduct a systematic review.

The main disadvantage of the evaluation of the articles of this work is the high heterogeneity of the confounding factors collected as the different materials, lengths, diameters, or dynamic loading conditions. There is no homogeneity in the design criteria established by each study in this regard.

With the objective of obtaining more clinically relevant information, future studies should incorporate and analyze the same parameters. Otherwise, there heterogeneity will continue creating doubt in the scientific literature in the field of dentistry [2]. Therefore and, in view of the results obtained in this systematic review, future in vitro analysis should have the same implants dimensions (diameter and length) and cyclic loading conditions (number of cycles, magnitude of force, angle, and medium). The no homogeneity found in these studies contribute to realize that it is necessary to standardize the criteria for carrying out the studies in order to make more concise and reliable comparisons.

Some studies [39,45,46,48,49] use ISO 14801 [24] which specifies the conditions that each type of implant must support to obtain certification. However, these conditions indicate that they should exceed a minimum number of cycles but say nothing about how to conduct experimental studies of implants that are already on the market, and therefore have already been certified.

In systematic reviews, a qualitative analysis of the included studies is required. This is done through the risk assessment of bias. This ensures that the data collected and analyzed have been managed in a controlled manner, avoiding all possible methodological errors in clinical trials. When the data are homogeneous, in addition, a quantitative analysis can be carried out, through meta-analysis [31].

This means that each study conditions to decide which subjects its implants, without any specific criteria, which makes it very difficult to know which of the implants have better mechanical behavior under certain conditions. This fact is evident in Tables 3 and 4 in which it is possible to observe the great variety of conditions used in the articles included in this review.

We used the Cochrane Handbook tool to assess the risk of bias in the studies, noting that in most domains, no data are given that give transparency to the studies. Generally, the criteria for randomization and allocation masking, and blinding of staff and data assessors are not indicated. Together with these detected defects and under recommendations some authors [70] in vitro studies should be treated to promote the quality of the tests: simple size calculation, meaningful difference between groups, sample preparation and handling, allocation sequence, randomization and blinding, statistical analysis.

The heterogeneity of data available in the scientific literature does not allow a meta-analysis in the field of in vitro fatigue and fracture of dental implants. As in the design of clinical trials in humans with the CONSORT (Consolidated Standards of Reporting Trials) guidelines [71], we consider it advisable to follow guidelines for in vitro studies such as the CRIS (Checklist for Reporting in vitro Studies) guidelines [70]. Also, define the criteria and conditions of applied loads (magnitude, angle, frequency, cycles ...) and are contained in an ISO standard.

Nevertheless, we have found, in the present review that the internal connections [42,44], and those based on the morse taper system [34,39,49,72] show a better performance against resistance to fracture in the dental implant compared to the external connection [34,48]. Moreover, the results revealed that the implant and the abutment have better behavior if both materials are the same.

In addition, these studies assessed a range of materials, but the most frequently used materials are still in order of use, titanium and zirconium, with a behavior similar to fatigue.

5. Conclusions

The limitations found in this review do not allow us to report consistent evidence. The results suggest that the internal and morse connections are the best for resisting the fracture of the dental implants and the most commonly used materials are titanium and zirconium. However, it is necessary to unify criteria in the methodological design of this type of in vitro studies.

Author Contributions: R.R. developed the main part of the review, created the search strategy, selected the articles, performed the statistical analysis and wrote part of the paper. M.P.-P. selected the articles, evaluated the methodological quality of the studies and wrote part of the paper. A.J.R. evaluated the methodological quality of the studies. J.C.P.-F. provided critical analysis and interpretation of data.

Acknowledgments: The authors received financial support of grants A-274 (Instradent Iberia® S.A., Alcobendas, Spain) and A-285 (Proclinic® S.A, Zaragoza, Spain).

References

1. Tian, K.; Chen, J.; Han, L.; Yang, J.; Huang, W.; Wu, D. Angled abutments result in increased or decreased stress on surrounding bone of single-unit dental implants: A finite element analysis. *Med. Eng. Phys.* **2012**, *34*, 1526–1531. [CrossRef] [PubMed]
2. Coray, R.; Zeltner, M.; Özcan, M. Fracture strength of implant abutments after fatigue testing: A systematic review and a meta-analysis. *J. Mech. Behav. Biomed. Mater.* **2016**, *62*, 333–346. [CrossRef] [PubMed]
3. Elias, C.N.; Fernandes, D.J.; Resende, C.R.S.; Roestel, J. Mechanical properties, surface morphology and stability of a modified commercially pure high strength titanium alloy for dental implants. *Dent. Mater.* **2015**, *31*, e1–e13. [CrossRef] [PubMed]
4. Osman, R.B.; Swain, M.V. A Critical Review of Dental Implant Materials with an Emphasis on Titanium versus Zirconia. *Materials* **2015**, *8*, 932–958. [CrossRef] [PubMed]
5. Ottria, L.; Lauritano, D.; Andreasi Bassi, M.; Palmieri, A.; Candotto, V.; Tagliabue, A.; Tettamanti, L. Mechanical, chemical and biological aspects of titanium and titanium alloys in implant dentistry. *J. Biol. Regul. Homeost. Agents* **2018**, *32*, 81–90. [PubMed]
6. Kirmanidou, Y.; Sidira, M.; Drosou, M.-E.; Bennani, V.; Bakopoulou, A.; Tsouknidas, A.; Michailidis, N.; Michalakis, K. New Ti-Alloys and Surface Modifications to Improve the Mechanical Properties and the Biological Response to Orthopedic and Dental Implants: A Review. *Biomed Res. Int.* **2016**, *2016*, 2908570. [CrossRef] [PubMed]
7. Kent, D.; Wang, G.; Dargusch, M. Effects of phase stability and processing on the mechanical properties of Ti-Nb based beta Ti alloys. *J. Mech. Behav. Biomed. Mater.* **2013**, *28*, 15–25. [CrossRef] [PubMed]
8. Hatamleh, M.M.; Wu, X.; Alnazzawi, A.; Watson, J.; Watts, D. Surface characteristics and biocompatibility of cranioplasty titanium implants following different surface treatments. *Dent. Mater.* **2018**, *34*, 676–683. [CrossRef] [PubMed]
9. Wennerberg, A. The importance of surface roughness for implant incorporation. *Int. J. Mach. Tools Manuf.* **1998**, *38*, 657–662. [CrossRef]
10. Komasa, S.; Taguchi, Y.; Nishida, H.; Tanaka, M.; Kawazoe, T. Bioactivity of nanostructure on titanium surface modified by chemical processing at room temperature. *J. Prosthodont. Res.* **2012**, *56*, 170–177. [CrossRef] [PubMed]
11. Kubasiewicz-Ross, P.; Dominiak, M.; Gedrange, T.; Botzenhart, U.U. Zirconium: The material of the future in modern implantology. *Adv. Clin. Exp. Med.* **2017**, *26*, 533–537. [CrossRef] [PubMed]
12. Freitas, G.; Hirata, R.; Bonfante, E.; Tovar, N.; Coelho, P. Survival Probability of Narrow and Standard-Diameter Implants with Different Implant-Abutment Connection Designs. *Int. J. Prosthodont.* **2016**, *29*, 179–185. [CrossRef] [PubMed]
13. Bartlett, D. Implants for life? A critical review of implant-supported restorations. *J. Dent.* **2007**, *35*, 768–772. [CrossRef] [PubMed]
14. Pardal-Pelaez, B.; Montero, J. Preload loss of abutment screws after dynamic fatigue in single implant-supported restorations. A systematic review. *J. Clin. Exp. Dent.* **2017**, *9*, e1355–e1361. [CrossRef] [PubMed]
15. Gehrke, S.A.; Delgado-Ruiz, R.A.; Prados Frutos, J.C.; Prados-Privado, M.; Dedavid, B.A.; Granero Marin, J.M.; Calvo Guirado, J.L. Misfit of Three Different Implant-Abutment Connections Before and After Cyclic Load Application: An In Vitro Study. *Int. J. Oral Maxillofac. Implant.* **2017**, *32*, 822–829. [CrossRef] [PubMed]
16. Prados-Privado, M.; Bea, J.A.; Rojo, R.; Gehrke, S.A.; Calvo-Guirado, J.L.; Prados-Frutos, J.C. A New Model to Study Fatigue in Dental Implants Based on Probabilistic Finite Elements and Cumulative Damage Model. *Appl. Bionics Biomech.* **2017**, *2017*. [CrossRef] [PubMed]
17. Schmitt, C.M.; Nogueira-Filho, G.; Tenenbaum, H.C.; Lai, J.Y.; Brito, C.; Döring, H.; Nonhoff, J. Performance of conical abutment (Morse Taper) connection implants: A systematic review. *J. Biomed. Mater. Res. Part A* **2014**, *102*, 552–574. [CrossRef] [PubMed]

18. Sailer, I.; Sailer, T.; Stawarczyk, B.; Jung, R.E.; Hämmerle, C.H.F. In vitro study of the influence of the type of connection on the fracture load of zirconia abutments with internal and external implant-abutment connections. *Int. J. Oral Maxillofac. Implant.* **2009**, *24*, 850–858. [CrossRef]
19. Kitagawa, T.; Tanimoto, Y.; Odaki, M.; Nemoto, K.; Aida, M. Influence of implant/abutment joint designs on abutment screw loosening in a dental implant system. *J. Biomed. Mater. Res. Part B Appl. Biomater.* **2005**, *75B*, 457–463. [CrossRef] [PubMed]
20. Dittmer, M.P.; Dittmer, S.; Borchers, L.; Kohorst, P.; Stiesch, M. Influence of the interface design on the yield force of the implant–abutment complex before and after cyclic mechanical loading. *J. Prosthodont. Res.* **2012**, *56*, 19–24. [CrossRef] [PubMed]
21. Ritter, J.E. Critique of test methods for lifetime predictions. *Dent. Mater.* **1995**, *11*, 147–151. [CrossRef]
22. Marx, R.; Jungwirth, F.; Walter, P.-O. Threshold intensity factors as lower boundaries for crack propagation in ceramics. *Biomed. Eng. Online* **2004**, *3*, 41. [CrossRef] [PubMed]
23. Alqahtani, F.; Flinton, R. Postfatigue fracture resistance of modified prefabricated zirconia implant abutments. *J. Prosthet. Dent.* **2014**, *112*, 299–305. [CrossRef] [PubMed]
24. Organization, I.S. *ISO 14801: Dentistry—Implants—Dynamic Fatigue Test for Endosseous Dental Implants*; ISO: Geneve, Switzerland, 2007.
25. Marchetti, E.; Ratta, S.; Mummolo, S.; Tecco, S.; Pecci, R.; Bedini, R.; Marzo, G. Mechanical Reliability Evaluation of an Oral Implant-Abutment System According to UNI EN ISO 14801 Fatigue Test Protocol. *Implant Dent.* **2016**, *25*, 613–618. [CrossRef] [PubMed]
26. Lee, C.K.; Karl, M.; Kelly, J.R. Evaluation of test protocol variables for dental implant fatigue research. *Dent. Mater.* **2009**, *25*, 1419–1425. [CrossRef] [PubMed]
27. Snauwaert, K.; Duyck, J.; van Steenberghe, D.; Quirynen, M.; Naert, I. Time dependent failure rate and marginal bone loss of implant supported prostheses: A 15-year follow-up study. *Clin. Oral Investig.* **2000**, *4*, 0013–0020. [CrossRef]
28. Hasan, I.; Bourauel, C.; Mundt, T.; Stark, H.; Heinemann, F. Biomechanics and load resistance of small-diameter and mini dental implants: A review of literature. *Biomed. Tech. Eng.* **2014**, *59*, 1–5. [CrossRef] [PubMed]
29. Moher, D.; Altman, D.G.; Liberati, A.; Tetzlaff, J. PRISMA statement. *Epidemiology* **2011**, *22*, 128. [CrossRef] [PubMed]
30. Centre for Rewies and Dissemination, University of York. *Systematic Reviews: CRD Guidance for Undertaking Reviews in Health Care*; University of York: York, UK, 2009.
31. Higgins, J.P.T.; Altman, D.G. Assessing risk of bias in included studies. In *Cochrane Handbook for Systematic Reviews of Interventions*; Higgins, J.P.T., Green, S., Eds.; Wiley: Hoboken, NJ, USA, 2008; pp. 187–241.
32. Landis, J.R.; Koch, G.G. The measurement of observer agreement for categorical data. *Biometrics* **1977**, *33*, 159–174. [CrossRef] [PubMed]
33. Balfour, A.; O'Brien, G.R. Comparative study of antirotational single tooth abutments. *J. Prosthet. Dent.* **1995**, *73*, 36–43. [CrossRef]
34. Khraisat, A.; Stegaroiu, R.; Nomura, S.; Miyakawa, O. Fatigue resistance of two implant/abutment joint designs. *J. Prosthet. Dent.* **2002**, *88*, 604–610. [CrossRef] [PubMed]
35. Cehreli, M.C.; Akca, K.; Iplikcioglu, H.; Sahin, S. Dynamic fatigue resistance of implant-abutment junction in an internally notched morse-taper oral implant: Influence of abutment design. *Clin. Oral Implant. Res.* **2004**, *15*, 459–465. [CrossRef] [PubMed]
36. Butz, F.; Heydecke, G.; Okutan, M.; Strub, J.R. Survival rate, fracture strength and failure mode of ceramic implant abutments after chewing simulation. *J. Oral Rehabil.* **2005**, *32*, 838–843. [CrossRef] [PubMed]
37. Gehrke, P.; Dhom, G.; Brunner, J.; Wolf, D.; Degidi, M.; Piattelli, A. Zirconium implant abutments: Fracture strength and influence of cyclic loading on retaining-screw loosening. *Quintessence Int.* **2006**, *37*, 19–26. [PubMed]
38. Kohal, R.-J.; Finke, H.C.; Klaus, G. Stability of prototype two-piece zirconia and titanium implants after artificial aging: An in vitro pilot study. *Clin. Implant Dent. Relat. Res.* **2009**, *11*, 323–329. [CrossRef] [PubMed]
39. Scarano, A.; Sacco, M.L.; Di Iorio, D.; Amoruso, M.; Mancino, C. Valutazione della resistenza a fatica ciclica di una connessione impianto-abutment cone morse e avvitata. *Ital. Oral Surg.* **2010**, *9*, 173–179. [CrossRef]

40. Magne, P.; Oderich, E.; Boff, L.L.; Cardoso, A.C.; Belser, U.C. Fatigue resistance and failure mode of CAD/CAM composite resin implant abutments restored with type III composite resin and porcelain veneers. *Clin. Oral Implant. Res.* **2011**, *22*, 1275–1281. [CrossRef] [PubMed]
41. Seetoh, Y.L.; Tan, K.B.; Chua, E.K.; Quek, H.C.; Nicholls, J.I. Load fatigue performance of conical implant-abutment connections. *Int. J. Oral Maxillofac. Implant.* **2011**, *26*, 797–806.
42. Stimmelmayr, M.; Edelhoff, D.; Güth, J.-F.; Erdelt, K.; Happe, A.; Beuer, F. Wear at the titanium–titanium and the titanium–zirconia implant–abutment interface: A comparative in vitro study. *Dent. Mater.* **2012**, *28*, 1215–1220. [CrossRef] [PubMed]
43. Foong, J.K.W.; Judge, R.B.; Palamara, J.E.; Swain, M.V. Fracture resistance of titanium and zirconia abutments: An in vitro study. *J. Prosthet. Dent.* **2013**, *109*, 304–312. [CrossRef]
44. Pintinha, M.; Camarini, E.T.; Sábio, S.; Pereira, J.R. Effect of mechanical loading on the removal torque of different types of tapered connection abutments for dental implants. *J. Prosthet. Dent.* **2013**, *110*, 383–388. [CrossRef] [PubMed]
45. Marchetti, E.; Ratta, S.; Mummolo, S.; Tecco, S.; Pecci, R.; Bedini, R.; Marzo, G. Evaluation of an Endosseous Oral Implant System According to UNI EN ISO 14801 Fatigue Test Protocol. *Implant Dent.* **2014**, *23*, 665–671. [CrossRef] [PubMed]
46. Rosentritt, M.; Hagemann, A.; Hahnel, S.; Behr, M.; Preis, V. In vitro performance of zirconia and titanium implant/abutment systems for anterior application. *J. Dent.* **2014**, *42*, 1019–1026. [CrossRef] [PubMed]
47. Mitsias, M.E.; Thompson, V.P.; Pines, M.; Silva, N.R.F.A. Reliability and failure modes of two Y-TZP abutment designs. *Int. J. Prosthodont.* **2015**, *28*, 75–78. [CrossRef] [PubMed]
48. Spies, B.C.; Nold, J.; Vach, K.; Kohal, R.-J. Two-piece zirconia oral implants withstand masticatory loads: An investigation in the artificial mouth. *J. Mech. Behav. Biomed. Mater.* **2016**, *53*, 1–10. [CrossRef] [PubMed]
49. Guilherme, N.M.; Chung, K.-H.; Flinn, B.D.; Zheng, C.; Raigrodski, A.J. Assessment of reliability of CAD-CAM tooth-colored implant custom abutments. *J. Prosthet. Dent.* **2016**, *116*, 206–213. [CrossRef] [PubMed]
50. Preis, V.; Kammermeier, A.; Handel, G.; Rosentritt, M. In vitro performance of two-piece zirconia implant systems for anterior application. *Dent. Mater.* **2016**, *32*, 765–774. [CrossRef] [PubMed]
51. Sahin, S.; Cehreli, M.C.; Yalcin, E. The influence of functional forces on the biomechanics of implant-supported prostheses—A review. *J. Dent.* **2002**, *30*, 271–282. [CrossRef]
52. Flanagan, D. Bite force and dental implant treatment: A short review. *Med. Devices* **2017**, *10*, 141–148. [CrossRef] [PubMed]
53. Chrcanovic, B.R.; Kisch, J.; Albrektsson, T.; Wennerberg, A. Factors Influencing Early Dental Implant Failures. *J. Dent. Res.* **2016**, *95*, 995–1002. [CrossRef] [PubMed]
54. Hoyer, S.A.; Stanford, C.M.; Buranadham, S.; Fridrich, T.; Wagner, J.; Gratton, D. Dynamic fatigue properties of the dental implant-abutment interface: Joint opening in wide-diameter versus standard-diameter hex-type implants. *J. Prosthet. Dent.* **2001**, *85*, 599–607. [CrossRef] [PubMed]
55. Piattelli, A.; Scarano, A.; Piattelli, M.; Vaia, E.; Matarasso, S. Hollow implants retrieved for fracture: A light and scanning electron microscope analysis of 4 cases. *J. Periodontol.* **1998**, *69*, 185–189. [CrossRef] [PubMed]
56. Tolman, D.E.; Laney, W.R. Tissue-integrated prosthesis complications. *Int. J. Oral Maxillofac. Implants* **1992**, *7*, 477–484. [CrossRef] [PubMed]
57. Steinebrunner, L.; Wolfart, S.; Ludwig, K.; Kern, M. Implant-abutment interface design affects fatigue and fracture strength of implants. *Clin. Oral Implant. Res.* **2008**, *19*, 1276–1284. [CrossRef] [PubMed]
58. Pita, M.S.; Anchieta, R.B.; Barao, V.A.; Garcia, I.R., Jr.; Pedrazzi, V.; Assuncao, W.G. Prosthetic platforms in implant dentistry. *J. Craniofac. Surg.* **2011**, *22*, 2327–2331. [CrossRef] [PubMed]
59. Gaviria, L.; Salcido, J.P.; Guda, T.; Ong, J.L. Current trends in dental implants. *J. Korean Assoc. Oral Maxillofac. Surg.* **2014**, *40*, 50–60. [CrossRef] [PubMed]
60. Finger, I.M.; Castellon, P.; Block, M.; Elian, N. The evolution of external and internal implant/abutment connections. *Pract. Proced. Aesthet. Dent.* **2003**, *15*, 625–632. [PubMed]
61. Binon, P.P. Implants and components: Entering the new millennium. *Int. J. Oral Maxillofac. Implant.* **2000**, *15*, 76–94.
62. Gracis, S.; Michalakis, K.; Vigolo, P.; Vult von Steyern, P.; Zwahlen, M.; Sailer, I. Internal vs. external connections for abutments/reconstructions: A systematic review. *Clin. Oral Implant. Res.* **2012**, *23* (Suppl. 6), 202–216. [CrossRef] [PubMed]

63. Macedo, J.P.; Pereira, J.; Vahey, B.R.; Henriques, B.; Benfatti, C.A.; Magini, R.S.; Lopez-Lopez, J.; Souza, J.C. Morse taper dental implants and platform switching: The new paradigm in oral implantology. *Eur. J. Dent.* **2016**, *10*, 148–154. [CrossRef] [PubMed]
64. Cheng, Y.C.; Lin, D.H.; Jiang, C.P.; Lee, S.Y. Design improvement and dynamic finite element analysis of novel ITI dental implant under dynamic chewing loads. *Biomed. Mater. Eng.* **2015**, *26* (Suppl. 1), S555–S561. [CrossRef] [PubMed]
65. Prados-Privado, M.; Prados-Frutos, J.C.; Calvo-Guirado, J.L.; Bea, J.A. A random fatigue of mechanize titanium abutment studied with Markoff chain and stochastic finite element formulation. *Comput. Methods Biomech. Biomed. Eng.* **2016**, *19*, 1583–1591. [CrossRef] [PubMed]
66. Anchieta, R.B.; Machado, L.S.; Hirata, R.; Bonfante, E.A.; Coelho, P.G. Platform-Switching for Cemented Versus Screwed Fixed Dental Prostheses: Reliability and Failure Modes: An In Vitro Study. *Clin. Implant Dent. Relat. Res.* **2016**, *18*, 830–839. [CrossRef] [PubMed]
67. Bordin, D.; Bergamo, E.T.P.; Fardin, V.P.; Coelho, P.G.; Bonfante, E.A. Fracture strength and probability of survival of narrow and extra-narrow dental implants after fatigue testing: In vitro and in silico analysis. *J. Mech. Behav. Biomed. Mater.* **2017**, *71*, 244–249. [CrossRef] [PubMed]
68. Kim, J.-W.; Covel, N.S.; Guess, P.C.; Rekow, E.D.; Zhang, Y. Concerns of hydrothermal degradation in CAD/CAM zirconia. *J. Dent. Res.* **2010**, *89*, 91–95. [CrossRef] [PubMed]
69. Kawai, Y.; Uo, M.; Wang, Y.; Kono, S.; Ohnuki, S.; Watari, F. Phase transformation of zirconia ceramics by hydrothermal degradation. *Dent. Mater. J.* **2011**, *30*, 286–292. [CrossRef] [PubMed]
70. Krithikadatta, J.; Gopikrishna, V.; Datta, M. CRIS Guidelines (Checklist for Reporting In-vitro Studies): A concept note on the need for standardized guidelines for improving quality and transparency in reporting in-vitro studies in experimental dental research. *J. Conserv. Dent.* **2014**, *17*, 301–304. [CrossRef] [PubMed]
71. Palmas, W. The CONSORT guidelines for noninferiority trials should be updated to go beyond the absolute risk difference. *J. Clin. Epidemiol.* **2017**, *83*, 6–7. [CrossRef] [PubMed]
72. Machado, L.S.; Bonfante, E.A.; Anchieta, R.B.; Yamaguchi, S.; Coelho, P.G. Implant-abutment connection designs for anterior crowns: Reliability and failure modes. *Implant Dent.* **2013**, *22*, 540–545. [CrossRef] [PubMed]

8

Corrosion Study of Implanted TiN Electrodes using Excessive Electrical Stimulation in Minipigs

Suzan Meijs [1,*], Kristian Rechendorff [2], Søren Sørensen [2] and Nico J.M. Rijkhoff [1]

[1] Department of Health, Science and Technology, Center for Sensory-Motor Interaction (SMI), Aalborg University, 9220 Aalborg, Denmark; nr@hst.aau.dk

[2] Materials Department, Danish Technological Institute, 8000 Århus, Denmark; krr@teknologisk.dk (K.R.); soren.steenfeldt.moller-sorensen@LEGO.com (S.S.)

* Correspondence: smeijs@hst.aau.dk

Abstract: (1) Background: Titanium nitride (TiN) electrodes have been used for implantable stimulation and sensing electrodes for decades. Nevertheless, there still is a discrepancy between the in vitro and in vivo determined safe charge injection limits. This study investigated the consequences of pulsing implanted electrodes beyond the in vivo safe charge injection limits. (2) Methods: The electrodes were implanted for a month and then pulsed at 20 mA and 50 mA and 200 Hz and 400 Hz. Afterwards, the electrodes were investigated using electrochemical and analytical methods to evaluate whether electrode degradation had occurred. (3) Results: Electrochemical tests showed that electrodes that pulsed at 20 mA and 200 Hz (lowest electrical dose) had a significantly lower charge injection capacity and higher impedance than the other used and unused electrodes. (4) Conclusions: The electrodes pulsed at the lowest electrical dose, for which no tissue damage was found, appeared to have degraded. Electrodes pulsed at higher electrical doses for which tissue damage did occur, on the other hand, show no significant degradation in electrochemical tests compared to unused implanted and not implanted electrodes. It is thus clear that the tissue surrounding the electrode has an influence on the charge injection properties of the electrodes and vice versa.

Keywords: implanted electrodes; electrical stimulation; corrosion

---

1. Introduction

Titanium nitride (TiN) has been used for implantable electrodes for many decades, starting with cardiac pacing electrodes [1]. The demands on cardiac pacing electrodes increased when it was desired to sense the heart rhythm, in order to provide rate-adaptive pacing [2]. Despite the high voltages applied during cardiac pacing, the electrode polarization should remain low so that the heart signal can reliably be recorded [1,3]. Porous electrodes were highly desirable for that purpose [4] but the electrodes should also be biocompatible [1] and corrosion resistant [3].

At the end of the previous century, TiN also received interest as a material for neural stimulation and recording electrodes [5]. Neural stimulation and recording applications within this field include, among others, visual prosthesis [6], brain implants [7,8] and cochlear implants [9]. Initially, studies reported conflicting results [5,6,9], which was likely due to differences in the fabrication method [4]. The majority of studies, however, reported very favourable properties of porous TiN [5,6], which were due to its large surface area rather than specific material properties [4].

The performance of stimulation and recording electrodes can be evaluated using their safe charge injection limits (Q_{inj}), charge storage capacity (CSC) and impedance. Q_{inj} is evaluated by comparing the electrode polarization under pulsing conditions to the safe potential limits established using slow sweep cyclic voltammetry (CV). The safe potential window is typically defined by the potentials at which water reduction and oxidation occurs. CSC is a measure of how much charge can be stored on the surface of the electrode and is measured using CV. The amount of charge available during fast pulsing, however, is typically much less than CSC. Impedance magnitude (typically at 1 kHz) can be used as a measure for battery consumption or recording performance. The lower the impedance, the better [10].

These properties are typically investigated under in vitro conditions in inorganic saline [5,6,9–20]. However, CSC, Q_{inj} and the impedance spectrum differ under acute and chronic in vivo circumstances [4,7,8,21–28]. Q_{inj} and electrode polarization have been reported to be significantly lower after implantation compared to in vitro measurements [7,8,23–27]. Moreover, they have been reported to decrease during the implanted period, when electrode failure does not occur [7,23–26].

TiN has long been known as a biocompatible [29–31] and corrosion resistant material [3,32,33], even under cathodic high voltage pulsing conditions [34]. Under anodic conditions, TiN oxidation reactions may occur, which primarily lead to passivation of these reactions until higher anodic voltages are reached [32]. At very high anodic voltages, TiN will eventually be degraded [34]. However, as Q_{inj} is lower when implanted compared to in vitro [7,8,21–27], unsafe voltages may be reached during pulsing. The aim of this study was therefore to investigate whether implanted TiN electrodes would degrade during pulsing when Q_{inj} measured in vivo was exceeded but Q_{inj} measured in saline was not.

2. Materials and Methods

Four Göttingen minipigs were implanted with four working electrodes (electrode pins) and four large surface area pseudo-reference disk electrodes. Minipigs were selected because the subcutaneous adipose tissue is similar to adipose tissue in humans. The number of electrodes and pseudo-reference electrodes was chosen in order not to cause excessive discomfort to the animals and thereby also to increase the homogeneity in the results. The electrodes were made from Ti6Al4V and coated with porous TiN. The animals recovered from anaesthesia and were monitored for one month before the corrosion experiments were conducted. The work was carried out according to Danish and European legislation.

2.1. Electrode Fabrication

TiN coatings were deposited on electrode pins (6 mm^2) made of a Ti6Al4V alloy and Ti disks (1000 mm^2) by reactive magnetron sputtering on a CC800/9 SiNOx coating unit (CemeCon AG, Würselen, Germany). The coatings were sputtered from four Ti targets (88 $\times$ 500 mm^2) with 99.5% purity in a Ar/N_2 mixture atmosphere. The purity of the gases was 99.999% and the Ar/N_2-flow was 300 sccm/350 sccm. The deposition time was 21,000 s. The electrodes underwent three-fold rotation during the coating process.

The electrodes coatings were investigated after deposition using samples taken from the same batch. Scanning electron microscopy (SEM) (Nova 600, FEI Company, Hillsboro, OR, USA) images were taken at magnifications ranging from 450$\times$ to 25,000$\times$ to get an overview of the surface and to investigate the surface structure of the electrodes in detail. A silicium sample coated in the same process was used to study the thickness, homogeneity and porous structure of the TiN coating using SEM. Images were recorded at a magnification of 40,000$\times$.

An ethylene tetrafluoroethylene (ETFE) coated 35N LT wire (Heraeus, Yverdon, Switzerland) was crimped to the hollow end of the electrode pins. A polyether ether ketone (PEEK) body and silicone tines were produced using injection moulding to insulate the electrode pins. The tines were first glued to the PEEK body using a silicone adhesive. The PEEK body with tines was then glued to the electrode pins also using a silicone adhesive. Further details and figures of the electrode production can be found in Reference [24]. The electrodes were cleaned thoroughly before they were sterilized by an overdose of electron-beam processing.

2.2. Surgical Procedure

For electrode implantation, the animals were anesthetized using Propofol. Small incisions were made in the back and the electrodes were implanted using a custom-made implantation tool. This ensured that the electrodes were placed in tight pockets in subcutaneous adipose tissue, which promotes fast healing and ingrowth. Five electrodes were implanted into each pig, four of which were used for intense electrical stimulation and one electrode in each pig served as a control. Four counter electrode disks with a percutaneous wire were also implanted in each pig. The minipigs recovered from the procedure and were carefully monitored in order to detect and treat cases of infection. The electrodes were not used for one month until the pigs were anaesthetized again using sevoflurane to perform electrical stimulation. After the stimulation sessions were completed, the electrodes were carefully dissected from the tissue. The electrodes were extensively cleaned using demineralized water and alcohol, they were then rinsed and stored dry, so that they could be further investigated.

2.3. Electrical Stimulation

Electrical stimulation was performed using a DS5 (Digitimer, Hertfordshire, UK) per electrode. The device was shorted between the pulses using a custom-build set-up to prevent drifting of the baseline potential. Biphasic, charge balanced 200 μs square pulses were applied, cathodic first with an inter-phase interval of 40 μs during which no current was applied. Stimulation was performed for 6 h in total, divided into three 2-h sessions. Before, between and after these sessions, voltage transient measurements (VTM) were recorded for each electrode using a VersaSTAT 3 potentio-galvanostat (Princeton Applied Research, Oak Ridge, TN, USA).

Four stimulation paradigms were applied:

- Group 1: 20 mA, 200 Hz
- Group 2: 20 mA, 400 Hz
- Group 3: 50 mA, 200 Hz
- Group 4: 50 mA, 400 Hz

During pilot experiments, it was verified that the group 1 stimulation paradigm did not cause tissue damage after one week of implantation (see Figure S1). To cause electrode damage, we decided to increase 2 parameters: stimulation frequency and stimulation amplitude. The stimulation frequency was doubled, which was expected to cause electrode damage due to an increasing trend in the inter-pulse potential [35]. The stimulation amplitude was set to the maximum the DS5 can deliver, which was expected to cause electrode damage by increasing the electrode potential during stimulation. It was expected that group 2–4 protocols would result in tissue damage; therefore photographs were taken of the tissue surrounding the electrodes to document the amount of tissue damage. However, the focus of this study is corrosion and the electrodes were thus investigated more extensively using electrochemical and analytical methods.

During stimulation, the voltage transients were recorded every 30 min using an oscilloscope. From these voltage transients, the resistive drop after pulse cessation (IR-drop) was calculated as [10]:

$$\text{IR-drop} = E_{\text{pulse\_end}} - E_{\text{pulse\_end}+40} \quad (1)$$

where $E_{\text{pulse\_end}}$ is the recorded potential at the end of the cathodic pulse and $E_{\text{pulse\_end}+40}$ is the potential 40 μs after pulse cessation (see Figure 1) [10]. E_{mc} and E_{ma} are the maximum cathodic and anodic voltage excursions after IR-drop is subtracted from the original voltage transient.

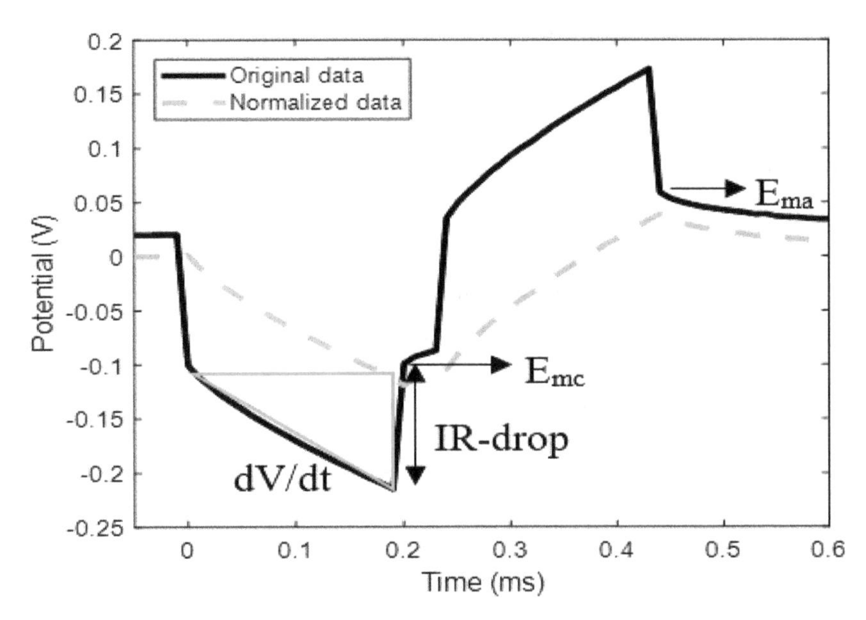

Figure 1. IR-drop, dV/dt, E_{mc} and E_{ma} are derived from the original data, while in the manuscript normalized data are presented. The data is normalized by subtracting the IR-drop and setting the pre-pulse potential to 0.

VTM before, between and after the stimulation blocks, was performed using the VersaSTAT 3 potentio-galvanostat using the same stimulation pulse.

The pulsing capacitance (C_{pulse}) was computed using the slope (dV/dt) of the voltage transient:

$$I_{stim} = C_{pulse} \cdot \frac{dV}{dt} \tag{2}$$

where I_{stim} is the stimulation current (1 mA while implanted and 5 mA in the electrochemical characterization). Q_{inj} was calculated using the current (I_{max}) at which E_{mc} or E_{ma} reached the safe potential limits (−0.6 and 0.9 V vs. open circuit potential, respectively) [10,24]:

$$Q_{inj} = \frac{I_{max} \cdot t}{A} \tag{3}$$

where t is the pulse duration (200 µs) and A is the geometrical surface area of the electrodes (6 mm^2). When voltage excursions exceeded machine limits (±10 V), I_{max} was extrapolated from the highest current assuming a linear relation.

$$V_{ext} = V_m \left(1 + \frac{I_{ext} - I_m}{I_m}\right) \tag{4}$$

where V_m and I_m were the measured potential and current, respectively, and V_{ext} and I_{ext} were the extrapolated potential and current. When V_{ex} reached the potential limits, I_{ext} was used as I_{max} in (3). This method provided accurate results using data for which I_{max} was measured.

2.4. Coating Characterization

After explantation, all electrodes were characterized using SEM and energy-dispersive x-ray spectroscopy (EDX). The electrochemical properties were investigated using electrochemical impedance spectroscopy (EIS), CV and VTM. Two electrode groups were added to the 4 groups of active implants, therefore these measurements were performed on six electrode groups:

- Group 1: implanted—20 mA, 200 Hz
- Group 2: implanted—20 mA, 400 Hz

- Group 3: implanted—50 mA, 200 Hz
- Group 4: implanted—50 mA, 400 Hz
- Group 5: implanted controls
- Group 6: not implanted controls

SEM (Nova 600, FEI Company, Hillsboro, OR, USA) images were recorded at magnifications varying from 450× to 10000× to obtain an overview of the surface and to investigate in detail the surface structure of the electrodes. SEM images of the not implanted control electrodes (group 6) were made, both to compare to the other electrode groups, as well as to investigate the uniformity of the coating after deposition. Further SEM analysis was carried out on a Si-wafer which was coated in the same process as the electrodes. The Si-wafer was mounted in a manner similar to the electrodes and the measured thickness is representative for the coating thickness on the electrodes. The advantage of using Si-wafer is that a cross-section analysis of the coating can be done easily. EDX (EDAX, AMETEK, Leicester, UK) spectra were made to investigate the chemical composition of the coatings after deposition and to determine whether the surface chemistry of the electrodes changed after having been implanted and after intense pulsing.

Electrochemical characterization measurements were performed in an electrochemical cell at room temperature using phosphate buffered saline as the electrolyte. The measurements were performed in a 3-electrode set-up, using the above mentioned porous TiN electrodes as working electrodes (0.06 cm^2), a Ag | AgCl reference electrode (1.6 cm^2) and a platinum foil counter electrode (50 cm^2).

Solartron, Model 1294 in conjunction with 1260 Impedance/gain-phase Analyzer (Solartron Analytical, Farnborough, UK) were used to perform EIS measurements. Accompanying SMaRT software was used to run the measurements. A sinusoidal current was used at frequencies from 0.1 Hz to 100 kHz, with 10 measurements per decade. Three different currents (5, 10 and 50 μA) were used to ensure that the measurement currents were in the linear operation range of the electrode [36]. An integration time of 10 s was used to obtain a reliable and noise-free signal.

Cyclic voltammetry (CV) was performed by cycling the electrode potential was cycled between the safe potential limits (−0.6 and 0.9 V vs. Ag | AgCl) previously established for similar electrodes [24]. The sweep rates used for CV were 0.05, 0.1, 0.5 and 1.0 V/s. Ten cycles were made at each sweep rate, the last cycle was used for data analysis. The cathodic charge storage capacity (CSC) was derived from the CV by taking the integral of the CV below the zero-current axis [10].

VTM were conducted in the same manner as described above for the implanted electrodes, except the 3-electrode setup and the electrochemical cell were employed. The maximum charge injection limit (Q_{inj}) and pulsing capacitance (C_{pulse}) were derived according to Equations (1)–(3).

2.5. Statistics

The data recorded during the 2-h pulsing sessions using an oscilloscope were filtered using a low-pass Butterworth filter (passband 5 kHz, stopband 15 kHz). E_{mc} and IR-drop were then normalized to the first measurement (session 1, start). E_{mc} was selected for statistical analysis to represent the electrode polarization and IR_{drop}, as a measure of the tissue resistance. Before, between and after the pulsing sessions, voltage transients were recorded using the VersaSTAT 3 (Princeton Applied Research, Oak Ridge, TN, USA). From these voltage transients Q_{inj} and C_{pulse} were used to further quantify the electrochemical performance of the electrodes. A linear mixed model was used to statistically analyse the data. Parameters (group, session, time and combinations thereof) were added stepwise to the model, until adding another parameter did not make a significant difference to the model.

One-way ANOVA was used to investigate the electrochemical properties of the electrodes after explantation (electrochemical cell setup). The following electrochemical properties of the 6 different electrode groups were used for statistical analysis:

- Cathodic CSC at 0.05 and 1.0 V/s
- Impedance magnitude at 0.1 Hz

- Q_{inj}

Significant findings are reported at p-values smaller than 0.05.

3. Results

All implantations were carried out without any complications. The animals recovered well from the surgery and no infections were observed during the month the electrodes were implanted.

3.1. General Coating Characteristics

In contrast to the well-known yellow-golden coloured TiN, the coatings on the electrodes had a brownish colour. To analyse the structure and chemical composition of the coating, the electrodes were studied in SEM and EDX. The overview SEM image (Figure 2a) shows a uniform coating on the electrode and in the corresponding EDX spectrum (Figure 2b) the expected peaks belonging to Ti and N are present. The quantification of the amounts of Ti and N from an EDX spectrum is difficult because the K-line of N and the L-line of Ti are very close. Here, numbers close to a 1:1 atomic ratio of Ti to N are found (note that weight-% is used in Figure 2b). The SEM images in Figure 3a clearly show a faceted structure, typical for TiN deposited at high pressure. The cross-section SEM image in Figure 3b shows the porous morphology of the coating as well. The coating thickness is approximately 6 μm.

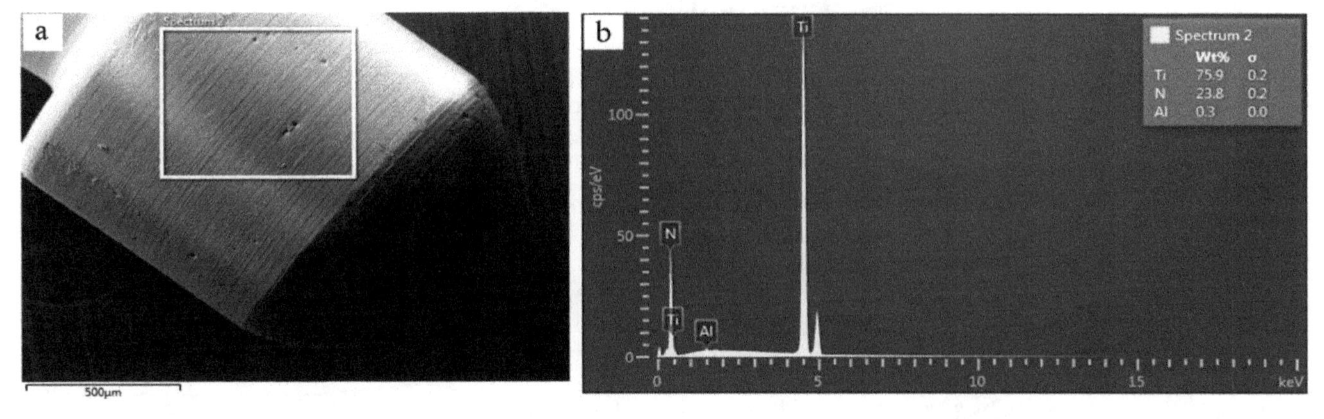

Figure 2. (**a**) Scanning electron microscope (SEM) image of the coated electrode. The indicated region is the area over which EDX was performed. The scale-bar is 500 μm. (**b**) Typical energy dispersive X-ray spectroscopy (EDX) spectrum corresponding to the area indicated in (**a**).

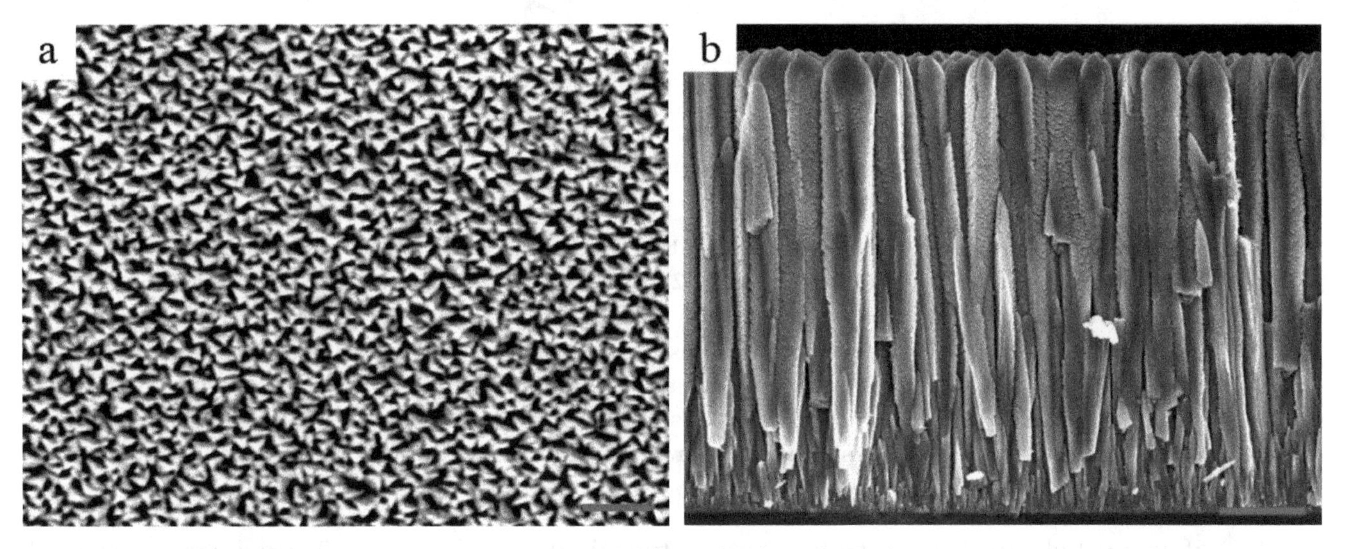

Figure 3. (**a**) Top-view (magnification: 25,000×) and (**b**) cross-section SEM images (magnification: 40,000×) of TiN coating on Si-wafer. Scale bar: 1 μm.

3.2. Changes in Electrochemical Properties during Intensive Pulsing

The shorting part of the setup broke down during the last series of measurements. The last two stimulation sessions could therefore not be completed with one of the electrodes in the 20 mA—00 Hz group. The data obtained with this electrode after the breakdown was not used in the analyses.

The significant parameters of the statistical model for IR-drop were: Time, Session, Group × Session and Time × Session. For E_{mc}, Group was an additional significant parameter of the statistical model. Figure 4 shows that the results for IR-drop and E_{mc} were similar. Both IR-drop and E_{mc} were significantly larger during session 1 compared to sessions 2 and 3 for electrode groups 2, 3 and 4. IR-drop and E_{mc} were only significantly larger during session 1 compared to sessions 2 and 3 at the 30 and 60 min measurements. Figure 4a,b also show that IR-drop (for groups 2, 3 and 4) and E_{mc} (all groups) were significantly larger after 30 and 60 min of pulsing compared to after 90 and 120 min of pulsing during session 1.

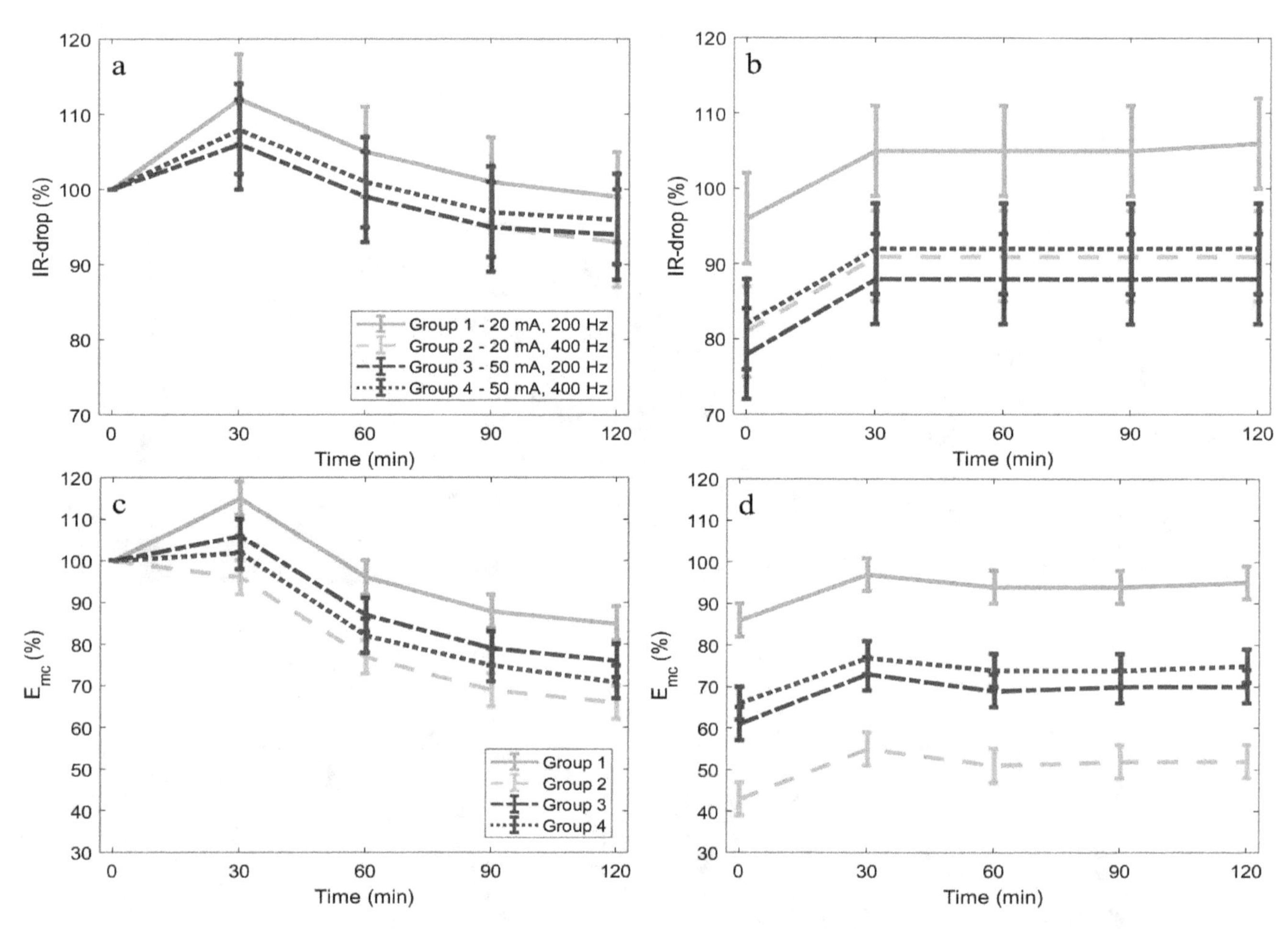

Figure 4. (**a**) IR-drop increased from baseline and then decreased during the first pulsing session for electrode groups 2, 3 and 4. (**b**) During session 2 and 3 (shown), an increase in IR-drop was seen from the start of stimulation, after which IR-drop remained stable. The IR-drop of group 2, 3 and 4 electrodes was significantly smaller during sessions 2 and 3 compared to session 1. (**c**) The same trend was observed for E_{mc} of all electrode groups but to an even greater extent (notice the axis). (**d**) An increase in E_{mc} was also observed from the start of stimulation during sessions 2 and 3 (shown). E_{mc} of group 2, 3 and 4 electrodes was also significantly smaller during sessions 2 and 3 compared to session 1 but notice again the difference in the axis of IR-drop and E_{mc}.

E_{mc} of group 1 electrodes was significantly larger than E_{mc} of group 2, 3 and 4 electrodes in all sessions. During session 2, E_{mc} of group 2 electrodes was significantly smaller than E_{mc} of group 1 and 3 electrodes and during session 3 E_{mc} of group 2 electrodes was significantly smaller than E_{mc} of group

1, 3 and 4 electrodes. Figure 4c,d show that IR-drop and E_{mc} of all electrode groups was significantly smaller at the start of stimulation compared to after 30, 60, 90 and 120 min of pulsing during session 3. The same was found for session 2.

3.3. Changes in Electrochemical Properties between Pulsing Sessions

For both C_{pulse} and Q_{inj}, the significant fixed effects were: Time, Group and Time*Group. The results of the statistical analysis for C_{pulse} and Q_{inj} were identical, except for a baseline difference between electrode groups observed for Q_{inj} ($p = 0.045$). Q_{inj} of group 2 electrodes (8.3 ± 2.4 μC/cm^2) was significantly smaller than Q_{inj} of group 3 and 4 electrodes (15.8 ± 2.4 μC/cm^2).

Figure 5a,b show that Q_{inj} and C_{pulse}, respectively, of group 1 electrodes did not change significantly. For group 2, 3 and 4 electrodes, Q_{inj} increased significantly to values of 26.45 ± 2.7 μC/cm^2, 50.00 ± 2.4 μC/cm^2 and 52.50 ± 2.4 μC/cm^2, respectively, after 6 h of intense pulsing. C_{pulse} of electrode groups 2, 3 and 4 increased significantly to capacitances of 54.1 ± 5.4 μF/cm^2, 97.1 ± 4.9 μF/cm^2 and 106.0 ± 4.9 μF/cm^2, respectively. Figure 5c,d show that the increase in C_{pulse} caused a decrease in electrode polarization. This decrease in electrode polarization led to an increased Q_{inj}.

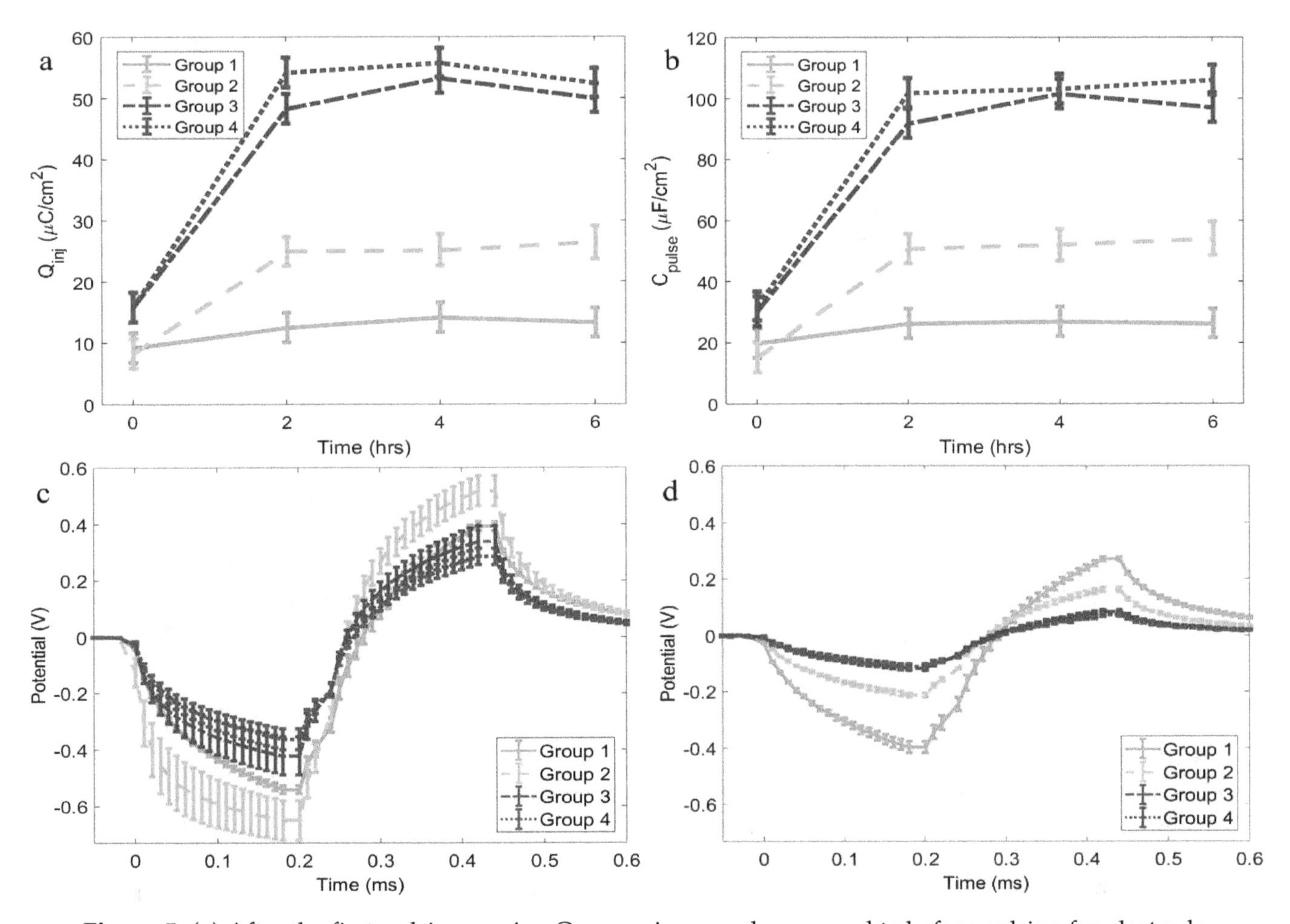

Figure 5. (**a**) After the first pulsing session Q_{inj} was increased compared to before pulsing for electrode groups 2, 3 and 4. (**b**) The same was observed for C_{pulse} of electrode groups 2, 3 and 4. (**c**) Normalized voltage transients recorded before pulsing at 3 mA. (**d**) Normalized voltage transients recorded at 3 mA after 2 h of pulsing.

Electrode group 1 had a significantly smaller Q_{inj} than all other electrode groups after the first pulsing session, which remained after the second and third pulsing session. Furthermore, electrode group 2 had a significantly smaller Q_{inj} than electrode groups 3 and 4 after the first pulsing session.

This difference also remained significant after pulsing sessions 2 and 3. The same group differences were observed for C_{pulse}.

3.4. Electrochemical Characteristics after Explantation

The results of the electrochemical characterization in phosphate buffered saline were largely consistent across measurements, as shown in Figure 6. Group 1 electrodes had a significantly smaller CSC at 0.05 and 1.0 V/s and a significantly larger impedance magnitude at 0.1 Hz compared to all other electrode groups. They also had a significantly smaller Q_{inj} compared to group 2 electrodes and the control electrodes in groups 5 and 6. But the Q_{inj} of group 1 electrodes was not significantly different from group 3 and 4 electrodes.

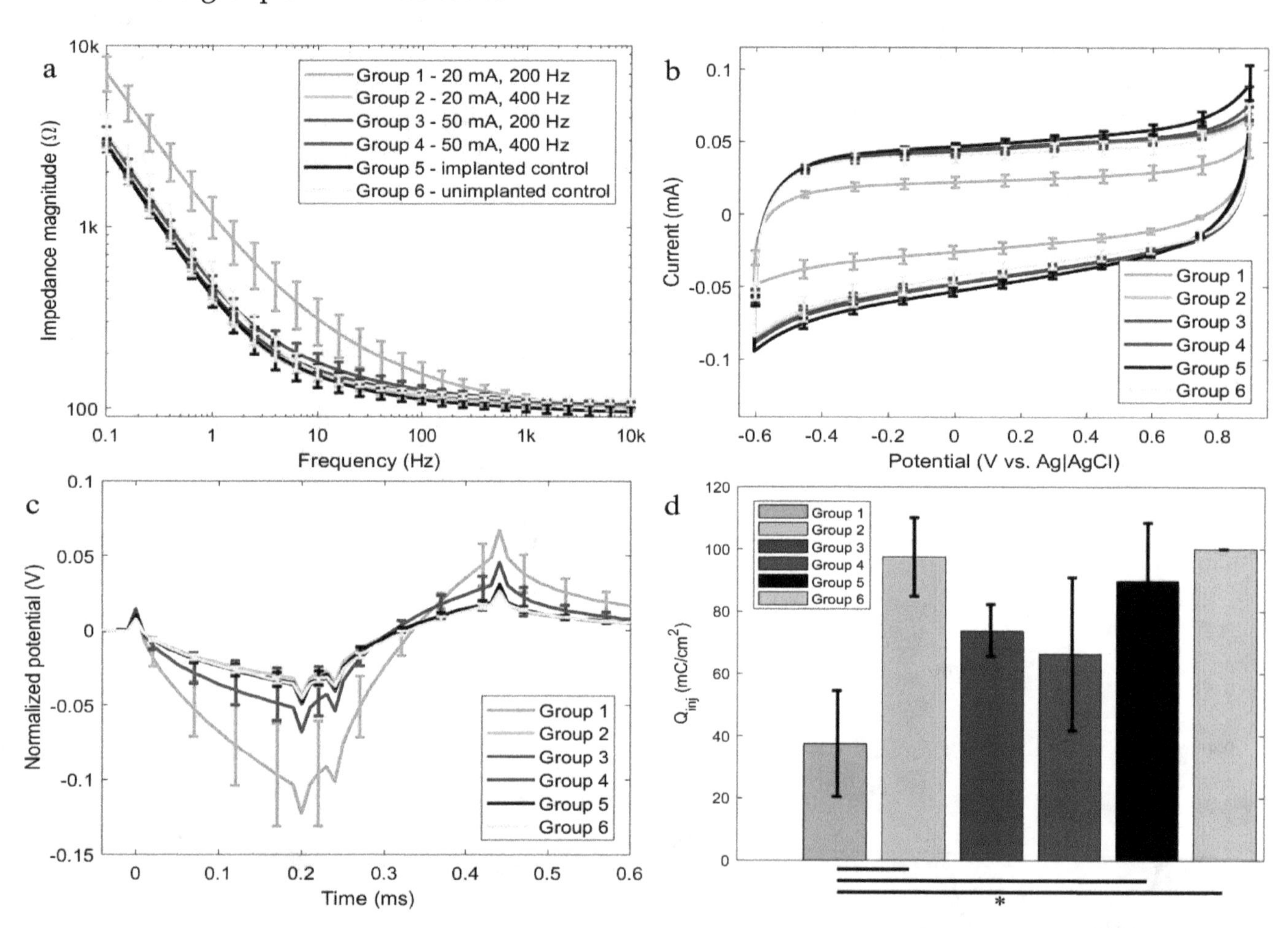

Figure 6. (**a**) The impedance magnitude was significantly larger for electrode group 1 compared to the other electrode groups, which was implanted and pulsed at the lowest electrical dose. (**b**) The cyclic voltammogram shows that the charge storage capacity of electrode group 1 was significantly smaller than the other electrode groups. (**c**) The normalized voltage transients at 5 mA show that the slope of the group 1 electrodes was larger than the slopes of the other electrode groups but no significant difference was found for C_{pulse}. (**d**) Q_{inj} of group 1 electrodes was significantly smaller than Q_{inj} of group 2, 5 and 6 electrodes.

3.5. Coating Properties after Explantation

SEM images (see Figure 7) showed that the electrode surfaces were intact after 6 h of intense stimulation. The coatings were all undamaged and had the same faceted structure, which is typical for TiN, as the electrodes had after deposition of the coating. Figure 7 indicates that there were no differences in chemical composition of group 1 and group 4 electrodes. The EDX spectra of all electrodes in all groups showed similar levels of titanium, nitrogen and oxide.

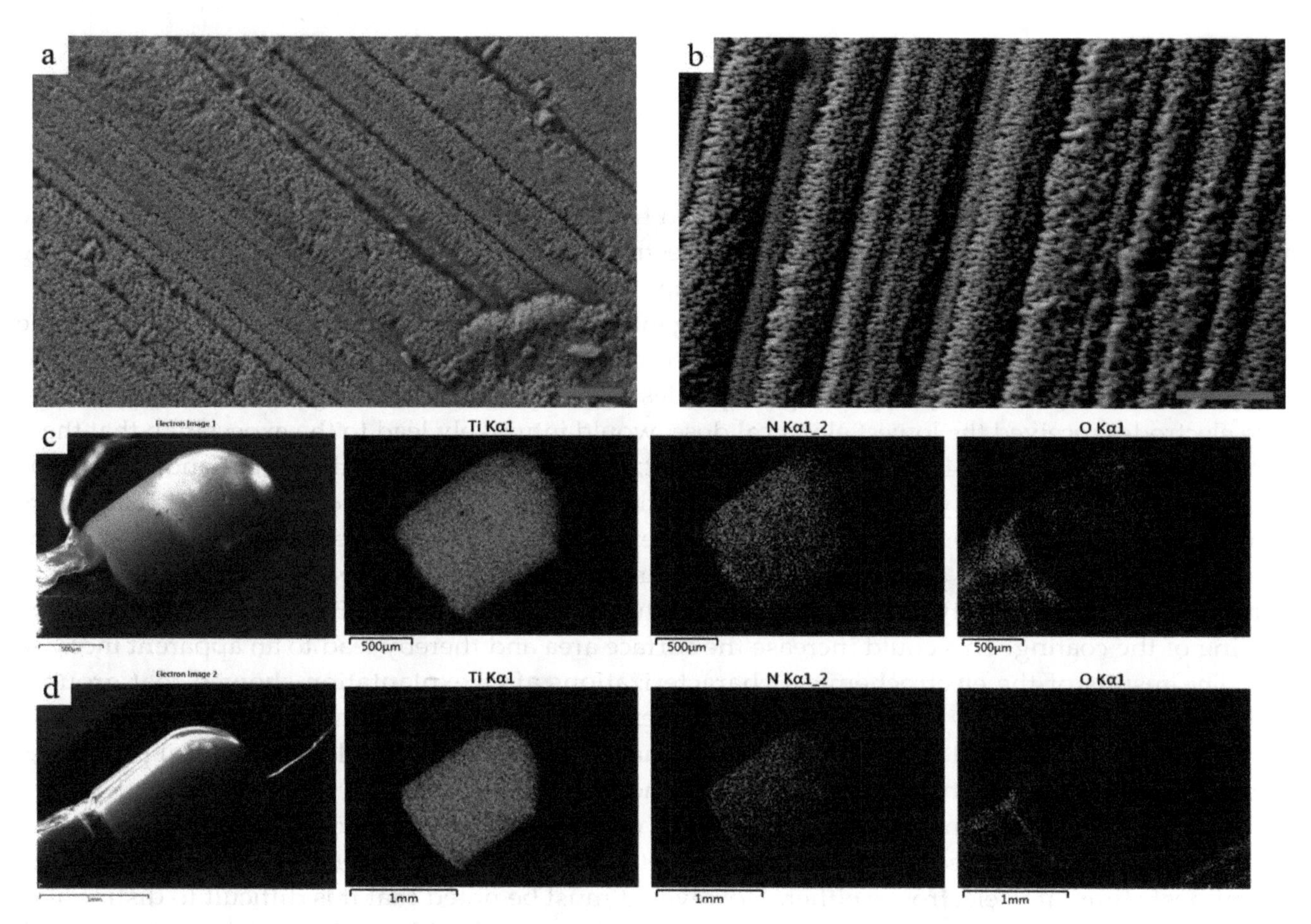

Figure 7. (**a**) Close-up SEM image (magnification: 10,000×) of an electrode in group 1 (20 mA—200 Hz), showing that the surface structure is still intact (scale bar: 10 μm). The same is true for (**b**), showing the detailed structure of an electrode in group 4 (50 mA—400 Hz, scale bar: 10 μm, magnification: 15,000×). (**c**) Overview SEM (magnification 450×) and corresponding EDX images of an electrode in group 1 show that the chemical composition of the electrode surface after pulsing was very similar to (**d**) the chemical composition of the electrode surface of an electrode in group 4. Also visualized by an overview SEM (magnification 450×) and corresponding EDX images of an electrode in group 4.

4. Discussion

Several previous studies [37–39] have shown that TiN coatings grown at high N-partial pressures become porous. These coatings have a brown colour that clearly distinguish them from the standard golden-coloured TiN used on tools and components. For use in electrode applications, it is essential to use the porous type of TiN coating as they have a high effective surface area, which in turn leads to a low impedance [3,4]. Cunha et al. [39] did a systematic study of the influence of N-content of the morphological structural and electrochemical properties of TiN coatings. In the case of high N-content (Ti:N ratio 1:1.34), these authors obtain results similar to ours regarding the coating morphology. The reason for the discrepancy in chemical composition (in our case we measure a Ti:N ratio 1:1 with EDX) could be that Cunha et al. [39] have determined the chemical composition using Rutherford Backscattering Spectrometry, which provides a more precise determination compared to EDX. Other studies have found that porous near-stoichiometric TiN coatings can be obtained just by adjusting the energy available during film growth [40].

Six groups of electrodes were used in this study, five of which were implanted and four of which were used for 6 h of intense pulsing at different electric doses:

- Group 1: 20 mA, 200 Hz
- Group 2: 20 mA, 400 Hz

- Group 3: 50 mA, 200 Hz
- Group 4: 50 mA, 400 Hz
- Group 5: implanted controls
- Group 6: un-implanted controls

During intense pulsing, IR-drop was derived from the voltage transients as a measure of tissue impedance and E_{mc} was used as a measure of electrode polarization. Between the three 2-h pulsing sessions, Q_{inj} was evaluated and from these voltage transients C_{pulse} was derived.

No significant changes were observed during intense pulsing for group 1 electrodes, receiving the lowest electrical dose. For group 2, 3 and 4 electrodes the IR-drop and E_{mc} decreased and Q_{inj} and C_{pulse} increased. The stability in pulsing properties of group 1 electrodes together with the fact that these electrodes received the lowest electrical dose, would intuitively lead to the expectation that these electrodes did not corrode [4,7,8,23–26]. For group 2, 3 and 4 electrodes, on the other hand, it could be expected that corrosion may have occurred, even though the observed electrochemical changes are favourable in the light of pulsing capability (higher charge injection, lower electrode polarization). Passivation at high anodic potentials was the main expected corrosive reaction [33,34], leading to decreased pulsing capability. Excessive bubbling due to water reduction [34], however, may lead to cracking of the coating. This could increase the surface area and thereby lead to an apparent increase Q_{inj}. The results of the electrochemical characterizations after explantation showed that group 1 electrodes had significantly deteriorated electrochemical properties compared to all other electrode groups. Group 2, 3 and 4 electrodes, on the other hand, had no different electrochemical properties after explantation than the two control groups (5 and 6).

Analytical investigations could neither confirm nor reject the electrochemical results. SEM images show that all coatings seemed to be intact. EDX spectra did not reveal differences between the harshest and mildest stimulated electrodes either. However, it must be noted that it is difficult to distinguish between the oxygen and nitrogen signal using EDX because the K-peaks of the two elements are quite close. As oxidation of the coating could have occurred, other analytical methods have been attempted (X-ray photoelectron spectroscopy and Time-of-Flight Secondary Ion Mass Spectrometry) in order to detect any differences in oxygen amounts. Preliminary results were unsuccessful, mainly because the geometry and size of the electrode is quite challenging in both the experimental set-ups. It is, however, very unlikely that the coating oxidized without showing any signs of damage. Norlin et al. [34] show SEM images porous TiN electrodes after anodic pulsing, which are severely damaged. For the current electrodes no signs of damage were found using SEM and EDX analysis.

Corrosion studies of stimulation electrodes have been performed extensively in saline [3,34,41–43]. Some studies found that the damage threshold is exactly at the water window limit [43], others suggest that the water window limit may be too conservative under pulsing conditions [41,44] and yet others suggest that corrosion may occur even within the limits of water reduction and oxidation [42]. TiN has been most extensively investigated by Norlin et al. [3,34,45]. Their pulsing study was carried out using 700V pulses of both anodic and cathodic polarity [34]. As expected, TiN showed severe corrosion upon anodic pulsing but was stable when cathodic pulses were applied. The voltages recorded in this study during constant current pulsing were, however, a factor 15–30 smaller. In a later study, the electrodes were aged using more conservative voltages (−3 and 1 V vs. Ag | AgCl), corresponding to 4 months of use based on the charge passed [3]. TiN proved very stable, which was expected, as high anodic voltages were avoided. In the current study, very high anodic voltages were also avoided by using a cathodic first stimulation paradigm. Anodic voltages between 1 and 3 V (vs. open circuit potential) were observed with the highest voltages in the 50 mA groups (groups 3 and 4). No signs of corrosion were observed for the electrodes in those groups, while corrosion of the anodically pulsed samples by Norlin et al. [34] was obvious in SEM images.

It has also been shown before that safe limits obtained in inorganic solutions do not necessarily apply to electrodes in protein containing solution [46] and implanted electrodes [47,48]. It was therefore concluded that proteins must protect the electrode surface against corrosion [46,48,49]. However,

Q_{inj} was never measured in vivo for these electrodes and it is therefore unclear whether or not it was exceeded [47,48]. The in vitro water window limits for platinum were not exceeded in either of the studies but in both studies corrosion was observed nevertheless [47,48]. Shepherd et al. [48] argue, however, that corrosion was not stimulation-induced but due to production failures. In the current study, Q_{inj} (measured in vivo) was exceeded for all stimulated electrode groups (1, 2, 3 and 4) during all stimulation sessions. However, group 1 and 2 electrodes were pulsed at approximately 20% of Q_{inj} as measured in saline and groups 3 and 4 electrodes were pulsed at approximately 50% of the in vitro Q_{inj}. Robblee et al. [47] stimulated their electrodes at approximately 5% and 30% of Q_{inj} in vitro. They observed platinum dissolution for all electrodes pulsed at 30% of Q_{inj} (in vitro) and less for electrodes pulsed at 5%. Shepherd et al. [48] stimulated the electrodes at 5-10% of Q_{inj} in vitro but concluded that the observed corrosion was not stimulation-induced. This makes it obvious that in vitro safe limits cannot be applied in vivo. However, it does not rule out that limits measured in vivo using techniques developed in vitro may be too conservative.

Interestingly, we found that corrosion most likely occurred for the electrodes pulsed at the lowest electrical dose (group 1 electrodes; 20% of Q_{inj} in vitro and 200 Hz). As group 2, 3 and 4 electrodes showed no signs of corrosion in the electrochemical characterizations after explantation, the occurrence of corrosion seems not only dependent on the electrode potentials or charge delivered. We suspect that the occurrence of corrosion is not only medium dependent (organic vs. inorganic, basic vs. acidic) but also tissue dependent. Although tissue damage was not the focus of this study, it seems to play an important role. No tissue damage seems to have occurred for electrode group 1, while tissue damage with increasing severity occurred for electrode groups 2–4 (see Figure S2). The lack of tissue damage for electrode group 1 is confirmed by the lack of change in IR-drop, which is representative of tissue impedance [10]. The same amount of charge density per phase was injected for electrode groups 1 and 2 but due to the increased frequency tissue damage is likely to have occurred in group 2 [50,51]. Tissue damage obviously occurred in electrode groups 3 and 4 (see Figure S2). Based on the IR-drop data, it appears that the tissue was damaged during the first hour of the first pulsing session for electrode groups 2, 3 and 4. It seems that a new electrode-tissue interface was formed that remained during pulsing session 2 and 3. This new electrode-tissue interface allowed for more charge injection, as Q_{inj} and C_{pulse} were significantly increased after the first pulsing session compared to before pulsing. And although the electrodes were still pulsed beyond the increased Q_{inj} in vivo, the formation of a new electrode-tissue interface and corresponding increase in Q_{inj} may thus have prevented corrosion.

The electrode-tissue interface appears to play a very important role with regards to the occurrence of corrosion. These results can therefore only be applied to stimulation electrodes implanted in adipose tissue, like ours [23,24,52] and like Bion [53] for example. They cannot be applied to implants in the brain [47], the cochlea [48] or the blood stream [1,2,4]. Furthermore, our electrode is a macro-electrode (0.06 cm^2). There are indications that different charge injection limits apply to smaller microelectrodes [51]. These results might therefore not apply to microelectrodes. Lastly, as it is challenging to work with larger animals, such as minipigs, the number of animals is low compared to rodent studies for example. The results, however, are consistent across measurements and rather homogeneous within the electrode groups and were thus statistically significant.

With the recent increase in investment in "electroceuticals," the development of novel, smaller and more sophisticated implants may be anticipated [54]. It is therefore more important than ever before to establish safe limits that apply to these specific implants [51]. We show that this is not only relevant in the light of tissue damage but also with respect to corrosion and long-term electrochemical performance of the implants. In the light of tissue damage due to corrosion, TiN appears to be a very suitable material for implants. There seems to occur no dissolution of the material [55], like with Pt [46] and IrOx [43]. As long as very high anodic potentials are avoided [34], we show that no corrosion occurs even after almost 9 million pulses. When corrosion does occur, its product (a passivation layer) remains attached to the electrodes and is not harmful to the tissue [33 55].

5. Conclusions

It was long suspected that in vitro safe limits established for implantable electrodes may not be applicable in vivo, which we confirm here. We also show that the type of tissue in which the electrode is implanted has an influence on safety limits. Biocompatibility and corrosion resistance cannot be viewed as two separate properties of implantable stimulation (and sensing) electrodes. Tissue responses influence the electrochemical behaviour of implanted electrodes and use of the electrodes influences the tissue surrounding the electrodes. It is therefore of great importance that safe limits are established for each electrode depending on the tissue in which it will be implanted.

Supplementary Materials: The following are available online at Figure S1: (a) Tissue around the tip of an electrode stimulated at 20 mA-200 Hz after 1 week of implantation. The tissue was stained using haematoxylin and eosin (H&E), which is the most commonly applied stain in medical diagnostics. Some inflammatory cells can still be observed but capsule formation has begun to take place. (b) Cells around the silicone part of the electrode appear very similar to those around the electrode tip, indicating no signs of stimulation-induced tissue damage. Figure S2: (a) The tissue around electrodes in group 1 showed no signs of tissue damage upon sacrifice. (b) The tissue around electrodes in group 2 showed some redness around the electrode tip, which likely is due to tissue damage. (c) The tissue around the electrode tips of electrodes in group 3 showed obvious tissue damage but the tissue around the insulated parts was unaffected. (d) The tissue around the electrode tips of electrodes in group 4 showed even more extensive tissue damage and bleeding. Nevertheless, the tissue around the insulated parts was unaffected.

Author Contributions: Conceptualization, S.M. and N.R.; methodology, S.M., K.R., S.S. and N.R.; writing—original draft preparation, S.M.; writing—review and editing, S.M., K.R., S.S and N.R.; project administration, N.R.; funding acquisition, N.R.

Acknowledgments: The authors thank Neurodan A/S, a member of the Ottobock group, for supplying the electrodes used in this study.

References

1. Saldach, M.; Hubmann, M.; Weikl, A.; Hardt, R. Sputter-deposited TiN electrode coatings for superior pacing and sensing performance. *Pacing Clin. Electrophysiol.* **1990**, *13*, 1891–1895. [CrossRef]
2. Lau, C.P.; Tse, H.F.; Camm, A.J.; Barold, S.S. Evolution of pacing for bradycardias: Sensors. *Eur. Heart J. Suppl.* **2007**, *9*, I11–I22. [CrossRef]
3. Norlin, A.; Pan, J.; Leygraf, C. Investigation of electrochemical behavior of stimulation/sensing materials for pacemaker electrode applications: I Pt, Ti and TiN coated electrodes. *J. Electrochem. Soc.* **2005**, *152*, J7–J15. [CrossRef]
4. Hubmann, M.; Bolz, A.; Hartz, R.; Saldach, M. Long term performance of stimulation and sensing behaviour of TiN and Ir coated pacemaker lead having a fractal surface structure. In Proceedings of the 1992 14th Annual International Conference of the IEEE Engineering in Medicine and Biology Society, Paris, France, 29 October–1 November 1992; Volume 6.
5. Janders, M.; Egert, U.; Stelzle, M.; Nisch, W. Novel thin film titanium nitride micro-electrodes with excellent charge transfer capability for cell stimulation and sensing applications. In Proceedings of the 1996 18th Annual International Conference of the IEEE Engineering in Medicine and Biology Society, Amsterdam, The Netherlands, 31 October–3 November 1 1996; Volume 1.
6. Zhou, D.M.; Greenberg, R.J. Electrochemical characterization of titanium nitride microelectrode arrays for charge-injection applications. In Proceedings of the 2003 25th Annual International Conference of the IEEE Engineering in Medicine and Biology Society, Cancun, Mexico, 17–21 September 2003; Volume 2.
7. Kane, S.R.; Cogan, S.F.; Ehrlich, J.; Plante, T.D.; McCreery, D.B.; Troyk, P.R. Electrical performance of penetrating microelectrodes chronically implanted in cat cortex. *IEEE Trans. Biomed. Eng.* **2013**, *60*, 2153–2160. [CrossRef] [PubMed]
8. Brunton, E.K.; Winther-Jensen, B.; Wang, C.; Yan, E.B.; Hagh Gooie, S.; Lowery, A.J.; Rajan, R. In vivo comparison of the charge densities required to evoke motor responses using novel annular penetrating microelectrodes. *Front. Neurosci.* **2015**, *9*, 265. [CrossRef]

9. Weiland, J.D.; Anderson, D.J.; Humayun, M.S. In vitro electrical properties for iridium oxide versus titanium nitride stimulating electrodes. *IEEE Trans. Biomed. Eng.* **2002**, *49*, 1574–1579. [CrossRef]
10. Cogan, S.F. Neural stimulation and recording electrodes. *Annu. Rev. Biomed. Eng.* **2008**, *10*, 275–309. [CrossRef] [PubMed]
11. Cogan, S.F.; Troyk, P.R.; Ehrlich, J.; Plante, T.D. In vitro comparison of the charge-injection limits of activated iridium oxide (AIROF) and platinum-iridium microelectrodes. *IEEE Trans. Biomed. Eng.* **2005**, *52*, 1612–1614. [CrossRef] [PubMed]
12. Cogan, S.F.; Troyk, P.R.; Ehrlich, J.; Plante, T.D.; Detlefsen, D.E. Potential-biased, asymmetric waveforms for charge-injection with activated iridium oxide (AIROF) neural stimulation electrodes. *IEEE Trans. Biomed. Eng.* **2006**, *53*, 327–332. [CrossRef] [PubMed]
13. Whalen, J.J.; Young, J.; Weiland, J.D.; Searson, P.C. Electrochemical characterization of charge injection at electrodeposited platinum electrodes in phosphate buffered saline. *J. Electrochem. Soc.* **2006**, *153*, C834–C839. [CrossRef]
14. Cogan, S.F.; Troyk, P.R.; Ehrlich, J.; Gasbarro, C.M.; Plante, T.D. The influence of electrolyte composition on the in vitro charge-injection limits of activated iridium oxide (AIROF) stimulation electrodes. *J. Neural Eng.* **2007**, *4*, 79–86. [CrossRef] [PubMed]
15. Cogan, S.F.; Ehrlich, J.; Plante, T.D.; Smirnov, A.; Shire, D.B.; Gingerich, M.; Rizzo, J.F. Sputtered iridium oxide films for neural stimulation electrodes. *J. Biomed. Mater. Res. B* **2009**, *89*, 353–361. [CrossRef]
16. Cogan, S.F.; Ehrlich, J.; Plante, T.D. The effect of electrode geometry on electrochemical properties measured in saline. In Proceedings of the 2014 36th Annual International Conference of the IEEE Engineering in Medicine and Biology Society (EMBC), Chicago, IL, USA, 26–30 August 2014.
17. Boehler, C.; Stieglitz, T.; Asplund, M. Nanostructured platinum grass enables superior impedance reduction for neural microelectrodes. *Biomaterials* **2015**, *67*, 346–353. [CrossRef] [PubMed]
18. Weremfo, A.; Carter, P.; Hibbert, D.B.; Zhao, C. Investigating the interfacial properties of electrochemically roughened platinum electrodes for neural stimulation. *Langmuir* **2015**, *31*, 2593–2599. [CrossRef]
19. Ghazavi, A.; Cogan, S.F. Electrochemical characterization of high frequency stimulation electrodes: Role of electrode material and stimulation parameters on electrode polarization. *J. Neural Eng.* **2018**, *15*, 036023. [CrossRef]
20. Deku, F.; Joshi-Imre, A.; Mertiri, A.; Gardner, T.J.; Cogan, S.F. Electrodeposited Iridium Oxide on Carbon Fiber Ultramicroelectrodes for Neural Recording and Stimulation. *J. Electrochem. Soc.* **2018**, *165*, D375–D380. [CrossRef]
21. Wei, X.F.; Grill, W.M. Impedance characteristics of deep brain stimulation electrodes in vitro and in vivo. *J. Neural Eng.* **2009**, *6*, 046008. [CrossRef] [PubMed]
22. Leung, R.T.; Shivdasani, M.N.; Nayagam, D.A.; Shepherd, R.K. In vivo and in vitro comparison of the charge injection capacity of platinum macroelectrodes. *IEEE Trans. Biomed. Eng.* **2015**, *62*, 849–857. [CrossRef]
23. Meijs, S.; Fjorback, M.; Jensen, C.; Sørensen, S.; Rechendorff, K.; Rijkhoff, N.J.M. Electrochemical properties of titanium nitride nerve stimulation electrodes: An in vitro and in vivo study. *Front. Neurosci.* **2015**, *9*, 268. [CrossRef] [PubMed]
24. Meijs, S.; Fjorback, M.; Jensen, C.; Sørensen, S.; Rechendorff, K.; Rijkhoff, N.J.M. Influence of fibrous encapsulation on electro-chemical properties of TiN electrodes. *Med. Eng. Phys.* **2016**, *38*, 468–476. [CrossRef] [PubMed]
25. Meijs, S.; Sørensen, C.; Sørensen, S.; Rechendorff, K.; Fjorback, M.; Rijkhoff, N.J.M. Influence of implantation on the electrochemical properties of smooth and porous TiN coatings for stimulation electrodes. *J. Neural Eng.* **2016**, *13*, 026011. [CrossRef] [PubMed]
26. Meijs, S.; Alcaide, M.; Sørensen, C.; McDonald, M.; Sørensen, S.; Rechendorff, K.; Gerhardt, A.; Nesladek, M.; Rijkhoff, N.J.; Pennisi, C.P. Biofouling resistance of boron-doped diamond neural stimulation electrodes is superior to titanium nitride electrodes in vivo. *J. Neural Eng.* **2016**, *13*, 056011. [CrossRef] [PubMed]
27. Black, B.J.; Kanneganti, A.; Joshi-Imre, A.; Rihani, R.; Chakraborty, B.; Abbott, J.; Pancrazio, J.J.; Cogan, S.F. Chronic recording and electrochemical performance of Utah microelectrode arrays implanted in rat motor cortex. *J. Neurophys.* **2018**, *120*, 2083–2090. [CrossRef] [PubMed]
28. Lempka, S.F.; Miocinovic, S.; Johnson, M.D.; Vitek, J.L.; McIntyre, C.C. In vivo impedance spectroscopy of deep brain stimulation electrodes. *J. Neural Eng.* **2009**, *6*, 046001. [CrossRef] [PubMed]

29. Cyster, L.A.; Grant, D.M.; Parker, K.G.; Parker, T.L. The effect of surface chemistry and structure of titanium nitride (TiN) films on primary hippocampal cells. *Biomol. Eng.* **2002**, *19*, 171–175. [CrossRef]
30. Cyster, L.A.; Parker, K.G.; Parker, T.L.; Grant, D.M. The effect of surface chemistry and nanotopography of titanium nitride (TiN) films on 3T3-L1 fibroblasts. *J. Biomed. Mater. Res. A* **2003**, *67*, 138–147. [CrossRef]
31. Cyster, L.A.; Parker, K.G.; Parker, T.L.; Grant, D.M. The effect of surface chemistry and nanotopography of titanium nitride (TiN) films on primary hippocampal neurones. *Biomaterials* **2004**, *25*, 97–107. [CrossRef]
32. Massiani, Y.; Medjahed, A.; Crousier, J.P.; Gravier, P.; Rebatel, I. Corrosion of sputtered titanium nitride films deposited on iron and stainless steel. In *Metallurgical Coatings and Materials Surface Modifications*; Hintermann, H.E., Spitz, J., Eds.; North-Holland: Amsterdam, The Netherlands, 1991; pp. 115–120.
33. Avasarala, B.; Haldar, P. Electrochemical oxidation behavior of titanium nitride based electrocatalysts under PEM fuel cell conditions. *Electrochim. Acta* **2010**, *55*, 9024–9034. [CrossRef]
34. Norlin, A.; Pan, J.; Leygraf, C. Investigation of Pt, Ti, TiN and nano-porous carbon electrodes for implantable cardioverter-defibrillator applications. *Electrochim. Acta* **2004**, *49*, 4011–4020. [CrossRef]
35. Merrill, D.R.; Bikson, M.; Jefferys, J.G. Electrical stimulation of excitable tissue: Design of efficacious and safe protocols. *J. Neurosci. Methods* **2005**, *141*, 171–198. [CrossRef]
36. Ragheb, T.; Geddes, L.A. Electrical properties of metallic electrodes. *Med. Biol. Eng. Comput.* **1990**, *28*, 182–186. [CrossRef]
37. Meng, L.J.; Santos, M.D. Characterization of titanium nitride films prepared by dc reactive magnetron sputtering at different nitrogen pressures. *Surf. Coat. Technol.* **1997**, *90*, 64–70. [CrossRef]
38. Chawla, V.; Jayaganthan, R.; Chandra, R. Structural characterizations of magnetron sputtered nanocrystalline TiN thin films. *Mater. Charact.* **2008**, *59*, 1015–1020. [CrossRef]
39. Cunha, L.T.; Pedrosa, P.; Tavares, C.J.; Alves, E.; Vaz, F.; Fonseca, C. The role of composition, morphology and crystalline structure in the electrochemical behaviour of TiNx thin films for dry electrode sensor materials. *Electrochim. Acta* **2009**, *55*, 59–67. [CrossRef]
40. Sánchez, G.; Rodrigo, A.; Bologna Alles, A. Titanium nitride pacing electrodes with high surface-to-area ratios. *Acta Mater.* **2005**, *53*, 4079. [CrossRef]
41. Brummer, S.B.; Turner, M.J. Electrical stimulation with Pt electrodes: II-estimation of maximum surface redox (theoretical non-gassing) limits. *IEEE Trans. Biomed. Eng.* **1977**, *5*, 440–443. [CrossRef]
42. McHardy, J.; Robblee, L.S.; Marston, J.M.; Brummer, S.B. Electrical stimulation with Pt electrodes. IV. Factors influencing Pt dissolution in inorganic saline. *Biomaterials* **1980**, *1*, 129–134. [CrossRef]
43. Negi, S.; Bhandari, R.; Rieth, L.; Van Wagenen, R.; Solzbacher, F. Neural electrode degradation from continuous electrical stimulation: Comparison of sputtered and activated iridium oxide. *J. Neurosci. Methods* **2010**, *186*, 8–17. [CrossRef]
44. Musa, S.; Rand, D.R.; Bartic, C.; Eberle, W.; Nuttin, B.; Borghs, G. Coulometric detection of irreversible electrochemical reactions occurring at Pt microelectrodes used for neural stimulation. *Anal. Chem.* **2011**, *83*, 4012–4022. [CrossRef]
45. Norlin, A.; Pan, J.; Leygraf, C. Investigation of interfacial capacitance of Pt, Ti and TiN coated electrodes by electrochemical impedance spectroscopy. *Biomol. Eng.* **2002**, *19*, 67–71. [CrossRef]
46. Robblee, L.S.; McHardy, J.; Marston, J.M.; Brummer, S.B. Electrical stimulation with Pt electrodes. V. The effect of protein on Pt dissolution. *Biomaterials* **1980**, *1*, 135–139. [CrossRef]
47. Robblee, L.S.; McHardy, J.; Agnew, W.F.; Bullara, L.A. Electrical stimulation with Pt electrodes. VII. Dissolution of Pt electrodes during electrical stimulation of the cat cerebral cortex. *J. Neurosci. Methods* **1983**, *9*, 301–308. [CrossRef]
48. Shepherd, R.K.; Murray, M.T.; Hougiton, M.E.; Clark, G.M. Scanning electron microscopy of chronically stimulated platinum intracochlear electrodes. *Biomaterials* **1985**, *6*, 237–242. [CrossRef]
49. Hibbert, D.B.; Weitzner, K.; Tabor, B.; Carter, P. Mass changes and dissolution of platinum during electrical stimulation in artificial perilymph solution. *Biomaterials* **2000**, *21*, 2177–2182. [CrossRef]
50. McCreery, D.B.; Agnew, W.F.; Yuen, T.G.H.; Bullara, L.A. Relationship between stimulus amplitude, stimulus frequency and neural damage during electrical stimulation of sciatic nerve of cat. *Med. Biol. Eng. Comput.* **1995**, *33*, 426–429. [CrossRef]
51. Cogan, S.F.; Ludwig, K.A.; Welle, C.G.; Takmakov, P. Tissue damage thresholds during therapeutic electrical stimulation. *J. Neural Eng.* **2016**, *13*, 021001. [CrossRef] [PubMed]

52. Meijs, S. The Influence of Tissue Responses on the Electrochemical Properties of Implanted Neural Stimulation Electrodes. Ph.D. Thesis, Aalborg Universitetsforlag, Aalborg, Denmark, 20 May 2016.
53. Loeb, G.E.; Richmond, F.J.; Baker, L.L. The BION devices: Injectable interfaces with peripheral nerves and muscles. *Neurosurg. Focus* **2006**, *20*, 1–9. [CrossRef]
54. Majid, A. *Electroceuticals*, 1st ed.; Springer: Basel, Switzerland, 2017.
55. Datta, S.; Das, M.; Balla, V.K.; Bodhak, S.; Murugesan, V.K. Mechanical, wear, corrosion and biological properties of arc deposited titanium nitride coatings. *Surf. Coat. Technol.* **2018**, *344*, 214–222. [CrossRef]

9

Hot Deformation Behavior and Processing Maps of Ti-6554 Alloy for Aviation Key Structural Parts

Qi Liu, Zhaotian Wang, Hao Yang and Yongquan Ning *

School of Materials Science and Engineering, Northwestern Polytechnical University, Xi'an 710072, China; liuqi2018@mail.nwpu.edu.cn (Q.L.); wzt199604043138@163.com (Z.W.); yryanghao@126.com (H.Y.)

* Correspondence: luckyning@nwpu.edu.cn

Abstract: With the development of the aviation industry, the performance requirements of materials for aviation large-scale structural parts are getting higher and higher. Ti-6554 alloy is the material of choice for aviation large-scale structural parts, but its forming process window is narrow and its microstructure is sensitive to process parameters, which affects the performance of the alloy. By adjusting the existing hot deformation process, it is of great significance to improve the properties of the alloy. Hot compression tests of Ti-6554 alloy were carried out at temperatures of 715–840 °C and strain rates of 0.001–1 s^{-1}. The results show that the flow stress and peak stress increased significantly with the increase of strain rate. At the same strain rate, the strain required for the stress to reach the peak point is smaller with the temperature increases. When the deformation temperature is below the phase transition point, the volume fraction and size of primary α phase gradually decrease with the increase of deformation temperature, while when the temperature is above the phase transition point, with the increase of deformation temperature, β grains grow up gradually, and the grain boundary bending effect is more obvious. The hyperbolic-sine Arrhenius constitutive equation was established. The correlation coefficient between experimental data and model calculated data reached 0.994. It indicates that the stress constitutive model proposed in this study can accurately reflect the stress characteristics of Ti-6554 alloy. Based on the dynamic material model, the processing maps of the alloy were established. The optimum hot deformation parameters range of the alloy was determined by analyzing the processing maps: the deformation temperature range of 800–830 °C, the strain rate range of 0.001–0.01 s^{-1}. Through the analysis of the processing maps, the instability regions in the process of cross-phase forging can be effectively avoided, and the performance of the forging can be effectively improved.

Keywords: Ti-6554 alloy; hot deformation; microstructure evolution; constitutive model; processing map

---

1. Introduction

Metastable β titanium alloy has high specific strength, deep hardenability, good corrosion resistance, and excellent processing performance, it is the material of choice for manufacturing large structural parts of aircraft [1–4]. Ti-6554 (Ti-6Cr-5Mo-5V-4Al) alloy is a new type of metastable beta high strength and high toughness titanium alloy, which is mainly used for manufacturing large-scale aviation structural parts, such as aircraft frames and landing gear. Its $[Mo]_{eq}$ is 13.3 and $[Al]_{eq}$ is 4 [5,6]. This alloy is a titanium alloy with high structural efficiency and excellent comprehensive performance. When the tensile strength of Ti-6554 alloy reaches 1270 MPa, its fracture toughness can reach more than 80 MPa.$m^{1/2}$. Its strength and toughness are better than BT-22 and Ti-1023 [7–9]. However, the alloy has large deformation resistance and poor thermal conductivity during thermal deformation. The flow stress and the microstructure are sensitive to the thermal deformation parameters [10,11]. In the actual production, various macroscopic and microscopic defects are easily generated, and the mechanical properties of the forgings are reduced. At present, there are few studies on the flow behavior and microstructure evolution of this alloy during thermal deformation. Hence, in order to obtain good performance of the alloy, it is necessary to study the flow behavior of the alloy.

During the thermal deformation of materials, the constitutive model can be used to describe the effects of deformation temperature, strain rate, and strain on the flow behavior, which is helpful to reduce the unnecessary forming defects and costs. In recent years, a large number of constitutive equations have been proposed or improved, which can be divided into three categories [12–14]: phenomenological, physically-based and artificial neural network (ANN). The phenomenological constitutive model is a model developed in the past to simulate the forming process of metals, alloys or composites at high temperature and high strain rate, Arrhenius constitutive relation is a typical representative of this constitutive model [15–17]. The model is widely used to describe the flow behavior of alloy materials such as titanium alloy [18,19], magnesium alloy [20], aluminum alloy [21], and nickel-base superalloy [22]. The physically-based constitutive model is based on the microscopic mechanism within the material to establish a model related to the deformation mechanism [23]. Due to the fact that the actual forming often involves multiple deformation mechanisms at the same time, the solution of the model parameters is relatively complicated, and the theory may deviate greatly from the reality, which leads to the difficulty in the test and application of the model. ANN model is composed of many simple and widely connected units. These units are arranged in layers, which can dynamically respond to input information and process information, and can describe complex nonlinear relationships [24]. However, ANN model has poor generality and requires long learning time. Therefore, for finite element simulation of engineering applications and macroscopic deformation, the phenomenological constitutive model is more practical.

The processing maps can be used as the basis to evaluate the machining performance of materials, and it is a powerful auxiliary tool for the process design of metal materials. At present, there are about two types of processing maps that are highly recognized. One is the Raj [25] processing maps based on the atomic model, and the other is the processing maps based on the dynamic material model (DMM). The Raj processing maps is only effective under steady-state conditions, and general complex alloys cannot be used directly, so it has great limitations in practical applications. The processing maps developed on the basis of DMM not only reflects the region where the specific deformation mechanism of specific microstructure is located, but also describes the unstable state that should be avoided during the hot processing [26]. Therefore, this paper chooses the processing maps based on DMM. The plastic deformation mechanism under different deformation conditions can be predicted by using the processing maps, and the unstable deformation parameters during the hot deformation process can be avoided. The cross-phase forging process can be guided by the processing maps.

In this study, the thermal deformation behavior of Ti-6554 alloy was studied by hot simulation compression test within the deformation temperatures of 715~840 °C and strain rates of 0.001 s^{-1}~1 s^{-1}. Through the analysis of the test results, the effects of deformation temperature and strain rates on the flow stress and microstructure evolution of the alloy were studied. The hyperbolic sinusoidal Arrhenius constitutive equation was established and the accuracy of the model was tested, which provides reference for the numerical simulation of the forming process of Ti-6554 alloy. Finally, the processing maps of the alloy was established, and the optimum range of hot processing parameters were determined. Through the analysis of the processing maps, the better forging process parameters of the cross-phase region were determined, which provides a reference for improving the service performance of forgings.

2. Materials and Methods

The test material was Ti-6554 alloy hot forging bar. It was prepared by three times of vacuum arc remelting, and the chemical composition is shown in Table 1. The phase transition point was 790 (±5) °C as measured by metallographic method. A sample having a diameter of 8 mm and a height of 12 mm was taken out from the original bar blank. Figure 1 shows the original microstructure of Ti-6554 alloy bar. From Figure 1, we can see that the microstructure of the sample is equiaxed α phase and β matrix, and its specific morphology is: along the grain boundary, there are slightly larger equiaxed α phase, while in the β matrix within the crystal, there are even smaller equiaxed α phases, and the number of equiaxed α phases in the crystal is obviously more. The volume fraction of the equiaxed α phase is about 13%, and the average diameter is about 5 μm.

The compression test was carried on a Gleebe-1500D thermal simulation tester (DSI, America) with six sets of deformation temperatures (715, 735, 755, 775, 810, 840 °C) and four sets of strain rates (0.001, 0.01, 0.1, 1 s^{-1}), the height was reduced by 60% to 4 mm. Before the experiment, tantalum pieces coated with glass lubricant were placed on both ends of the sample to reduce the friction between the table and the sample. During the experiment, the heating rate of the sample was set to 10 °C/s. When the temperature reached the set value, the sample was kept warm for 5 min in order to make the temperature of the center of the sample consistent with that of the surface. After the experiment, the samples were quenched immediately to retain the high temperature microstructure of the deformed samples. The system can automatically collect the relevant data in the process of compression. The sample was mechanically polished and then etched with a solution of 2 mL HF:10 mL HNO_3:88 mL H_2O. Observe and take metallographic photographs under the OLYMPUS GX-71 optical microscope (Olympus, Tokyo, Japan).

Table 1. Chemical composition of Ti-6554 alloy (mass fraction, %).

Cr	Mo	V	Al	Fe	Si	C	Ti
5.7	4.7	4.81	3.93	0.080	0.028	0.025	Bal.

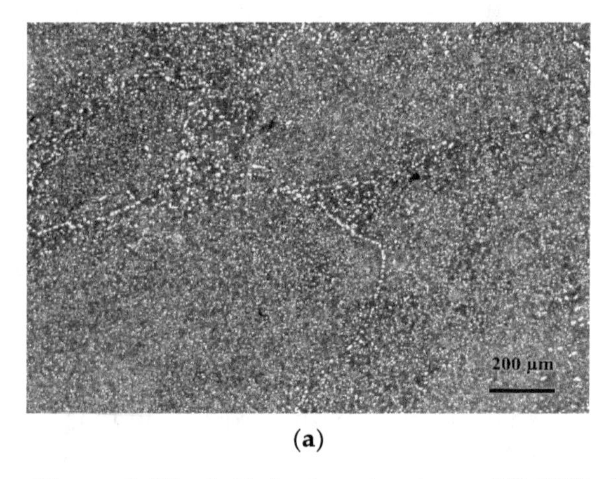

(a)

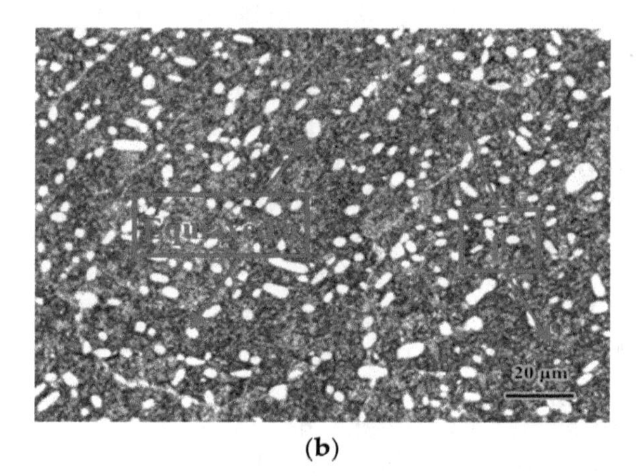

(b)

Figure 1. The initial microstructure of Ti-6554 alloy bar. (**a**) Microstructure at 50× magnification; (**b**) Microstructure at 500× magnification.

3. Results and Discussion

3.1. Flow Stress Characteristics

The true stress–strain curves of Ti-6554 titanium alloy obtained through the thermal compression test are depicted in Figure 2. It can be seen from the figure that Ti-6554 titanium alloy has similar true stress–strain curve variation rule under different strain rates. In the early deformation stage, the flow stress rises rapidly with increasing strain. The reason is that at the early stage of deformation, dislocations proliferate and accumulate at the grain boundary through slip. The increase of dislocation density makes the critical shear stress of slip increase rapidly [27]. Therefore, in the early deformation stage, the overall work hardening rate of the deformed alloy is relatively high. When the strain reaches the critical strain of dynamic recrystallization (generally less than 0.1 [28]). With the beginning of dynamic recrystallization (DRX), the dynamic softening effect gradually increases, and the work hardening rate of the alloy decreases. The dynamic softening and work hardening reached the first dynamic equilibrium at the peak strain [29]. After that, it enters the dynamic softening stage, and the dislocation network generated in the previous stage evolves to form a large number of new crystal nuclei, which absorb the surrounding dislocation and gradually grow into dynamic recrystallized grains, reducing the dislocation density in the alloy and weakening the work hardening effect. At last, it entered the steady state flow stage. The dislocation growth rate in the alloy was basically equal to the consumption rate of the growth of new grain nucleation with the further increase of strain, and the work hardening and dynamic softening reached the second dynamic equilibrium.

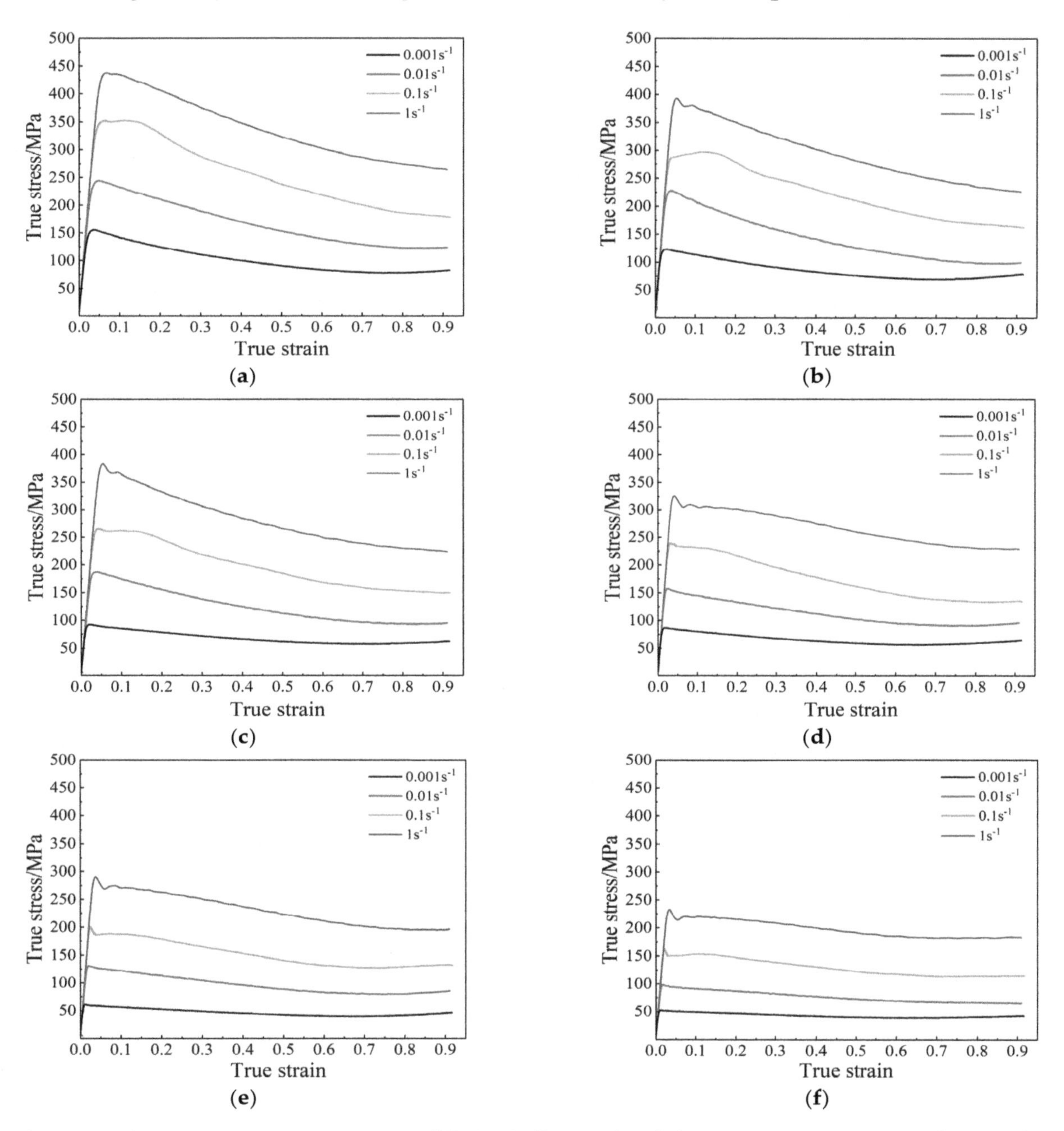

Figure 2. The true stress–strain curves of Ti-6554 alloy under deformation temperatures of (**a**) 715 °C; (**b**) 735 °C; (**c**) 755 °C; (**d**) 775 °C; (**e**) 810 °C; (**f**) 840 °C.

Thermal deformation process parameters have an important effect on the microstructure and properties of the material after hot forming. It can be seen from Figure 2 that the flow stress of Ti6554 alloy has high strain rate sensitivity, and its main characteristics are as follows:

(1) Under the same temperature conditions, the greater the strain rate, the greater the stress corresponding to the same strain. This is because when the strain rate is large, the time of unit strain is short, a large number of dislocations move simultaneously in a short time, which will increase the distortion degree of Ti-6554 titanium alloy, leading to the increase of critical shear strain force, the decrease of grain boundary slip momentum, and leading to the increase of work-hardening rate.

(2) Under the condition of high strain rate, the stress–strain curves have a sharp initial stress peak, discontinuous yield occurs, and then continues to soften. This is mainly because: in the elastic deformation stage, the dislocation density in Ti-6554 alloy is low, but when the plastic deformation occurs, the dislocations starts to move and multiply, when the dislocation movement to the grain boundary will be blocked and produce plug accumulation, so that the stress increases rapidly. As the dislocation accumulates until the critical value is exceeded, dynamic recovery (DRV) will occur in the β phase, and the dislocation of the plugging product will move through climbing, cross slip, and other means, and a large number of dissimilar dislocations will annihilate each other, so that the dislocation density in the grain will decrease, which shows that the stress will drop significantly in a short time.

(3) At the same strain rate, the strain required for the stress to reach the peak point is smaller with the temperature increases. This is mainly because: when the alloy is deformed at a higher temperature, a small amount of recrystallization occurs under a smaller strain, which causes the alloy to enter the softening stage in advance and presents the characteristics of an earlier peak point in the curve.

3.2. Establishment of Constitutive Model

Arrhenius equation is widely used to describe the relationship between material flow stress and deformation parameters [30]. Its specific equation is:

$$\dot{\varepsilon} = AF(\sigma)\exp\left(-\frac{Q}{RT}\right) \tag{1}$$

$F(\sigma)$ is the stress function, which has the following three forms under different stress states:

$$F(\sigma) = \begin{cases} \sigma^{n'} & \alpha\sigma < 0.8 \\ \exp(\beta\sigma) & \alpha\sigma > 1.2 \\ [\sinh(\alpha\sigma)]^{n} & \text{for all } \sigma \end{cases} \tag{2}$$

where $\dot{\varepsilon}$ is deformation strain rate (s^{-1}), T is deformation temperature (K), σ is the flow stress (MPa), R is gas constant (8.31 $J{\cdot}mol^{-1}{\cdot}K^{-1}$), Q is activation energy ($KJ{\cdot}mol^{-1}$), $\alpha = \beta/n'$, A, β, n, and n' are material constants. The power exponent-type is usually the preferred method for describing creep, but it cannot be used when the stress is large. Under the conditions of relatively low temperature or high strain rate, the exponential type is applicable. The hyperbolic-sine type can be applied to various temperatures and strain rates [31]. Therefore, the hyperbolic sine is more general expression. In order to obtain the above material constants, substitute Equation (2) into Equation (1), and then take natural logarithm for both sides at the same time to get the following formula:

$$\ln\dot{\varepsilon} = n'\ \ln\sigma + \left(\ln A_1 - \frac{Q}{RT}\right) \tag{3}$$

$$\ln\dot{\varepsilon} = \beta\sigma + \left(\ln A_2 - \frac{Q}{RT}\right) \tag{4}$$

$$\ln\dot{\varepsilon} = n\ln[\sinh(\alpha\sigma)] + \left(\ln A - \frac{Q}{RT}\right) \tag{5}$$

In addition, the thermal deformation condition can be expressed by the Zener-Hollomon [32] parameter as follows:

$$Z = \dot{\varepsilon}\exp\left(\frac{Q}{RT}\right) = AF(\sigma) = A[\sinh(\alpha\sigma)]^{n} \tag{6}$$

Through the transformation of Equation (6), the expression of stress can be obtained:

$$\sigma = \frac{1}{\alpha}\ln\left\{\left(\frac{Z}{A}\right)^{\frac{1}{n}} + \left[\left(\frac{Z}{A}\right)^{\frac{2}{n}} + 1\right]^{\frac{1}{2}}\right\} \tag{7}$$

The results of thermal simulation compression test show that the stress value of Ti-6554 alloy varies greatly under different thermal deformation parameters. However, according to Figure 2, most of the stress values tend to be stable or only change in a small range as the strain increases when the true strain reaches 0.6. It can be considered that the Ti-6554 alloy has entered the steady state flow stage. Therefore, stress $\sigma_{0.6}$ is selected to establish the constitutive equation of Ti-6554 alloy. Since the large difference between the deformation microstructure of the titanium alloy in different phase region, in order to further understand its phase region characteristics, the constitutive equations of the $\alpha + \beta$ phase region and the β region are solved separately.

According to Equations (3) and (4), the constants $n' = \partial \ln \dot{\varepsilon} / \partial \ln \sigma$ and $\beta = \partial \ln \dot{\varepsilon} / \partial \sigma$ are the coefficients of $\ln \sigma - \ln \dot{\varepsilon}$ and $\sigma - \ln \dot{\varepsilon}$ at a certain strain, respectively. When the true strain is 0.6, the relationship diagrams of $\ln \sigma - \ln \dot{\varepsilon}$ and $\sigma - \ln \dot{\varepsilon}$ at different temperatures are drawn respectively, as depicted in Figure 3. The least square method is used to fit the data points linearly, and the values of n' and β in different phase regions can be obtained, the specific values are shown in Table 2. Hence, we can derive the value of α in different phase regions according to formula $\alpha = \beta / n'$, and the specific α value is shown in Table 2.

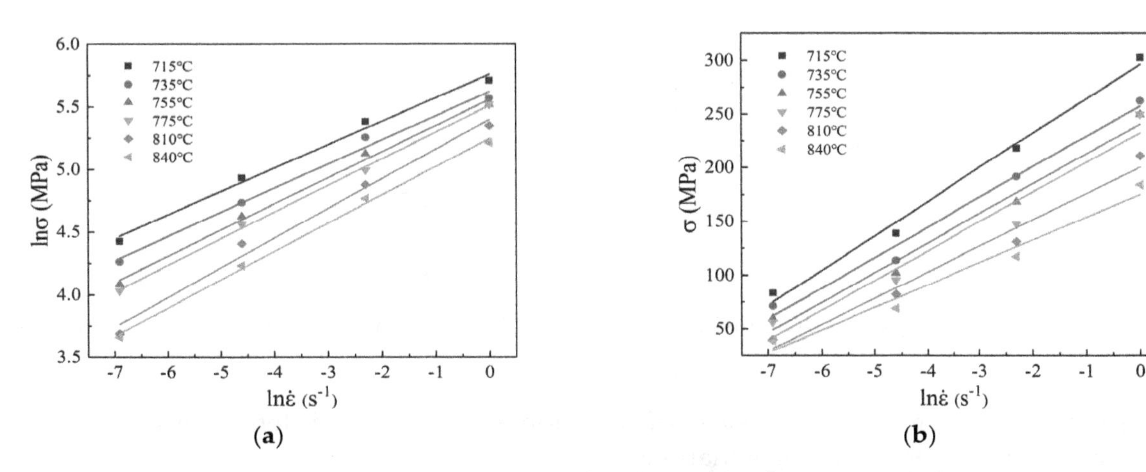

Figure 3. The relationships of (**a**) $\ln \sigma$ and $\ln \dot{\varepsilon}$; (**b**) σ and $\ln \dot{\varepsilon}$.

Table 2. Constitutive model parameters for Ti-6554 alloy.

Model Parameters	Parameters Values	
	A + β Phase	**β Phase**
n'	5.003402	4.331211
β	0.034779	0.04428
α	0.006951	0.010223
n	3.706985	3.197382
Q	244.8959	183.815
lnA	24.82687	16.08101

Find the partial differential of Equation (5) about $\ln[\sinh(\alpha\sigma)]$, and sort it out as follows:

$$n = \frac{\partial \ln \dot{\varepsilon}}{\partial \ln[\sinh(\alpha\sigma)]} \tag{8}$$

The variation of $\ln[\sinh(\alpha\sigma)]$ with $\ln\dot{\varepsilon}$ as shown in Figure 4a. The data points at different temperatures are fitted into a straight line, and the inverse of the slope is the approximate value of n at that temperature. Finding the average of the inverse of the slope of the straight line at different phase region temperature is the value of n, as shown in Table 2.

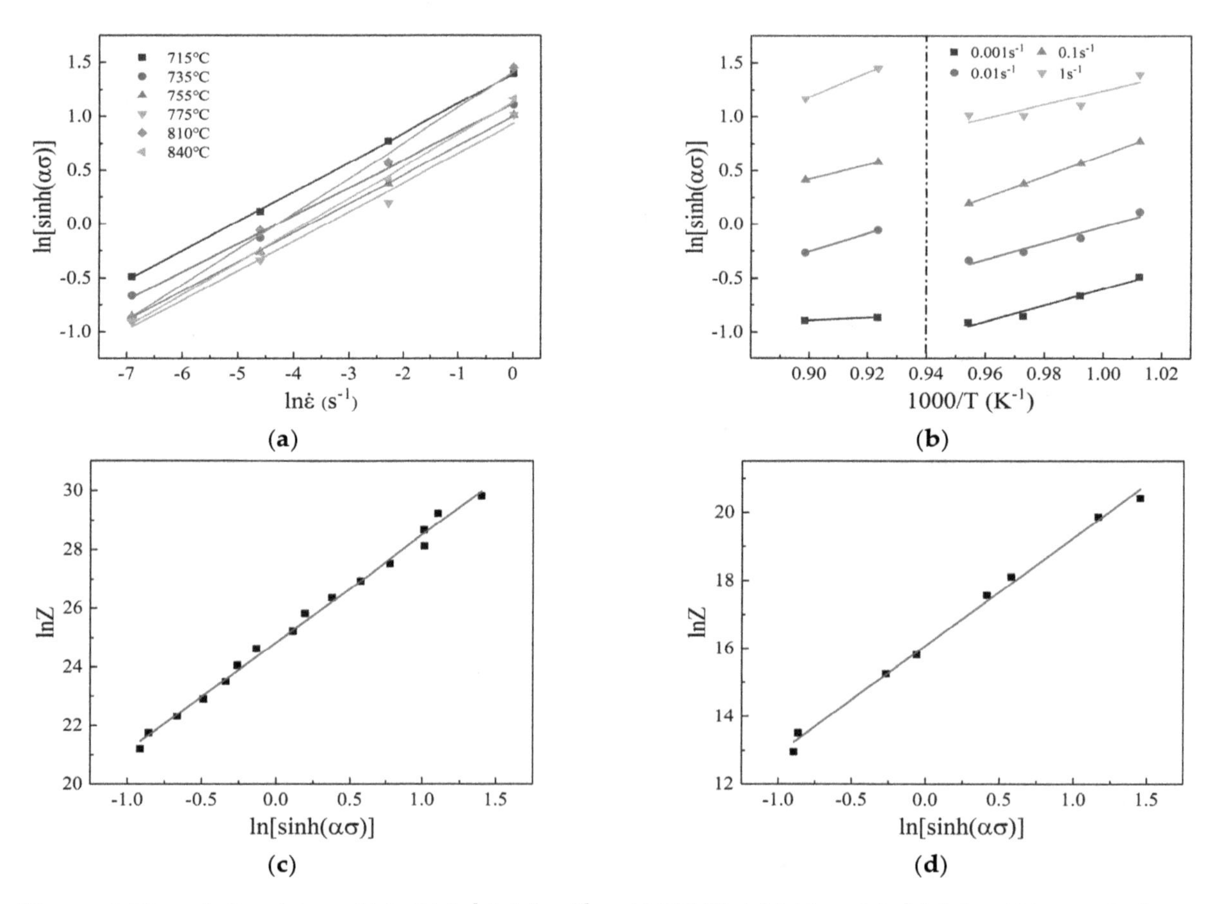

Figure 4. The relationships of (**a**); (**b**) $\ln[\sinh(\alpha\sigma)]$and 1000/T; (**c**) lnZ and $\ln[\sinh(\alpha\sigma)]$ in $\alpha + \beta$ phase; (**d**) lnZ and $\ln[\sinh(\alpha\sigma)]$ in β phase.

At a given strain rate, solve both sides of Equation (6) for partial differentials about T^{-1} at the same time, and sort out the following equation:

$$Q = nR\frac{\partial \ln[\sinh(\alpha\sigma)]}{\partial\left(\frac{1}{T}\right)}\bigg|_{\varepsilon,\dot{\varepsilon}} \tag{9}$$

The theory of plastic deformation believes that the deformation activation energy is an energy barrier that must be overcome in the process of atomic transition of the material during thermal deformation, and it can reflect the difficulty of material deformation. According to Figure 4b, Q-value is further determined by the linear fitting of $\ln[\sinh(\alpha\sigma)]$ and 1000/T. Table 3 lists the Q values of several typical high strength-toughness titanium alloys during processing in different phase regions. During the β-phase region deformation, the Q values of all the listed alloys are small, which may be related to the difference in the crystal structure of the α-phase and the β-phase. The Q of the pure titanium β phase is 153 KJ/mol, and the Q of the α phase is 242 KJ/mol [33]. Under different deformation conditions, the thermal deformation activation energy obtained reflects the thermal deformation mechanism of the material to a certain extent. The Q value of the Ti-6554 alloy when deformed in the β phase region is 184 KJ/mol, which is between the Q values of α phase and β phase, indicating that the main deformation mechanism may be the DRV caused by dislocation slip and climbing. The Q value during deformation in the $\alpha + \beta$ phase region is 245 KJ/mol, which is slightly higher than the pure titanium α phase, indicating that the main deformation mechanism may be dynamic and metastable recrystallization.

Table 3. Activation energy for several typical high strength-toughness titanium alloys.

Alloy	Transformation Point $T_\beta/^\circ C$	Deformation Temperature $T/^\circ C$	Strain Rate $\dot{\varepsilon}/s^{-1}$	$Q/kJ\cdot mol^{-1}$
Ti-1300 [34]	875	860~890 800~860	10^{-2}~10	178 216
Ti-55511 [35]	845	850~950 700~800	10^{-3}~10	137 288
Ti-55531 [36]	803	823~843 763~783	10^{-2}~1	148 275
Ti-6554	790	810~840 715~775	10^{-3}~10	184 245

Substituting Q, n, α, T, and $\dot{\varepsilon}$ into Equation (6), the Z values of different thermal deformation parameters can be obtained. As shown in Figure 4c,d, draw the relation diagram of $\ln Z - \ln[\sinh(\alpha\sigma)]$ and perform linear fitting with the least square method. The intercept of the line is the value of lnA.

Substitute all the above data into Equation (5), the flow stress model of Ti-6554 alloy under different process parameters can be obtained as follows:

$\alpha + \beta$ two-phase region:

$$\dot{\varepsilon} = 6.06 \times 10^{10}[\sinh(0.006951\sigma)]^{3.70}\exp\left(-\frac{2.45 \times 10^5}{RT}\right)$$

β single phase region:

$$\dot{\varepsilon} = 9.64 \times 10^{6}[\sinh(0.010223\sigma)]^{3.20}\exp\left(-\frac{1.84 \times 10^5}{RT}\right)$$

3.3. *Verification of Different Strain Constitutive Models*

So far, the constitutive equation of a certain true strain of 0.6 has been obtained. However, during thermal deformation of metallic materials, the effect of strain on flow stress is very obvious, so in order to describe the flow stress of materials more accurately, it is necessary to verify the constitutive model under different strains. Hence, the material constants (a, lnA, n, Q) corresponding to different strains can be obtained by the above method. Then select Equations (10)–(13) to perform fifth-order polynomial regression on the four material constants. The polynomial fitting curves of the material constants to the true-stress as depicted in Figure 5. The polynomial fitting results are shown in Tables 4 and 5.

$$\alpha = X_0+X_1\varepsilon + X_2\varepsilon^2+X_3\varepsilon^3+X_4\varepsilon^4+X_5\varepsilon^5 \quad (10)$$

$$n = N_0+N_1\varepsilon + N_2\varepsilon^2+N_3\varepsilon^3+N_4\varepsilon^4+N_5\varepsilon^5 \quad (11)$$

$$Q = Q_0+Q_1\varepsilon + Q_2\varepsilon^2+Q_3\varepsilon^3+Q_4\varepsilon^4+Q_5\varepsilon^5 \quad (12)$$

$$\ln A = Y_0+Y_1\varepsilon + Y_2\varepsilon^2+Y_3\varepsilon^3+Y_4\varepsilon^4+Y_5\varepsilon^5 \quad (13)$$

Through the above method, we can build constitutive models under different strains. In order to verify the reliability of the established Arrhenius constitutive model, as shown in Figure 6, the model calculated stress value is compared with the experimentally stress value. The observation shows that the deviation between the theoretical value and the measured value is not large, which indicates that the established Arrhenius model is reliable. To further quantitatively test the accuracy of the proposed stress constitutive model of Ti-6554 alloy, the correlation coefficient R and the average absolute relative error (AARE) are used to estimate the model accuracy. The mathematical expression is as follows:

$$R = \frac{\sum_{i=1}^{N}(E_i - \overline{E})(P_i - \overline{P})}{\sqrt{\sum_{i=1}^{N}(E_i - \overline{E})^2 \sum_{i=1}^{N}(P_i - \overline{P})^2}} \tag{14}$$

$$AARE(\%) = \frac{1}{N}\sum_{i=1}^{N}\left|\frac{E_i - P_i}{E_i}\right| \times 100 \tag{15}$$

where E_i is the measured stress, P_i is the model calculated stress, $\overline{E}$ is the measured stress average. As shown in Figure 7, the values of R and AARE are 0.994 and 3.80%, respectively. This indicated that the established constitutive model has good precision. From the above qualitative and quantitative analysis, it can be seen that the constitutive model proposed in this study can accurately reflect the stress characteristics of Ti-6554 alloy, and has good precision, thus verifying the reliability of the model.

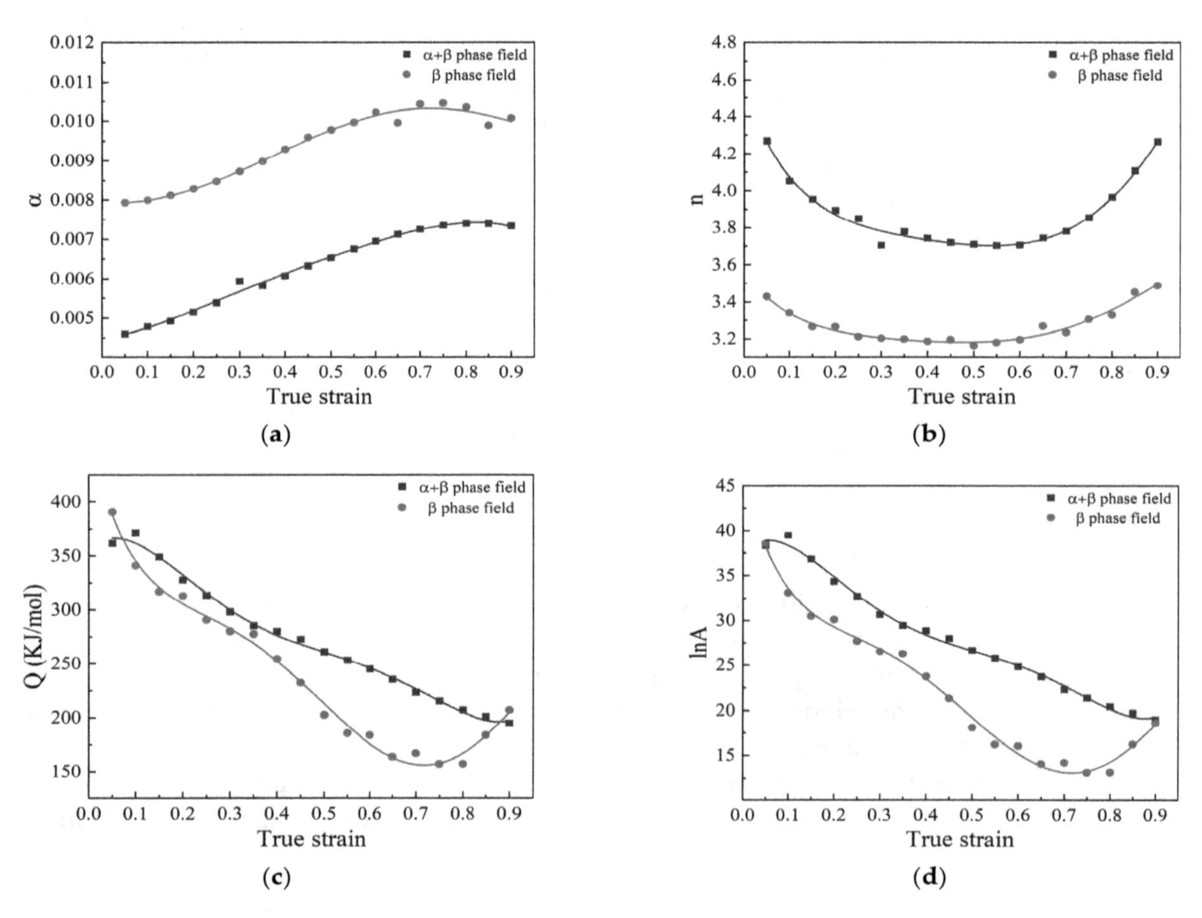

Figure 5. The fitted curves of (**a**) α; (**b**) n; (**c**) Q; (**d**) lnA.

Table 4. Polynomial fitting results of a, lnA, n, Q in α + β phase region.

	α		n		Q		lnA
X_0	0.00449	N_0	4.53506	Q_0	356.55303	Y_0	37.73193
X_1	0.00117	N_1	−6.57274	Q_1	390.72639	Y_1	46.276
X_2	0.01945	N_2	23.81874	Q_2	−4473.36504	Y_2	−531.88745
X_3	−0.04781	N_3	−45.73538	Q_3	12113.23612	Y_3	1449.71709
X_4	0.05363	N_4	42.39728	Q_4	−13845.55598	Y_4	−1663.14237
X_5	−0.02422	N_5	−13.74889	Q_5	5700.25554	Y_5	685.98485

Table 5. Polynomial fitting results of a, lnA, n, Q in β phase region.

α		n		Q		lnA	
X_0	0.00794	N_0	3.56092	Q_0	457.62985	Y_0	42.82754
X_1	−0.000679	N_1	−3.21573	Q_1	−1759.56953	Y_1	−191.89453
X_2	0.01243	N_2	12.24894	Q_2	8148.57959	Y_2	890.00338
X_3	0.00287	N_3	−25.02692	Q_3	−19516.71413	Y_3	−2131.70623
X_4	−0.2847	N_4	25.43849	Q_4	20285.7594	Y_4	2216.20461
X_5	0.01556	N_5	−9.37857	Q_5	−7368.92828	Y_5	−805.47528

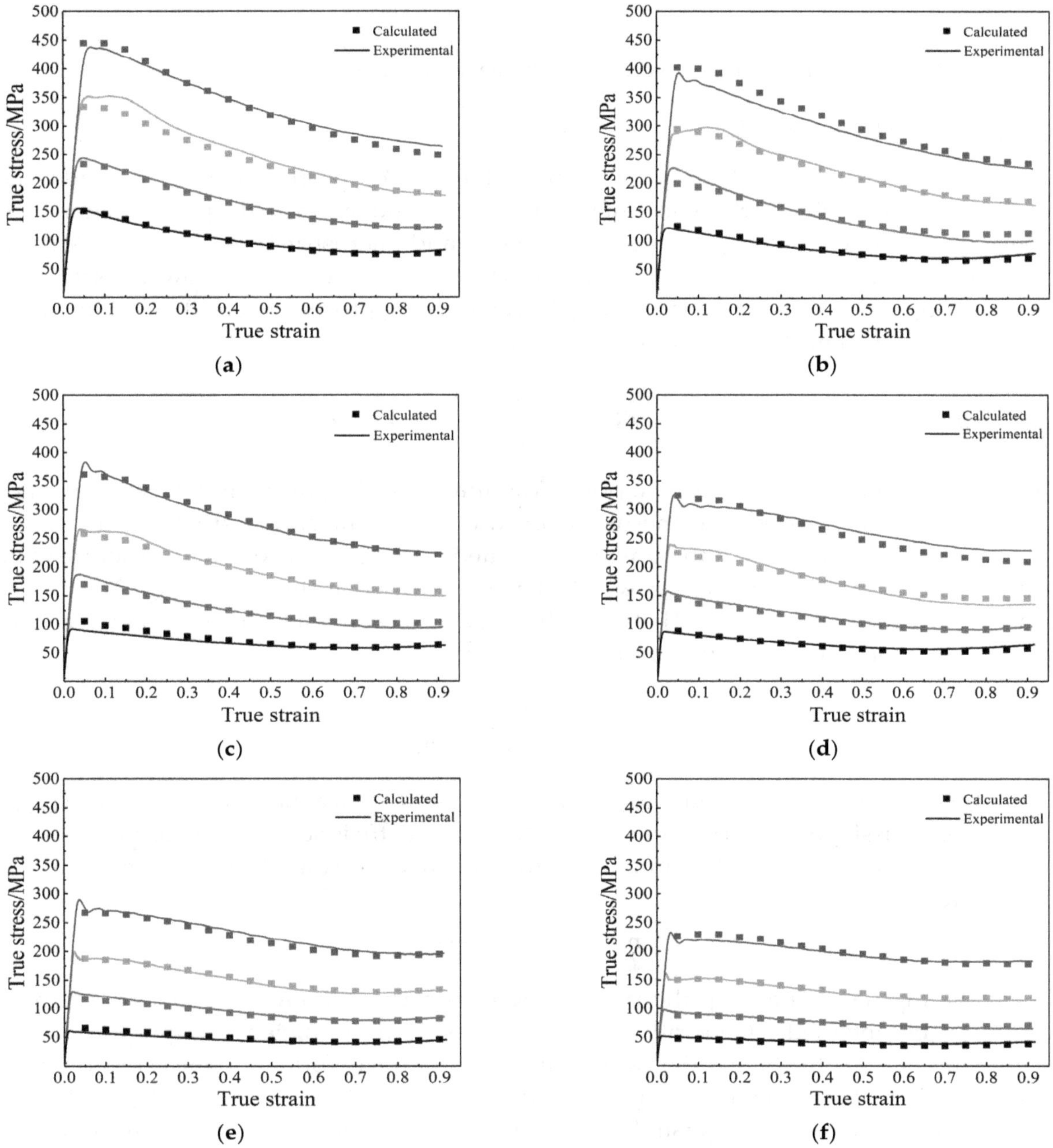

(a)

(b)

(c)

(d)

(e)

(f)

Figure 6. Comparison between experimental and model calculated flow stress at deformation temperatures of (**a**) 715 °C; (**b**) 735 °C; (**c**) 755 °C; (**d**) 775 °C; (**e**) 810 °C; (**f**) 840 °C.

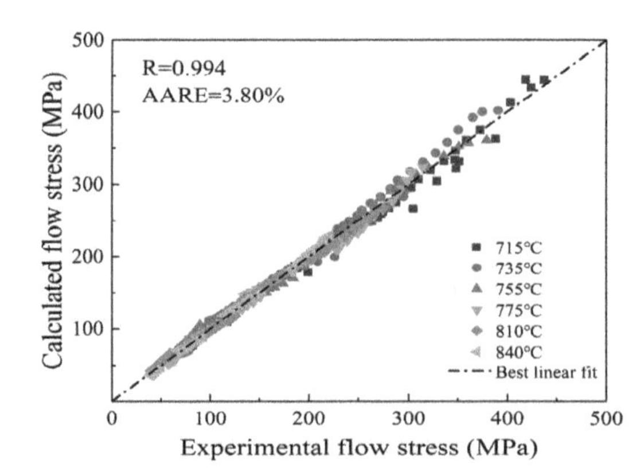

Figure 7. Correlations between experimental and model calculated flow stress.

3.4. Establishment of Processing Map

The processing map is mainly established through the DMM proposed by Prasad and Gegel [37]. It can evaluate the processing performance of metal materials and optimize the process parameters, and can achieve the role of controlling product performance and microstructure evolution. According to the DMM, the workpiece undergoing thermal deformation is a non-linear power dissipation unit, and the power input P of the workpiece during the hot forming process can be divided into two parts: dissipation content (G) and dissipation co-content (J), as follows:

$$P = \sigma\dot{\varepsilon} = G + J = \int_0^{\dot{\varepsilon}} \sigma d\dot{\varepsilon} + \int_0^{\sigma} \dot{\varepsilon}\, d\sigma \tag{16}$$

Dissipation content (G) refers to the power dissipated during thermal deformation, and most of the dissipation is converted into thermal energy, and a small part is stored in the workpiece; dissipation co-content (J) refers to the power consumed in the process of microstructure evolution during thermal deformation of materials, which is finally released as thermal energy. The distribution relationship of the total input power P between the dissipation content G and dissipation co-content J is determined by strain rate sensitivity index (m), as follows:

$$m = \frac{\partial J}{\partial G} = \frac{\dot{\varepsilon}\partial\sigma}{\sigma\partial\dot{\varepsilon}} = \frac{\partial \ln \sigma}{\partial \ln \dot{\varepsilon}} \tag{17}$$

For an ideal linear dissipative unit, the m value is 1, and J has a maximum value $J = J_{max} = \frac{\sigma\dot{\varepsilon}}{2}$. For the nonlinear dissipation process, the power dissipation efficiency η can be used to describe the efficiency of the energy consumed by microstructural changes during plastic deformation, which is shown as the following:

$$\eta = \frac{J}{J_{max}} = \frac{2m}{m+1} \tag{18}$$

By calculating the value of η, the power dissipation maps under different hot deformation conditions can be obtained. Generally, the power dissipation efficiency is proportional to the deformation mechanism, the higher the value of η, the better the deformation mechanism. In order to deal with the deformation instability of materials during hot working, Prasad et al. [38] proposed an instability criterion that can distinguish the thermal deformation instability region from the stability region of materials. The conditions of deformation instability of materials are as follows:

$$\xi(\dot{\varepsilon}) = \frac{\partial \ln\left(\frac{m}{m+1}\right)}{\partial \ln \dot{\varepsilon}} + m < 0 \tag{19}$$

The region with a negative parameter $\xi(\dot{\varepsilon})$ is the instability region, which is the processing risk region, and may produce adiabatic shear band, crack, mechanical twinning, and so on. By changing the instability criterion, the instability maps of the material can be drawn, and the processing maps of the material can be obtained by superposing the instability maps with the power dissipation maps. Based on the above method, the power dissipation maps under different strains are shown in Figure 8. It can be seen from Figure 8 that the peak value of power dissipation efficiency of the alloy varies greatly under different strains, and the peak value of power dissipation efficiency decreases with the increase of strain. When the strain is 0.3, at the deformation temperature of 790–840 °C and the strain is less than 0.01 s^{-1}, the peak power dissipation efficiency can reach 0.56, which indicates that the better microstructure can be obtained by hot working in this range. With the increase of strain, the peak area of power dissipation efficiency is concentrated in the deformation temperature of 800–830 °C, and the strain rate is 0.01–0.001 s^{-1}, which indicates that this range is the better processing parameter range of Ti-6554 alloy. At this time, the material's DRX is more sufficient, the microstructure is fine, and it has good mechanical properties. Conversely, the shaded area in the processing maps (as shown in Figure 9) shows the characteristic of instability. The instability region is basically the same under different strains, mainly concentrated in the low temperature (715–750 °C) and high strain rate (0.1–1 s^{-1}) area. However, when the strain reached 0.9, a small part of the instability region was added near the medium temperature (770–795 °C) and high strain rate (0.001–0.003 s^{-1}) region. Therefore, in the process of hot working, processing in these regions should be avoided, otherwise it will lead to uneven distribution of material structure, crack, and mechanical property decline, which will seriously deteriorate the mechanical properties of the material.

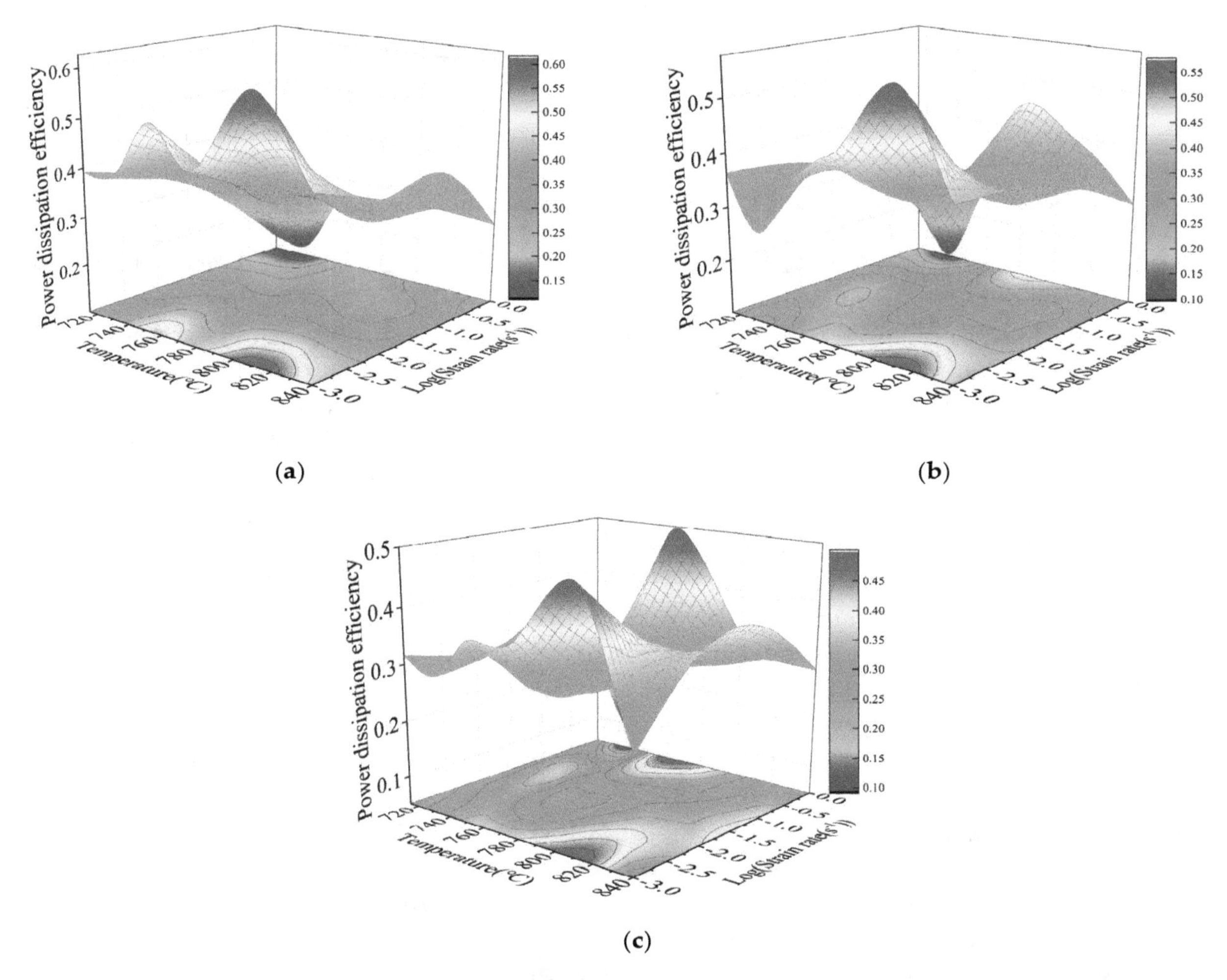

Figure 8. The power dissipation map of Ti-6554 alloy at the true strain of: (**a**) 0.3; (**b**) 0.6; (**c**) 0.9.

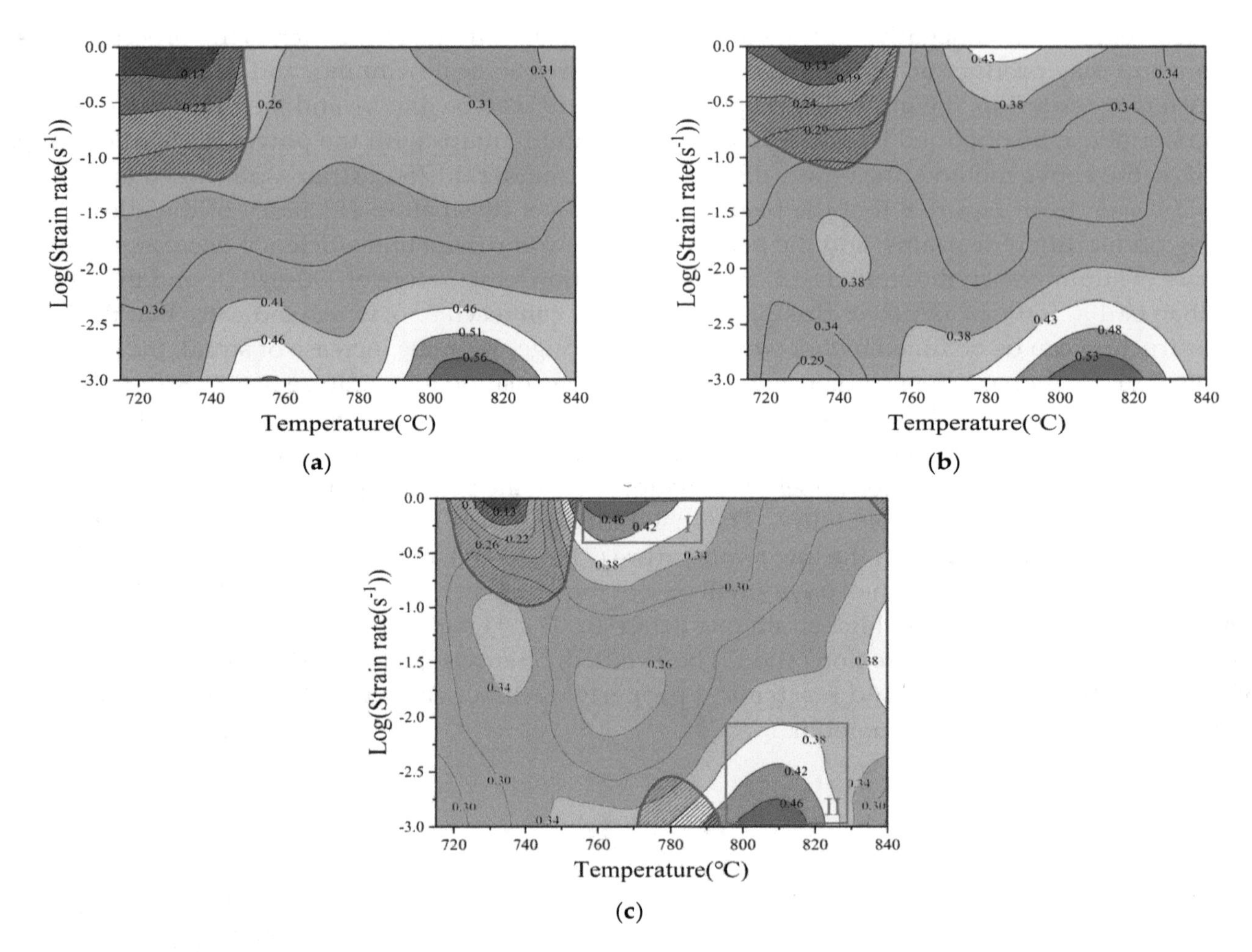

Figure 9. Processing maps of Ti-6554 alloy at the true strain of: (**a**) 0.3; (**b**) 0.6; (**c**) 0.9.

As shown in Figure 9c, when the true strain is 0.9, the stable deformation region is composed of two regions: domain I with the deformation temperatures of 760–790 °C and strain rates of 0.178–1 s^{-1}, the η-value is 0.34–0.46; domain II with the deformation temperatures of 800–830 °C and strain rates of 0.001–0.01 s^{-1}, the η-value is 0.38–0.46. However, the range of effective deformation temperature and strain rate in domain I is narrow, which is not conducive to large-scale processing of the alloy. Therefore, the optimal hot processing area determined by the processing map are the deformation temperatures of 800–830 °C and the strain rates of 0.001–0.01 s^{-1}. The following shows a typical microstructure of instability region and safety region, as shown in Figure 10. Among them, Figure 10a shows the microstructure in the instability area. It can be seen from the figure that the α phase is elongated along the direction perpendicular to the compression direction, and the aspect ratio of some α phases reaches 10:1, which is easy to be broken into fine α phase. Moreover, due to the high strain rate, the center area of the alloy is prone to a large temperature rise effect, and the thermal conductivity of Ti6554 alloy is poor, which is easy to cause unstable tissue properties. Figure 10b shows the microstructure in the safety area, mainly composed of β grains, and contains a small amount of equiaxed β grains, indicating that DRX has occurred under this condition, and the alloy has better overall performance.

Through the above analysis, we have determined the hot workability of the alloy for an area of temperatures and strain rates. During the forging process of the titanium alloy in the cross-phase region, since forging needs to be performed in different phase regions, it is particularly important to ensure that the hot processing process parameters selected in the different phase regions are not in the instability region. According to the processing maps, we can determine the better cross-phase forging process parameters: in the β single-phase region, the forging temperatures of 800–830 °C and the strain rates of 0.001–0.01 s^{-1}; in the α + β phase region, the forging temperatures of 740–765 °C and the strain rates of 0.001–0.01 s^{-1}. The distribution of the deformation of different phase regions in the cross-phase forging can also be determined by reference to the processing maps. Therefore, the processing maps can be used to guide the cross-phase forging and provide reference for determining the best cross-phase forging process parameters.

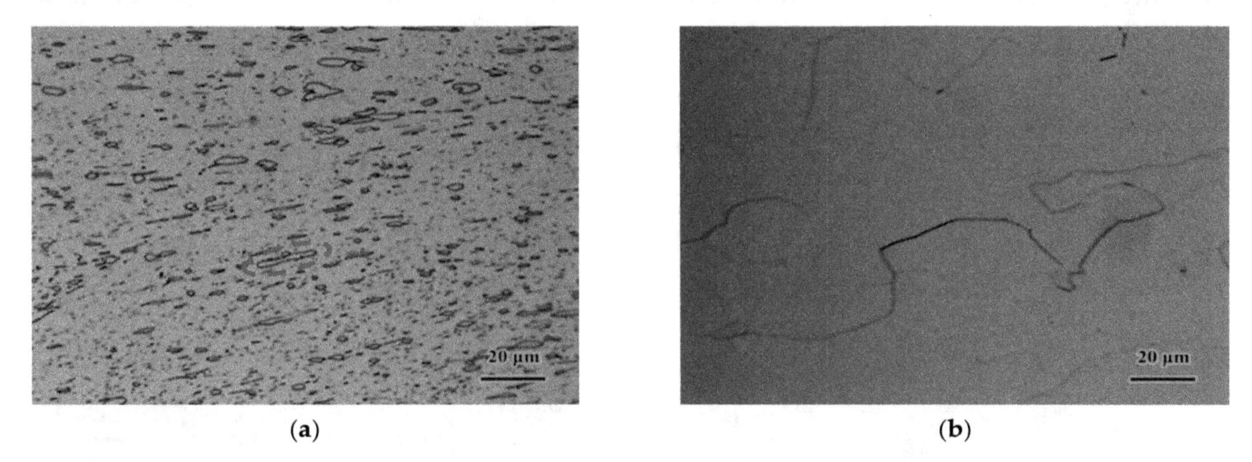

Figure 10. Typical microstructure of instability region and safety region. (**a**) 735 °C/1 s^{-1}; (**b**) 810 °C/0.001 s^{-1}.

3.5. Microstructure Evolution

Figure 11 shows the microstructure of Ti-6554 alloy after a 60% thermal deformation at different deformation temperatures under the condition of strain rate 0.001 s^{-1}. During thermal deformation below the phase transition point (790 °C), the deformed structure is mainly composed of equiaxed α phases evenly distributed on the β matrix. As shown in Figure 11a,b, when the deformation temperature rises from 715 °C to 735 °C, the shape of the primary α phase is equiaxed, and the volume fraction and size of the primary α phase does not change much. When the deformation temperature rises to 755 °C, as shown in Figure 11c, the shape of the primary α phase is basically unchanged, but its volume fraction and size decrease. When the deformation temperature reaches 775 °C, as shown in Figure 11d, the equiaxed α phase gradually dissolves, and its volume fraction and size decrease amplitude increases. This is because the higher the deformation temperature, the greater the degree of transformation of the α phase into the β phase within the alloy. On the other hand, as the deformation temperature increases, the α phase will swallow the surrounding fine α phase, thereby increasing the size of the α phase. During the thermal deformation of the alloy, these two mechanisms coexist, but when the temperature rises to a certain level, the former mechanism dominates, resulting in a rapid decline in the volume fraction and size of the primary α phase. Below the phase change point, the general change trend is as the deformation temperature increases, the volume fraction and size of the primary α phase in the microstructure of Ti-6554 alloy decrease simultaneously.

During thermal deformation above the phase transition point, the α phase in the microstructure is completely dissolved and the microstructure of the alloy consists mainly of flattened β grains, as depicted in Figure 11e,f. At this point, the β grain is severely deformed, elongated, and forms distinct streamlines, the grain boundary is large serrated, and a small amount of relatively small equiaxed β grains are observed around the flattened β grain. Among them, the flattened β grains are the remains of the original microstructure, while the equiaxed β grains are the new grains produced by dynamic recrystallization.

Above the phase transition point, with the deformation temperature increases, the β grain grows gradually, and the bending effect of grain boundary is more obvious. This is because: on the one hand, at higher temperatures, the ability of atoms to move is enhanced due to thermal activation, the faster the atom diffusion rate, the faster the migration rate, the smaller the hindrance effect of defects such as vacancy on dislocation, and the enhanced grain boundary slip ability. On the other hand, as the temperature increases, the large β grains are more likely to swallow the small β grains, resulting in the β grain coarsening. At the same time, although with the deformation temperature increases, the ability of atom migration and diffusion is stronger, the occurrence of DRV in Ti-6554 alloy reduces the distortion energy, thus reducing the driving force of DRX nucleation, but the high temperature promotes the growth of recrystallized grains, which makes the size of DRX grains larger and the volume fraction increased.

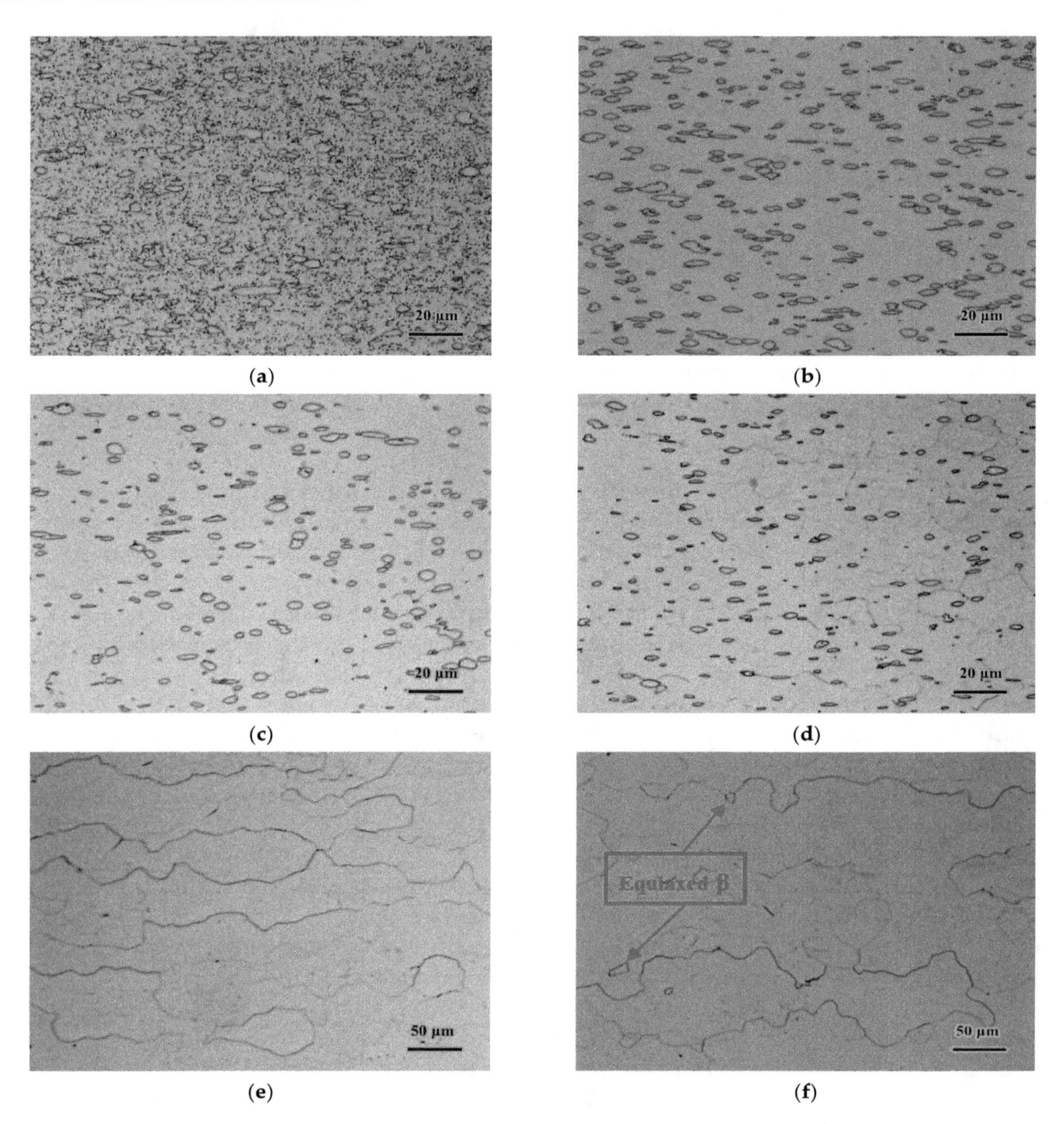

Figure 11. The microstructure of different deformation temperatures at 0.001 s^{-1}: (**a**) 715 °C; (**b**) 735 °C; (**c**) 755 °C; (**d**) 775 °C; (**e**) 810 °C; (**f**) 840 °C.

Figure 12 shows the microstructures of Ti-6554 titanium alloy in single phase region after hot deformation. According to Figure 12a,b, when the strain rates were 0.001 s^{-1} and 0.1 s^{-1}, the β grain boundary is relatively curved, and some of the β grains present an equiaxed shape. This is because with the acceleration of strain, the deformation duration is shortened, the mechanisms of grain boundary sliding and diffusion creep are gradually replaced by dislocation movement, and the shape of the grain changes into a band. At a low strain rate, the dynamic recovery consumes only a small amount of dislocations in the alloy grain, resulting in a large distortion energy. The driving force of dynamic recrystallization is large, so there are many nucleation points and a relatively large amount of recrystallization. The softening mechanism inside the alloy is mainly DRX.

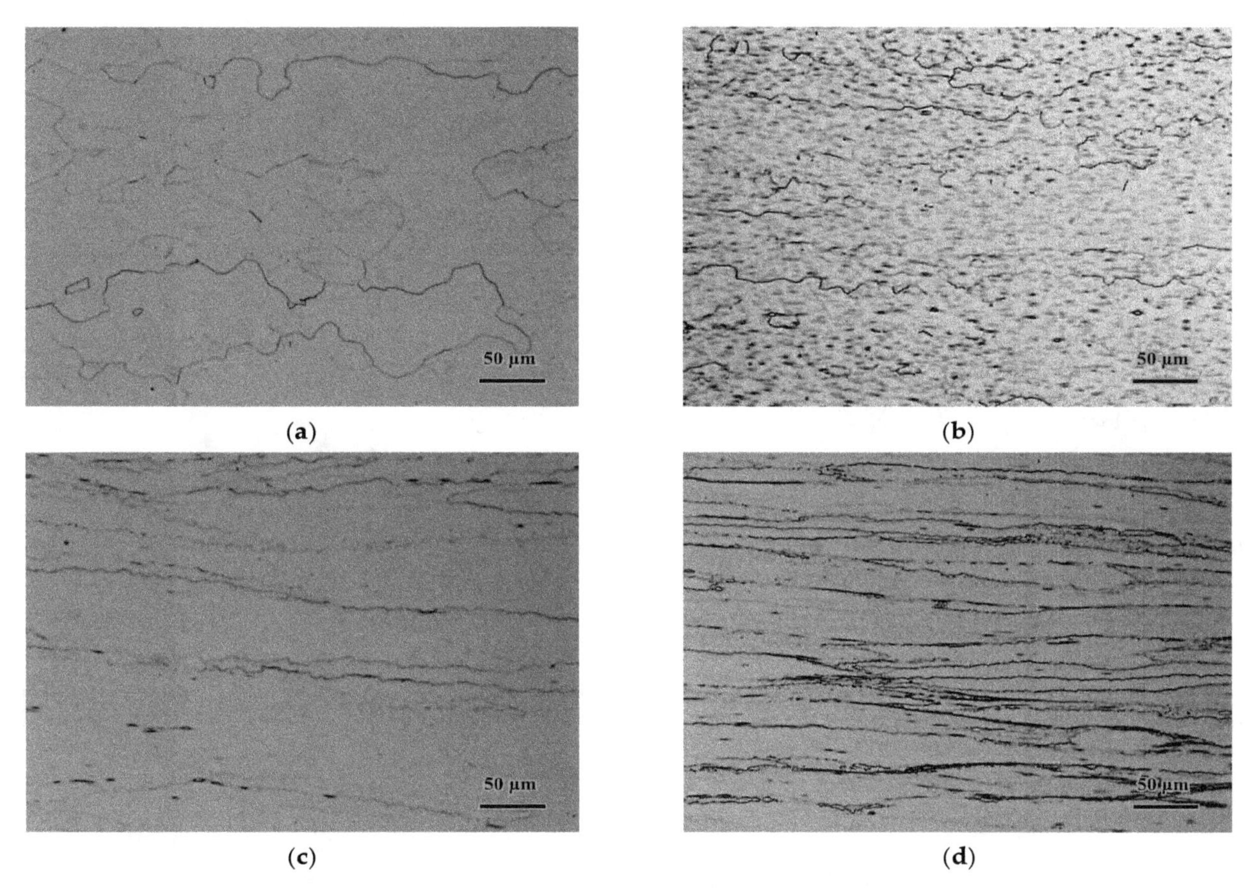

Figure 12. The microstructure of different strain rate at 840 °C: (**a**) 0.001 s^{-1}; (**b**) 0.01 s^{-1}; (**c**) 0.1 s^{-1}; (**d**) 1 s^{-1}.

However, under the higher strain rate (0.1 s^{-1}, 1 s^{-1}), it can be observed from Figure 12c,d that β grains are seriously elongated and strip like. The β grain boundary is relatively smooth, has a fine zigzag shape, and no recrystallized grains appear. When the strain rate is high, the deformation time is shortened, the atom has no time to diffuse, the grain boundary slip and diffusion creep are difficult, and the cross slip has no time to complete, the dislocation density increases rapidly in a short time, resulting in dislocation entanglement, and the softening mechanism in the alloy is mainly dynamic recovery. By means of dynamic recovery and local migration of grain boundaries, the surface tension of grain boundaries can reach a dynamic balance with the ever-changing dislocation density, thus forming a fine jagged grain boundary [39].

4. Conclusions

In this study, the thermal deformation behavior of Ti-6554 alloy under the deformation temperatures of 715–840 °C and the strain rates of 0.001 s^{-1}–1 s^{-1} was studied by thermal simulation compression test. Some important conclusions are summarized as follows:

(1) The flow stress is sensitive to the technological parameters of thermal deformation. The flow stress and peak stress increased with the increase of strain rate. At the same strain rates, the strain required for the stress to reach the peak point is smaller with the temperature increases.

(2) The constitutive equations of Ti-6554 alloy in different phase regions were established, respectively. $\alpha + \beta$ two-phase region:

$$\dot{\varepsilon} = 6.06 \times 10^{10}[\sinh(0.006951\sigma)]^{3.70} \exp\left(-\frac{2.45 \times 10^5}{RT}\right)$$

β single-phase region:

$$\dot{\varepsilon} = 9.64 \times 10^{6}[\sinh(0.010223\sigma)]^{3.20} \exp\left(-\frac{1.84 \times 10^5}{RT}\right)$$

(3) According to the processing maps, the best range of hot working process parameters were determined as follows: the deformation temperature range of 800–830 °C, the strain rate range of 0.001–0.01 s^{-1}.

(4) Below the phase transition point, with the deformation temperature increases, the volume fraction and size of primary α phase gradually decrease due to the gradual isomeric transformation of primary α phase.

(5) Above the phase transition point, with the deformation temperature increases, β grains grow up gradually, and the grain boundary bending effect is more obvious. With the increase of strain rate, the β grains deformation becomes more serious, and grain boundary changes from big ripple shape to fine zigzag shape. Dynamic recrystallization mechanism is gradually replaced by dynamic recovery mechanism.

Author Contributions: Methodology, Q.L.; investigation, Q.L. and H.Y.; validation, Y.N.; writing—original draft preparation, Q.L. and Z.W.; writing—reviewing and editing, Y.N.; funding acquisition, Y.N. All authors have read and agreed to the published version of the manuscript.

References

1. Xiao, J.F.; Nie, Z.H.; Tan, C.W.; Zhou, G.; Chen, R.; Li, M.R.; Yu, X.D.; Zhao, X.C.; Hui, S.X.; Ye, W.J.; et al. The dynamic response of the metastable β titanium alloy Ti-2Al-9.2Mo-2Fe at ambient temperature. *Mater. Sci. Eng. A* **2019**, *751*, 191–200. [CrossRef]
2. Zhao, Q.Y.; Yang, F.; Torrens, R.; Bolzoni, L. Evaluation of the hot workability and deformation mechanisms for a metastable beta titanium alloy prepared from powder. *Mater. Charact.* **2019**, *146*, 226–238. [CrossRef]
3. Fan, J.; Li, J.; Li, J.S.; Zhang, Y.D.; Kou, H.C.; Germain, L.; Esling, C. Formation and crystallography of nano/ultrafine-trimorphic structure in metastable β titanium alloy Ti-5Al-5Mo-5V-3Cr-0.5Fe processed by dynamic deformation at low temperature. *Mater. Charact.* **2017**, *130*, 149–155. [CrossRef]
4. Banerjee, D.; Williams, J.C. Perspective on Titanium science and technology. *Acta. Mater.* **2013**, *61*, 844–879. [CrossRef]
5. Long, S.; Xia, Y.F.; Wang, P.; Zhou, Y.T.; Gong, F.J.; Zhou, J.; Zhang, J.S.; Cui, M.L. Constitutive modelling, dynamic globularization behavior and processing map for Ti-6Cr-5Mo-5V-4Al alloy during hot deformation. *J. Alloys Compd.* **2015**, *796*, 65–76. [CrossRef]
6. Li, C.L.; Mi, X.J.; Ye, W.J.; Hui, S.X.; Yu, Y.; Wang, W.Q. Effect of solution temperature on microstructures and tensile properties of high strength Ti-6Cr-5Mo-5V-4Al alloy. *Mater. Sci. Eng. A* **2013**, *578*, 103–109. [CrossRef]
7. Kumar, J.; Singh, V.; Ghosal, P.; Kumar, V. Characterization of fracture and deformation mechanism in a high strength beta titanium alloy Ti-10-2-3 using EBSD technique. *Mater. Sci. Eng. A* **2015**, *623*, 49–58. [CrossRef]
8. Wang, W.Q.; Yang, Y.L.; Zhang, Y.Q.; Li, F.L.; Yang, H.L.; Zhang, P.H. The microstructure and mechanical properties of high-strength and high-toughness titanium alloy BTi-6554 bar. *Mater. Sci. Forum* **2009**, *618–619*, 173–176. [CrossRef]
9. Boyer, R.R.; Briggs, R.D. The use of β titanium alloys in the aerospace industry. *J. Mater. Eng. Perform.* **2005**, *14*, 681–685. [CrossRef]

10. Zhan, H.; Kent, D.; Wang, G.; Wang, G.; Dargusch, M. The dynamic response of a β titanium alloy to high strain rates and elevated temperatures. *Mater. Sci. Eng. A* **2014**, *607*, 417–426. [CrossRef]
11. Zhan, H.; Zeng, W.; Wang, G.; Kent, D.; Dargusch, M. Microstructural characteristics of adiabatic shear localization in a metastable beta titanium alloy deformed at high strain rate and elevated temperatures. *Mater. Charact.* **2015**, *102*, 103–113. [CrossRef]
12. Trimble, D.; O'Donnell, G.E. Constitutive modelling for elevated temperature flow behaviour of AA7075. *Mater. Des.* **2015**, *76*, 150–168. [CrossRef]
13. Lin, Y.C.; Chen, X.M. A critical review of experimental results and constitutive descriptions for metals and alloys in hot working. *Mater. Des.* **2011**, *32*, 1733–1759. [CrossRef]
14. Rusinek, A.; Rodríguez-Martínez, J.A.; Arias, A. A thermo-viscoplastic constitutive model for FCC metals with application to OFHC copper. *Int. J. Mech. Sci.* **2010**, *52*, 120–135. [CrossRef]
15. He, A.; Xie, G.; Zhang, H.; Wang, X.T. A comparative study on Johnson-Cook, modified Johnson-Cook and Arrhenius-type constitutive models to predict the high temperature flow stress in 20CrMo alloy steel. *Mater. Des.* **2013**, *52*, 677–685. [CrossRef]
16. Senthilkumar, V.; Balaji, A.; Narayanasamy, R. Analysis of hot deformation behavior of Al 5083-TiC nanocomposite using constitutive and dynamic material models. *Mater. Des.* **2012**, *37*, 102–110. [CrossRef]
17. Liu, Y.H.; Ning, Y.Q.; Yao, Z.K.; Guo, H.Z. Hot deformation behavior of Ti-6.0Al-7.0Nb biomedical alloy by using processing map. *J. Alloys Compd.* **2014**, *587*, 183–189. [CrossRef]
18. Pilehva, F.; Zarei-Hanzaki, A.; Ghambari, M.; Abedi, H.R. Flow behavior modeling of a Ti-6Al-7Nb biomedical alloy during manufacturing at elevated temperatures. *Mater. Des.* **2013**, *51*, 457–465. [CrossRef]
19. Qin, C.; Yao, Z.K.; Ning, Y.Q.; Shi, Z.F.; Guo, H.Z. Hot deformation behavior of TC11/Ti-22Al-25Nb dual-alloy in isothermal compression. *Trans. Nonferrous Met. Soc. China* **2015**, *25*, 2195–2205. [CrossRef]
20. Jia, W.T.; Xu, S.; Le, Q.C.; Fu, L.; Ma, L.F.; Tang, Y. Modified Fields-Backofen model for constitutive behavior of as-cast AZ31B magnesium alloy during hot deformation. *Mater. Des.* **2016**, *106*, 120–132. [CrossRef]
21. Haghdadi, N.; Zarei-Hanzaki, A.; Abedi, H.R. The flow behavior modeling of cast A356 aluminum alloy at elevated temperatures considering the effect of strain. *Mater. Sci. Eng. A* **2012**, *535*, 252–257. [CrossRef]
22. Ning, Y.Q.; Fu, M.W.; Chen, X. Hot deformation behavior of GH4169 superalloy associated with stick δ phase dissolution during isothermal compression process. *Mater. Sci. Eng. A* **2012**, *540*, 164–173. [CrossRef]
23. Zhang, H.M.; Chen, G.; Chen, Q.; Han, F.; Zhao, Z.D. A physically-based constitutive modelling of a high strength aluminum alloy at hot working conditions. *J. Alloys Compd.* **2018**, *743*, 283–293. [CrossRef]
24. Sun, Y.; Zeng, W.D.; Zhao, Y.Q.; Zhang, X.M.; Shu, Y.; Zhou, Y.G. Modeling constitutive relationship of Ti40 alloy using artificial neural network. *Mater. Des.* **2011**, *32*, 1537–1541. [CrossRef]
25. Raj, R. Development of a processing map for use in warm-forming and hot-forming processes. *Metall. Trans. A* **1981**, *12*, 1089–1097. [CrossRef]
26. Zhang, B.Y.; Liu, X.M.; Yang, H.; Ning, Y.Q.; Wen, S.F.; Wang, Q.D. The deformation behavior, microstructural mechanism, and process optimization of PM/Wrought dual superalloys for manufacturing the dual-property turbine disc. *Metals* **2019**, *9*, 1127. [CrossRef]
27. Chen, X.M.; Lin, Y.C.; Wen, D.X.; Zhang, J.L.; He, M. Dynamic recrystallization behavior of a typical nickel-based superalloy during hot deformation. *Mater. Des.* **2014**, *57*, 568–577. [CrossRef]
28. Wu, K.; Liu, G.Q.; Hu, B.F.; Wang, C.Y.; Zhang, Y.W.; Tao, Y.; Liu, J.T. Effect of processing parameters on hot compressive deformation behavior of a new Ni-Cr-Co based P/M superalloy. *Mater. Sci. Eng. A* **2011**, *528*, 4620–4629. [CrossRef]
29. Chen, F.; Liu, J.; Ou, H.G.; Liu, B.; Gui, Z.S.; Long, H. Flow characteristics and intrinsic workability of IN718 superalloy. *Mater. Sci. Eng. A* **2015**, *642*, 279–287. [CrossRef]
30. Sellars, C.M.; Mctegart, W.J. On the mechanism of deformation. *Acta Metall.* **1966**, *14*, 1136–1138. [CrossRef]
31. Mcqueen, H.J.; Ryan, N.D. Constitutive analysis in hot working. *Mater. Sci. Eng. A* **2002**, *322*, 43–63. [CrossRef]
32. Zener, C.; Hollomon, J.H. Effect of strain rate upon plastic flow of steel. *J. Appl. Phys.* **1944**, *15*, 22–32. [CrossRef]
33. Sargent, P.M.; Ashby, M.F. Deformation maps for titanium and zirconium. *Scr. Metall.* **1982**, *16*, 1415–1422. [CrossRef]
34. Zhao, H.Z.; Xiao, L.; Ge, P.; Sun, J.; Xi, Z.P. Hot deformation behavior and processing maps of Ti-1300 alloy. *Mater. Sci. Eng. A* **2014**, *604*, 111–116. [CrossRef]

35. Nie, X.A.; Hu, Z.; Liu, H.Q.; Yi, D.Q.; Chen, T.X.; Wang, B.F.; Gao, Q.; Wang, D.C. High temperature deformation and creep behavior of Ti-5Al-5Mo-5V-1Fe-1Cr alloy. *Mater. Sci. Eng. A* **2014**, *613*, 306–316. [CrossRef]
36. Warchomicka, F.; Poletti, C.; Stockinger, M. Study of the hot deformation behaviour in Ti-5Al-5Mo-5V-3Cr-1Zr. *Mater. Sci. Eng. A* **2011**, *528*, 8277–8285. [CrossRef]
37. Prasad, Y.V.R.K.; Gegel, H.L.; Doraivelu, S.M.; Malas, J.C.; Morgan, J.T.; Lark, K.A.; Barker, D.R. Modeling of dynamic material behavior in hot deformation: Forging of Ti-6242. *Metall. Trans. A* **1984**, *15*, 1883–1892. [CrossRef]
38. Prasad, Y.V.R.K.; Rao, K.P. Processing maps and rate controlling mechanisms of hot deformation of electrolytic tough pitch copper in the temperature range 300–950 °C. *Mater. Sci. Eng. A* **2005**, *391*, 141–150. [CrossRef]
39. Dikovits, M.; Poletti, C.; Warchomicka, F. Deformation mechanisms in the near-β titanium alloy Ti-55531. *Metall. Mater. Trans. A* **2014**, *45*, 1586–1596. [CrossRef]

10

Insights into Machining of a β Titanium Biomedical Alloy from Chip Microstructures

Damon Kent [1,2,3,*], Rizwan Rahman Rashid [4,5], Michael Bermingham [2,3], Hooyar Attar [2], Shoujin Sun [4] and Matthew Dargusch [2,3]

1 School of Science and Engineering, University of the Sunshine Coast, Maroochydore DC 4558, Australia
2 Queensland Centre for Advanced Materials Processing and Manufacturing (AMPAM), The University of Queensland, St. Lucia 4072, Australia; m.bermingham@uq.edu.au (M.B.); h.attar@uq.edu.au (H.A.); m.dargusch@uq.edu.au (M.D.)
3 ARC Research Hub for Advanced Manufacturing of Medical Devices, St. Lucia 4072, Australia
4 School of Engineering, Faculty of Science, Engineering and Technology, Swinburne University of Technology, Victoria 3122, Australia; rrahmanrashid@swin.edu.au (R.R.R.); ssun@swin.edu.au (S.S.)
5 Defence Materials Technology Centre, Victoria 3122, Australia
* Correspondence: dkent@usc.edu.au

Abstract: New metastable β titanium alloys are receiving increasing attention due to their excellent biomechanical properties and machinability is critical to their uptake. In this study, machining chip microstructure has been investigated to gain an understanding of strain and temperature fields during cutting. For higher cutting speeds, $\geq$60 m/min, the chips have segmented morphologies characterised by a serrated appearance. High levels of strain in the primary shear zone promote formation of expanded shear band regions between segments which exhibit intensive refinement of the β phase down to grain sizes below 100 nm. The presence of both α and β phases across the expanded shear band suggests that temperatures during cutting are in the range of 400–600 °C. For the secondary shear zone, very large strains at the cutting interface result in heavily refined and approximately equiaxed nanocrystalline β grains with sizes around 20–50 nm, while further from the interface the β grains become highly elongated in the shear direction. An absence of the α phase in the region immediately adjacent to the cutting interface indicates recrystallization during cutting and temperatures in excess of the 720 °C β transus temperature.

Keywords: machining; titanium; temperature; strain; grain refinement; ultrafine; nanocrystalline

---

1. Introduction

β titanium alloys possess high strength to weight, excellent toughness, corrosion resistance and biocompatibility and so have excellent potential for a wide range of biomedical applications [1]. In the last decade, there has been significant focus on the development of a variety of new metastable β titanium alloys with lower Young's moduli approaching that of human bone. These alloys employ various combinations of elements to stabilise the body-centred cubic β titanium phase and can exhibit both shape memory and pseudoelastic behaviours [2,3]. A metastable Ti-Nb based β titanium alloy (Ti-25Nb-3Mo-3Zr-2Sn wt.%) with excellent mechanical and biological compatibility has recently been the subject of extensive research and development by the authors [4–6].

Biomedical components manufactured from titanium alloys typically require machining to achieve their required form, size and surface finish. However, machining can be problematic, particularly at high speeds, due to issues with build-up of heat at the cutting zone associated with titanium's relatively low thermal conductivity and high levels of chemical affinity which lead to reaction with and 'sticking' to the cutting tool materials [7,8]. Typically, more than 70% of the heat generated during machining is delivered to the cutting tool, intensifying the degree of chemical interaction between the tool and workpiece [9,10]. For these reasons, most titanium alloys are considered difficult to machine and much of the fabrication cost for geometrically complex components may be due to machining. Hence, there is a strong incentive to better understand the machining process to improve material removal for these alloys. A further driver to study these processes comes from observations that the plastic deformation which takes place at the cutting interface can also significantly influence cell viability and adhesion on metallic implant materials [11].

Previously, Rashid et al. studied the machinability of the Ti-25Nb-3Mo-3Zr-2Sn alloy including cutting forces, temperatures and macroscopic chip characteristics [12,13]. For the solution treated and aged Ti-25Nb-3Mo-3Zr-2Sn alloy, the main cutting force decreases from around 600 N at low cutting speeds to around 430 N for speeds above 30 m/min, remaining constant at this level for speeds up to almost 200 m/min. Measurements of the external chip surface temperatures in the cutting zone using infrared thermography revealed that the temperatures are $\leq$300 °C for low surface cutting speeds (below 10 m/min), increasing markedly to more than 700 °C for high surface cutting speeds approaching 200 m/min [12]. The machining chips transition from a continuous form at low cutting speeds to a segmented saw-tooth morphology for surface cutting speeds $\geq$60 m/min [13]. The frequency of shear regions between individual sawtooth segments is associated with significant fluctuations in component forces during machining which exacerbate tool wear.

The shear regions between the sawtooth segments are subject to localised, high strain rate, severe plastic deformation at elevated temperatures. Due to the relatively small chip mass, the metal is effectively quenched as it leaves the cutting zone, preserving the as-machined microstructures. The extreme deformation conditions may result in the formation of nano-crystalline and/or ultrafine-grain microstructures of interest from the perspective of improving fundamental knowledge of severe plastic deformation processes and their associated microstructures, as well as understanding the cutting process. Schneider et al. used focussed ion beam (FIB) specimens from machining chips in conjunction with transmission electron microscopy (TEM) to study the fine microstructural features within the secondary deformation zone from cutting of the $\alpha + \beta$, Ti-6Al-4V (wt.%) alloy [14]. A layered microstructure with fine grains near the cutting interface transitioning to coarse grains toward the free surface was observed. A 10 nm thick recrystallised layer was present at the cutting interface which adjoined a 20 nm thick amorphous layer. To the best of the author's knowledge, similar high level characterisation of the fine scale deformation features formed during machining of the increasingly important β titanium class of alloys has not yet been undertaken.

Recently, the deformation behaviours of the Ti-25Nb-3Mo-3Zr-2Sn (wt.%) alloy under high strain rates ($\approx$1000 s^{-1}), in the order of those encountered in machining, were studied using Split Hopkinson Pressure Bar testing [15]. High strain rates alone did not significantly alter the deformation mechanisms from those occurring under quasistatic strain conditions which involve twinning ({332} <113> and {112} <111> twinning systems) as well as stress induced formation of the α'' and ω phases. The strain hardening behaviour of the alloy was also strain rate insensitive under these conditions due to limited adiabatic heating. However, at elevated deformation temperatures ($\geq$300 °C), the preferred deformation mechanism shifts to dislocation slip due to an increased relative stability of the β phase promoting textural changes in the β grain orientation to those favouring slip, i.e., the <001> and <111> fibre textures [16]. At elevated temperatures the yield stress also significantly reduces due to the cessation of mechanical twinning in association with significant thermal softening.

These observations can inform the interpretation of the deformation processes taking place in the chips during machining of the Ti-25Nb-3Mo-3Zr-2Sn alloy. Hence, the aim of this study is to investigate deformation microstructures preserved in the machining chips and to relate these to the strain and temperature fields present during machining. This will assist to better predict, model and optimize machining operations involving β titanium biomedical alloys.

2. Materials and Methods

The investigated alloy has a nominal alloy composition of Ti-25Nb-3Mo-3Zr-2Sn (wt.%). A 25 kg ingot was produced by alloying commercially pure Ti sponge (99.5 wt.% purity), pure Zr bars (99.7 wt.%), pure Sn bars (99.9 wt.%), pure Mo powder (99.8 wt.%) and an intermediate Nb-47 wt.% Ti alloy. The alloy was melted twice by non-consumable arc melting to ensure chemical homogeneity and low levels of impurities. The ingots were forged and then hot rolled to produce cylindrical bars 33 mm in diameter. The bars were solution treated at 750 °C followed by air cooling and ageing at 450 °C for 2 h followed by air cooling.

The machining operation was performed on 3.5 hp Hafco Metal Master lathe (Brisbane, QLD, Australia), Model AL540. A carbide tool CNMX1204A2-SMH13A provided by Sandvik with +15° rake angle, −6° inclination angle and entry angle of 45° was used to machine the Ti-25Nb-3Mo-3Zr-2Sn alloy under dry machining conditions. The machining operation took place under a constant feed rate of 0.19 mm/rev and a constant depth of cut of 1 mm. Microstructural examination was performed on machining chips from cutting with surface cutting speeds around 90 m/min.

The machining chips were mounted and polished with the width direction of the chip perpendicular to the polished surface to reveal the serrated chip cross-section. The chips were etched with Kroll's reagent for observation with scanning electron microscopy (SEM) performed on a JEOL 6460 instrument (Sydney, Australia) equipped with backscatter detector. Hardness testing was conducted on polished specimens using a Struers Vickers microhardness tester (Brisbane, Australia). X-ray diffraction (XRD) was conducted using a Bruker D8 Advance X-ray Diffractometer (Melbourne, Australia) operated at 40 KV and 30 mA, equipped with a graphite monochromator, a Ni-filtered Cu Kα (λ = 1.5406 nm) source and a scintillation counter. Specimens for transmission electron microscopy (TEM) were prepared from transverse mounted sections of the machining chips by dual focused ion beam (FIB) milling using a Zeiss Auriga FIB-SEM (Adelaide, Australia). Sections approximately 100 nm in thickness were milled using a Ga+ beam with typical dimensions of 5 μm × 12 μm. The sections were attached to a C-section copper grid. The TEM was performed using a Philips Tecnai 20 FEG instrument (Brisbane, Australia).

3. Results

3.1. Workpiece Material

The solution treated and aged Ti-25Nb-3Mo-3Zr-2Sn alloy shown in Figure 1a with XRD phase analysis in b consists of β grains with grain sizes in the order of 50 μm and lath shaped α precipitates (the dark phase in the SEM image) located primarily around the grain boundaries and protruding into the β grains. Some α laths are also present within the interior of the β grains. The solution treated and aged alloy has a hardness of 265 ± 5 HV, an ultimate tensile strength of approximately 800 MPa with typical tensile elongation of around 8% [12].

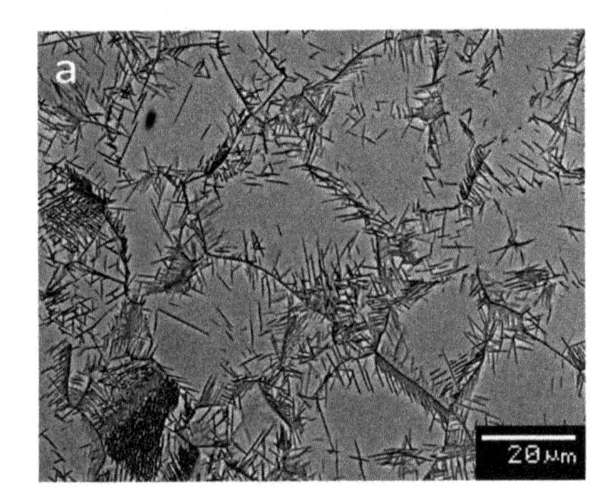

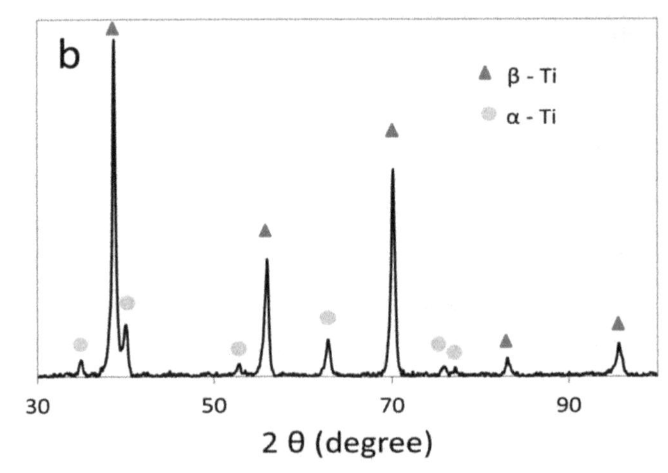

Figure 1. Scanning electron microscopy (SEM) image (**a**) and X-ray diffraction (XRD) spectrum (**b**) from the solution treated and aged workpiece material.

3.2. Machining Chip Characteristics

Cross-sections of Ti-25Nb-3Mo-3Zr-2Sn chips produced for cutting surface speeds of approximately 90 m/min are shown in Figure 2. Within this cutting regime the chips have segmented morphologies characterised by a serrated appearance with bands of severe plastic deformation, referred to herewith as expanded shear band regions, with more limited deformation in adjoining regions. According to previous research by the authors', at these cutting speeds the undeformed chip surface length, i.e., the distance between the regions of severe deformation measured from the top surface of the chip, are approximately 0.08 mm while the average chip thickness is approximately 0.15 mm and the chip roughness ratio is around 0.2 [12,13].

The expanded shear band regions between the sawtooth chips feature extensive deformation as indicated in Figure 2a, while the regions outside these bands are subject to more limited deformation and the original β grain structure of the workpiece material remains discernable, examples of which are indicated in Figure 2b. Typical microhardness values measured from within the expanded shear band regions were 398 $\pm$ 15 HV, while adjoining regions with more limited deformation had hardness of around 292 $\pm$ 8 HV which is still substantially harder than the initial starting material (265 $\pm$ 5 HV). As is also the case for formation of continuous chips, during formation of segmented chips deformation of the material occurs ahead of the tool in the region referred to as the primary shear zone. The transition from continuous to segmented chip morphologies, featuring thermoplastic instability-induced shear banding, emerges once the smooth chip flow becomes insufficient to dissipate the energy through homogeneous plastic flow [17]. The other significant area of deformation in the chip is the secondary shear zone shown in Figure 2c, which is the region adjoining the tool rake face during cutting. Hardness measurements from within the secondary shear zone were around 363 $\pm$ 8 HV. However, it should be noted that the measurements were approximately 20 μm from the outer edge of the chip which was as near to the cutting interface as could be reliably tested through microhardness measurements.

To gain a higher level understanding of the deformation microstructures formed during machining, TEM investigations were undertaken on samples from the expanded shear band regions formed in the primary deformation zone and from the secondary shear zone, shown in Figures 3 and 4, respectively.

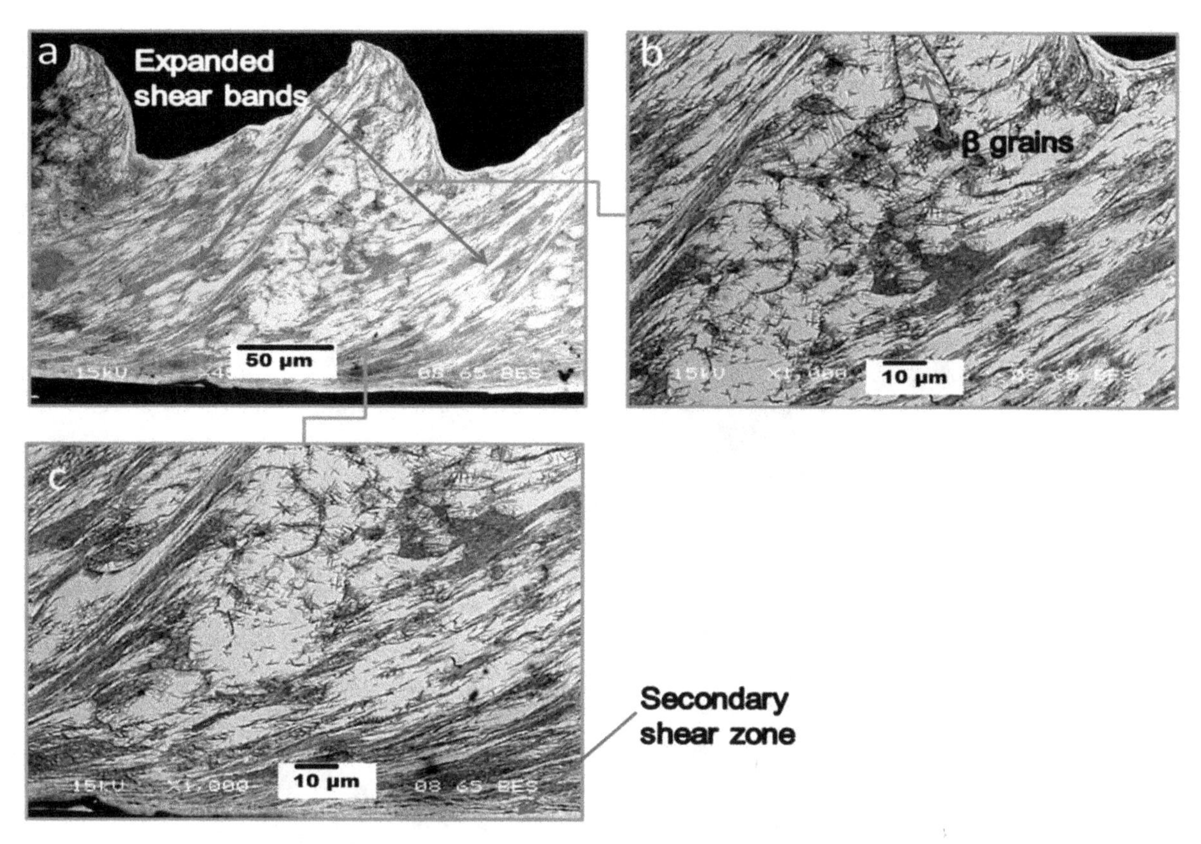

Figure 2. SEM of machining chip microstructure for surface cutting speed around 90 m/min: (**a**) Low magnification image of segmented chip morphology. (**b**) Undeformed chip region with original β grain structure of the workpiece material with in-tact β grains highlighted. (**c**) Higher magnification image of the secondary shear zone which adjoins the tool rake face during cutting.

3.3. Transmission Electron Microscopy (TEM) Analysis of Expanded Shear Band Region

A montage of bright field (BF) TEM images showing the typical microstructure within the expanded shear band region is presented in Figure 3a and a selected area diffraction pattern (SADP) from this region is included inset. They reveal that the microstructure in the bands is highly refined, consisting of fine elongated β grains as the matrix phase interspersed with lath-like α precipitates (lighter contrast). The SAPD, indexed to the β and α phases, exhibits a continuous ring pattern characteristic of very fine and randomly oriented grains with large grain boundary misorientations. There is a gradient of deformation from the highly refined, smaller grains at the centre to more coarse grains at the left and right extremities. Some individual β grains can be identified (arrowed in Figure 3a) due to dark strain contrast arising from high levels of dislocation activity. The arrowed β grains reveal the progression of refinement from the extremities to the interior. The β grains at the extremities have a diamond-like shape with grain sizes in the order of 400–500 nm with α phase laths often sitting along their diagonal boundaries. Closer to the heavily refined central region, the β grains transition to an elongated form with grains around 20–50 nm in width and 300–400 nm in length, while the α laths with their long axis closely aligned to the length-wise axis of the β grains are 5–10 nm in width and 300–400 nm in length. In the most highly refined region, the β grain size is less than 100 nm.

Higher magnification BF and corresponding hollow cone dark field (HCDF) images from this highly refined region are presented in Figure 3b,c, respectively. They further reveal the fine β grains with considerable internal deformation structure as well as the lath-like α phase. A high magnification BF image showing the coarse diamond shaped β grains at the extremities of the deformation region is shown in Figure 3d.

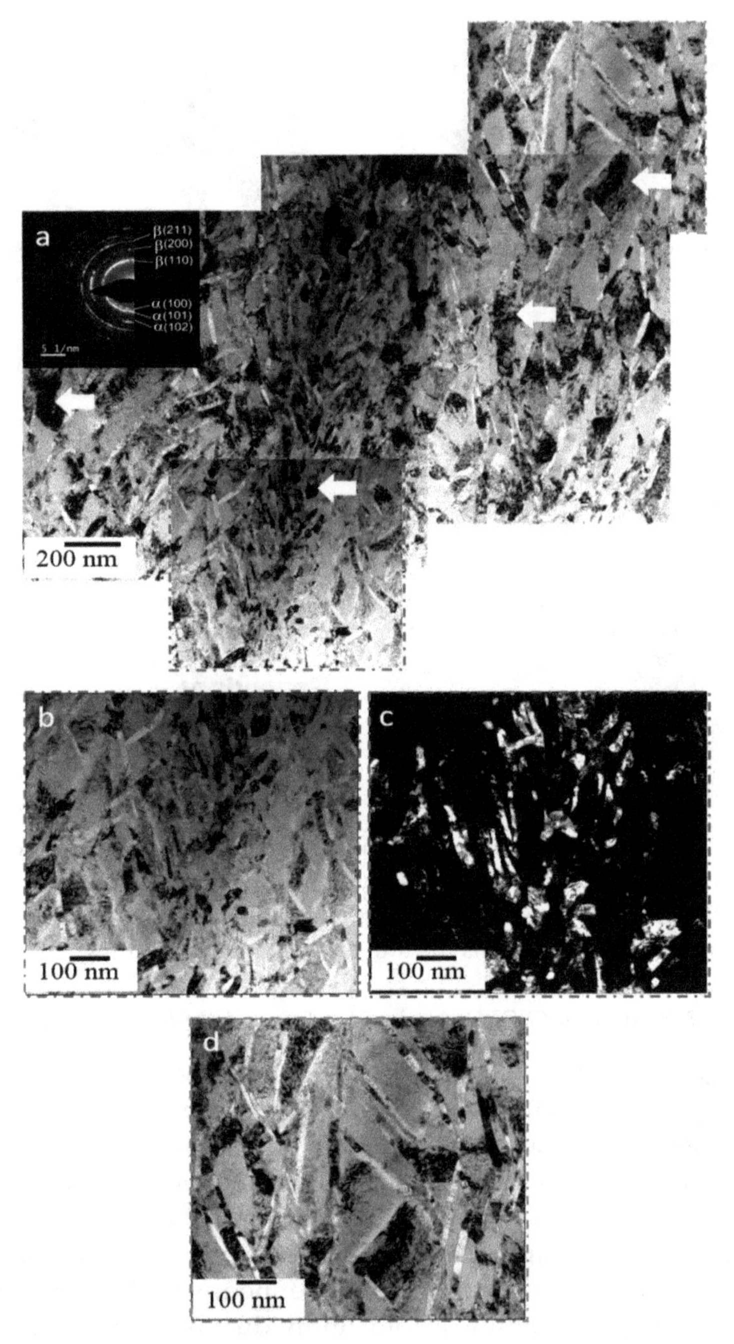

Figure 3. Transmission electron microscopy (TEM) from the expanded shear band region: (**a**) Montage of bright field (BF) images with selected area diffraction pattern (SADP) inset indexed to the β and α phases. (**b**) BF image and (**c**) corresponding hollow cone dark field (HCDF) image from the red dashed region in (**a**) formed from β (110) and α (100), (002), (101) diffraction rings. (**d**) Higher magnification BF image from the blue dashed region in (**a**).

3.4. TEM Analysis of Secondary Shear Zone

A series of TEM images from the secondary shear zone are presented in Figure 4. They reveal the chip microstructure in the region immediately adjacent to the rake face (a), i.e., at the chip extremity, and then moving incrementally into the chip interior, (b) and (c).

Figure 4a shows a BF TEM image with SADP inset and corresponding HCDF image from the region immediately adjacent to the rake face. The SADP exhibits a continuous ring pattern characteristic of very fine and randomly oriented grains with reflections from the β phase only. The BF image reveals a gradient of deformation from left to right from fine equiaxed grains immediately adjacent to the cutting interface at the left with grain sizes around 30–50 nm to more elongated grains 50 to 100 nm in width and several hundreds of nm in length at the right further from the cutting interface. A BF TEM image showing the significantly elongated β grains formed further from the cutting interface (approximately 1–2 μm) and a corresponding HCDF formed from the β (110) diffraction ring are shown in Figure 4b. Figure 4c reveals the microstructure in the secondary shear zone 5–10 μm from the cutting interface becomes less refined and consists of significantly larger, elongated β grains with long lath-like α precipitates (lighter contrast). Again, some β grains exhibit dark contrast arising from high levels of dislocation activity within their interiors. The SAPD from this region exhibits a discontinuous ring pattern indicative of a less refined structure with reflections from both the β and α phases.

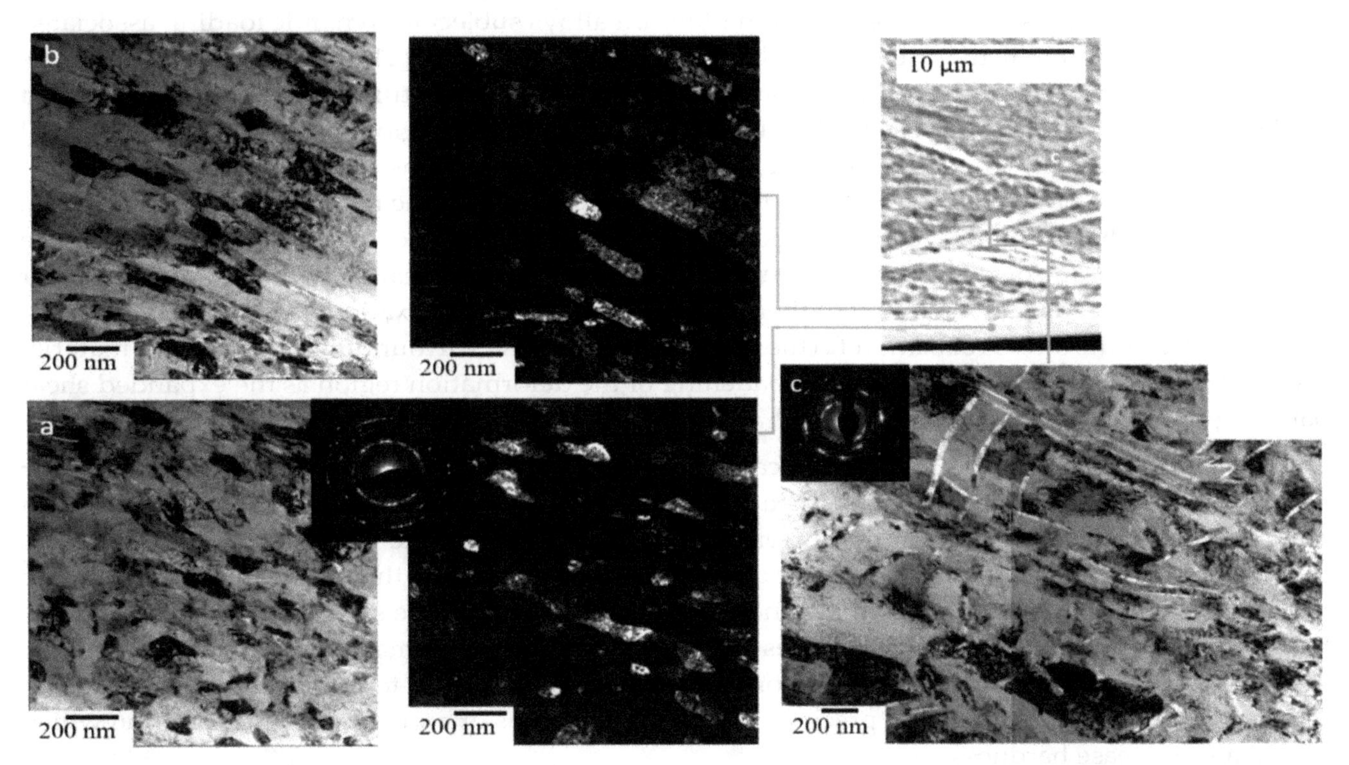

Figure 4. Series of TEM images from the secondary shear zone region immediately adjacent to the rake face (**a**), i.e., the chip extremity, and incrementally further into the chip interior (**b**,**c**). (**a**) BF image with SADP inset showing reflections almost entirely from the β phase and corresponding HCDF image formed from the β (110) diffraction ring. (**b**) BF image and corresponding HCDF image formed from β (110) diffraction ring. (**c**) Montage of BF images with SADP inset with reflections from the β and α phases.

4. Discussion

The TEM characterisation revealed important details of the machining chip microstructures which are linked to the deformation processes taking place in the chips during machining of the Ti-25Nb-3Mo-3Zr-2Sn alloy. Significant differences were identified between the expanded shear bands regions and the secondary shear zone which are discussed in respect to the influence of strain, heat generation and temperatures encountered during cutting.

High levels of deformation in the expanded shear bands formed within the primary deformation zone promote extensive localised refinement of the β phase to almost equiaxed grains with sizes below 100 nm. Additionally, α phase laths frequently occupy the β grain boundaries, aligned to the shear direction. This grain refinement consequently led to an approximate 50% increase in hardness from that of the starting material, primarily due to locally enhanced Hall-Petch strengthening [18]. In regions adjacent to the fine equiaxed zone, there is a transition to firstly larger elongated β grains, which according to the work of Zhan et al. [16] can be inferred to be the <001> and <111> fibre textures, with α laths aligned to the shear direction, and then to larger diamond shaped β grains with α phase often located on their diagonal axes. Across the entire width of the expanded shear band region, this variation in refinement suggests a cyclic process of formation such that the expanded shear band consists of an accumulation of individual localised shear events. The presence of both β and α phases across the entire expanded shear band region indicates that temperatures associated with adiabatic heating during cutting do not exceed the β transus temperature of around 720 °C for the Ti-25Nb-3Mo-3Zr-2Sn alloy [19]. The high density of α phase laths in this region suggest that dynamic precipitation may also take place in conjunction with the deformation. This suggests that temperatures during cutting are likely in the range of 400–600 °C [20].

Shear localisations are often observed in titanium alloys subject to dynamic loading associated with their low heat conductivity and high adiabatic shearing sensitivity [21]. Chip formation occurs by concentrated shear within the deformation band and the microstructural refinement is attributed to the large shear strains imposed. The formation of the shear regions is linked to substantial reductions in the yield stress which occur during cutting due to substantial thermal softening of the Ti-25Nb-3Mo-3Zr-2Sn alloy at elevated temperatures [16]. Under the momentum diffusion-based shear band evolution model, a highly localised primary shear band forms at the centre of the primary deformation zone and large deformation occurs inside the shear band as the localised shear deformation proceeds [22]. Subsequent thermal softening enables relaxation of the stress within the shear band and the stress relaxation further propagates into the surrounding undisturbed material giving rise to momentum diffusion and broadening of the deformation region as the expanded shear band regions evolve.

In comparison to the highly localised shear bands observed in cutting of other titanium alloys [23–26], the expanded shear band regions observed for the Ti-25Nb-3Mo-3Zr-2Sn alloy are relatively diffuse. Under the current cutting conditions, for the Ti-25Nb-3Mo-3Zr-2Sn alloy these zones are around 100 μm in width and of a similar magnitude to that of the regions with more limited deformation. An increase in hardness of around 10% from the starting material in regions between the expanded shear bands indicates that deformation in the primary shear zone is not entirely confined to the expanded shear band regions. Additionally, increased temperatures during cutting may also promote further α phase precipitation in these regions of more limited deformation which would also increase hardness.

For the secondary shear zone, strains at the cutting interface can be very large (>5) [27], typically leading to formation of ultrafine and/or nanocrystalline grain structures. The level of strain and hence refinement decays with increasing distance from the cutting interface across the secondary shear zone, which is in the order of 15–20 μm in width for the Ti-25Nb-3Mo-3Zr-2Sn alloy. At the cutting interface the β grains are heavily refined and approximately equiaxed with very fine nanocrystalline grain sizes around 20–50 nm while further from the interface, approximately 1–2 μm, the β grains become highly elongated in the shear direction with grains in the order of 100 nm in width and 0.5 to 1 μm in length. Again it can be inferred that the elongated grains are aligned to the <001> and <111> fibre textures [16]. An absence of the α phase in the equiaxed and elongated β grain regions of the secondary shear zone indicates that recrystallization takes place during cutting and temperatures are in excess of the alloy's 720 °C β transus temperature while subsequent rates of cooling are sufficiently high to preserve the single phase β microstructure. At 5–10 μm from the cutting interface a mixture of larger elongated β grains and long α laths are observed and the hardness is enhanced by more than 30% over that of the starting material. This can be attributed to the effects of significant β phase refinement resulting from deformation during cutting which enhances Hall-Petch strengthening as well as dynamic precipitation taking place due to the elevated temperatures during cutting. The α laths tend to have thinner, longer morphologies than those observed within the expanded shear band region which is potentially due to the influence of comparatively higher temperatures and strains in this region, which favour growth of precipitates over nucleation through dynamic precipitation effects.

As the temperature and intensity of strain decrease across the secondary shear zone as a function of the distance from the cutting interface, the microstructures observed reflect their differing thermomechanical histories. The variations observed in the microstructure across the secondary deformation zone for the Ti-25Nb-3Mo-3Zr-2Sn alloy are largely consistent with those reported for adiabatic shear bands formed in titanium alloys [21,28,29] involving a progression to finer, more equiaxed grains in the region of most intense deformation, in this case at the cutting interface [14].

Previously, Rashid et al. [12] used infrared thermography to measure the maximum temperatures at the back surface of the chips in the cutting region during machining of the Ti-25Nb-3Mo-3Zr-2Sn alloy. They reported the average temperatures to be 540–600 °C for surface cutting speeds of 80–100 m/min for the solution treated and aged alloy. However, as the cutting edge was covered by the chip, the temperatures reported were those from the cutting zone at the external surface of the chip, i.e., on the opposite side to the cutting interface. Therefore, the actual temperature at the interface between the cutting face of the tool and the chip may be substantially higher than those reported. Additionally, infrared thermography temperature measurements are acknowledged to be subject to substantial error [30,31] and temperatures in the cutting zone are far from uniform, being typically characterised by regions of high temperature gradient [32,33]. The microstructural characterisation of the chips gives some insight into the degree of these temperature gradients for machining of the solution treated and aged Ti-25Nb-3Mo-3Zr-2Sn alloy. For the chip, the significant sources of heat are in the primary deformation zone due to plastic work associated with shear and in the secondary deformation zone due to work done in deformation of the chip and in association with sliding friction at the tool-chip interface.

Based on the assumption that all mechanical work is converted to heat, the cutting forces can be applied to estimate heat generation during cutting. In this case, the heat generated in the primary deformation zone, Q_s, can be calculated from [32]:

$$Q_s = W_c = F_V \cdot V \tag{1}$$

where F_V is the tangential cutting force and V is the cutting velocity.

The amount of heat generated due to work done in the secondary deformation zone along the tool rake face is calculated from the frictional energy given by:

$$Q_r = \frac{F_{fr} \cdot V}{\lambda} \tag{2}$$

where F_{fr} is the total shear force acting on the rake face and λ is the chip thickness ratio. The shear force can be calculated from:

$$F_{fr} = F_V \sin(\alpha) + F_S \cos(\alpha) \tag{3}$$

where F_S is the feed force and α is the rake angle. An estimate of the heat generated in the primary and secondary deformation zone made from the above equations using the cutting forces reported by Rashid et.al. [12] with a tangential cutting force, $F_V = 450$ N, a feed force, $F_S = 240$ N, a cutting velocity, $V = 90$ m/min, a chip thickness ratio, $\lambda = \frac{0.19 \text{ mm}}{0.15 \text{ mm}}$ and a rake angle $\alpha = 9°$.yields heat generation of approximately 40 kW in the primary deformation zone and 22 kW in the secondary shear zone. From this analysis, the heat generated in the secondary shear zone for cutting of the Ti-25Nb-3Mo-3Zr-2Sn alloy is proportionately quite high at around 55% of that in the primary deformation zone. For context, Tay et al. observe that typically the total heat generation due to plastic deformation and frictional sliding in the secondary deformation zone for continuous chips from a non-abrasive material at medium cutting speeds is around 20% to 30% of the heat generated in the primary cutting zone [34].

While cutting heat is removed by the chip, the tool, the workpiece and some of the heat generated at the shear plane (primary shear zone) is transferred to the tool-chip interface. Hence, the temperature in the chip in the region adjacent to the tool rake face rises due to the combination of heat from the primary and secondary shear zones. Predictions of the temperature fields in the chip on the basis of heat generation are complex and have been the subject of various analytical and numerical investigations involving modelling of heat conduction, kinematics, geometries and energy aspects of the machining process. Others have attempted to measure temperature both at the cutting interface zone and across the chip, tool and workpiece through the methods including embedded thermocouples, radiation pyrometers and metallographic techniques [32,33]. In general, the highest temperatures are reportedly near the tool-chip interface in the secondary deformation zone. This is consistent with the microstructural analysis of the Ti-25Nb-3Mo-3Zr-2Sn alloy chips which indicated that temperatures in this region were in excess of 720 °C during cutting.

5. Conclusions

Ultrafine grain microstructures formed by severe plastic deformation during machining of the solution treated and aged Ti-25Nb-3Mo-3Zr-2Sn biomedical β titanium alloy have been investigated by TEM analyses of specimens obtained by FIB from transverse section of chips. The investigations have revealed that:

1. High levels of deformation in the primary shear zone promote extensive refinement of the β phase within expanded shear band regions approximately 100 μm in width to almost equiaxed grains with sizes below 100 nm in regions of intense deformation, while α phase laths frequently occupy the grain boundaries aligned to the shear direction. There is a transition to firstly elongated β grains and then to larger diamond-shaped β grains in adjoining regions of less intense deformation. The presence of a high density of α phase laths across the entire expanded shear band region suggests that temperatures in this region are likely in the range of 400–600 °C during cutting.
2. For the secondary shear zone, large strains at the cutting interface result in recrystallised, approximately equiaxed grains with nanocrystalline grain sizes around 20–50 nm, while further (1–2 μm) from the interface the β grains become highly elongated in the shear direction with grains in the order of 100 nm in width and 0.5 to 1 μm in length. At the cutting interface, an absence of the α phase indicates that the temperatures exceed the alloy's 720 °C β transus temperature. At 5–10 μm, from the cutting interface a mixture of large elongated β grains and long α phase laths are observed. The microstructural variation across the secondary shear zone reflects the decay of strain and temperature away from the cutting interface.
3. The microstructural characterisation of the chips infers information on the temperature fields present across the chips during cutting. The highest cutting temperatures occur within the secondary shear zone at the cutting interface, associated with proportionately high levels of heat generation due to deformation and friction.

Author Contributions: Conceptualization, D.K., M.D., S.S. and M.B.; Methodology, D.K.; Investigation, D.K., R.R.R. and H.A.; Writing-Original Draft Preparation, D.K.; Writing-Review & Editing, R.R.R., M.B. and S.S.; Project Administration, M.D.; Funding Acquisition, D.K. M.B. and M.D.

Acknowledgments: The authors acknowledge the facilities, and the scientific and technical assistance of the Australian Microscopy & Microanalysis Research Facility at the Centre for Microscopy and Microanalysis, The University of Queensland and at the Australian Centre for Microscopy & Microanalysis at the University of Sydney.

References

1. Long, M.; Rack, H.J. Titanium alloys in total joint replacement—A materials science perspective. *Biomaterials* **1998**, *19*, 1621–1639. [CrossRef]

2. Niinomi, M. Mechanical biocompatabilities of titanium alloys for biomedical applications. *J. Mech. Behav. Biomed. Mater.* **2008**, *1*, 30–42. [CrossRef] [PubMed]
3. Ping, D.H.; Mitarai, Y.; Yin, F.X. Microstructure and shape memory behavior of a Ti-30Nb-3Pd alloy. *Scr. Mater.* **2005**, *52*, 1287–1291. [CrossRef]
4. Yu, Z.; Wang, G.; Ma, X.Q.; Dargusch, M.S.; Han, J.Y.; Yu, S. Development of biomedical near beta titanium alloys. In *Materials Science Forum: 4th International Light Metals Technology Conference*; Trans Tech Publications: Zurich, Switzerland, 2009.
5. Kent, D.; Wang, G.; Yu, Z.; Dargusch, M.S. Pseudoelastic behaviour of a β Ti-25Nb-3Zr-3Mo-2Sn alloy. *Mater. Sci. Eng. A* **2010**, *527*, 2246–2252. [CrossRef]
6. Kent, D.; Wang, G.; Yu, Z.; Ma, X.; Dargusch, M.S. Strength enhancement of a biomedical titanium alloy through a modified accumulative roll bonding technique. *J. Mech. Behav. Biomed. Mater.* **2011**, *4*, 405–416. [CrossRef] [PubMed]
7. Rahman, M.; Wong, Y.S.; Zareena, A.R. Machinability of titanium alloys. *JSME Int. J. Ser. C* **2003**, *46*, 107–115. [CrossRef]
8. Yang, X.; Liu, C.R. Machining titanium and its alloys. *Mach. Sci. Technol.* **1999**, *3*, 107–139. [CrossRef]
9. Ezugwu, E.O.; Wang, Z.M. Titanium alloys and their machinability—A review. *J. Mater. Process. Technol.* **1997**, *68*, 262–274. [CrossRef]
10. Machado, A.R.; Wallbank, J. Machining of titanium and its alloys—A review. *Proc. Inst. Mech. Eng. Part B* **1990**, *204*, 53–60. [CrossRef]
11. Uzer, B.; Toker, S.M.; Cingoz, A.; Bagci-Onder, T.; Gerstein, G.; Maier, H.J.; Canadinc, D. An exploration of plastic deformation dependence of cell viability and adhesion in metallic implant materials. *J. Mech. Behav. Biomed. Mater.* **2016**, *60*, 177–186. [CrossRef] [PubMed]
12. Rashid, R.A.R.; Sun, S.; Wang, G.; Dargusch, M.S. Machinability of a near beta titanium alloy. *Proc. Inst. Mech. Eng. Part B* **2011**, *225*, 2151–2162. [CrossRef]
13. Rashid, R.A.R.; Sun, S.; Wang, G.; Dargusch, M.S. Experimental investigation of various chip parameters during machining of the Ti25Nb3Mo3Zr2Sn beta titanium alloy. *Adv. Mat. Res.* **2013**, *622*, 366–369. [CrossRef]
14. Schneider, J.; Dong, L.; Howe, J.Y.; Meyer, H.M. Microstructural characterization of Ti-6Al-4V metal chips by focused ion beam and transmission electron microscopy. *Metall. Mater. Trans. A* **2011**, *42*, 3527–3533. [CrossRef]
15. Zhan, H.; Zeng, W.; Wang, G.; Kent, D.; Dargusch, M. On the deformation mechanisms and strain rate sensitivity of a metastable β Ti-Nb alloy. *Scr. Mater.* **2015**, *107*, 34–37. [CrossRef]
16. Zhan, H.; Wang, G.; Kent, D.; Dargusch, M. The dynamic response of a metastable β Ti-Nb alloy to high strain rates at room and elevated temperatures. *Acta Mater.* **2016**, *105*, 104–113. [CrossRef]
17. Ye, G.G.; Xue, S.F.; Ma, W.; Dai, L.H. Onset and evolution of discontinuously segmented chip flow in ultra-high-speed cutting Ti-6Al-4V. *Int. J. Adv. Manuf. Technol.* **2017**, *88*, 1161–1174. [CrossRef]
18. Hughes, G.D.; Smith, S.D.; Pande, C.S.; Johnson, H.R.; Armstrong, R.W. Hall-petch strengthening for the microhardness of twelve nanometer grain diameter electrodeposited nickel. *Scr. Metall.* **1986**, *20*, 93–97. [CrossRef]
19. Zhentao, Y.; Lian, Z. Influence of martensitic transformation on mechanical compatibility of biomedical β type titanium alloy tlm. *Mater. Sci. Eng. A* **2006**, *438*, 391–394. [CrossRef]
20. Kent, D.; Pas, S.; Zhu, S.; Wang, G.; Dargusch, M.S. Thermal analysis of precipitation reactions in a Ti–25nb–3mo–3zr–2sn alloy. *Appl. Phys. A* **2012**, *107*, 835–841. [CrossRef]
21. Wang, B.; Wang, X.; Li, Z.; Ma, R.; Zhao, S.; Xie, F.; Zhang, X. Shear localization and microstructure in coarse grained beta titanium alloy. *Mater. Sci. Eng. A* **2016**, *652*, 287–295. [CrossRef]
22. Ye, G.G.; Xue, S.F.; Jiang, M.Q.; Tong, X.H.; Dai, L.H. Modeling periodic adiabatic shear band evolution during high speed machining Ti-6Al-4V alloy. *Int. J. Plast.* **2013**, *40*, 39–55. [CrossRef]
23. Arrazola, P.J.; Garay, A.; Iriarte, L.M.; Armendia, M.; Marya, S.; Le Maître, F. Machinability of titanium alloys (Ti6Al4V and Ti555.3). *J. Mater. Process. Technol.* **2009**, *209*, 2223–2230. [CrossRef]
24. Joshi, S.; Pawar, P.; Tewari, A.; Joshi, S.S. Effect of β phase fraction in titanium alloys on chip segmentation in their orthogonal machining. *CIRP J. Manuf. Sci. Technol.* **2014**, *7*, 191–201. [CrossRef]
25. Sun, S.; Brandt, M.; Dargusch, M.S. Characteristics of cutting forces and chip formation in machining of titanium alloys. *Int. J. Mach. Tool. Manuf.* **2009**, *49*, 561–568. [CrossRef]

26. Dargusch, M.S.; Sun, S.; Kim, J.W.; Li, T.; Trimby, P.; Cairney, J. Effect of tool wear evolution on chip formation during dry machining of ti-6al-4v alloy. *Int. J. Adv. Manuf. Tech.* **2018**, *126*, 13–17. [CrossRef]
27. Oxley, P.L.B. *Mechanics of Machining*; Ellis Horwood: New York, NY, USA, 1989.
28. Zhan, H.; Zeng, W.; Wang, G.; Kent, D.; Dargusch, M. Microstructural characteristics of adiabatic shear localization in a metastable beta titanium alloy deformed at high strain rate and elevated temperatures. *Mater. Charact.* **2015**, *102*, 103–113. [CrossRef]
29. Yang, Y.; Jiang, F.; Zhou, B.M.; Li, X.M.; Zheng, H.G.; Zhang, Q.M. Microstructural characterization and evolution mechanism of adiabatic shear band in a near beta-ti alloy. *Mater. Sci. Eng. A* **2011**, *528*, 2787–2794. [CrossRef]
30. Davies, M.A.; Ueda, T.; M'Saoubi, R.; Mullany, B.; Cooke, A.L. On the measurement of temperature in material removal processes. *CIRP Ann. Manuf. Technol.* **2007**, *56*, 581–604. [CrossRef]
31. Lane, B.; Whitenton, E.; Madhavan, V.; Donmez, A. Uncertainty of temperature measurements by infrared thermography for metal cutting applications. *Metrologia* **2013**, *50*, 637–653. [CrossRef]
32. Abukhshim, N.A.; Mativenga, P.T.; Sheikh, M.A. Heat generation and temperature prediction in metal cutting: A review and implications for high speed machining. *Int. J. Mach. Tool. Manuf.* **2006**, *46*, 782–800. [CrossRef]
33. Sutter, G.; Ranc, N. Temperature fields in a chip during high-speed orthogonal cutting—an experimental investigation. *Int. J. Mach. Tool. Manuf.* **2007**, *47*, 1507–1517. [CrossRef]
34. Tay, A.O.; Stevenson, M.G.; De Vahl Davis, G. Using the finite element method to determine temperature distributions in orthogonal machining. *Proc. Inst. Mech. Eng.* **1974**, *188*, 627–638. [CrossRef]

Diffraction Line Profile Analysis of 3D Wedge Samples of Ti-6Al-4V Fabricated using Four Different Additive Manufacturing Processes

Ryan Cottam †, Suresh Palanisamy [1,2,*], Maxim Avdeev [3], Tom Jarvis [4], Chad Henry [5], Dominic Cuiuri [6], Levente Balogh [7] and Rizwan Abdul Rahman Rashid [1,2]

1 School of Engineering, Faculty of Science, Engineering and Technology, Swinburne University of Technology, Hawthorn, VIC 3122, Australia; rrahmanrashid@swin.edu.au
2 Defence Materials Technology Centre, Hawthorn, VIC 3122, Australia
3 The Bragg Institute, Australian Nuclear Science and Technology Organisation (ANSTO), Lucas Heights, NSW 2234, Australia; maxim.avdeev@ansto.gov.au
4 Monash Centre for Additive Manufacturing, Monash University, Notting Hill, VIC 3168, Australia; tom.jarvis@monash.edu
5 Commonwealth Scientific and Industrial Research Organization (CSIRO), Clayton, VIC 3168, Australia; wchadry@yahoo.com
6 School of Mechanical, Materials, and Mechatronic Engineering, Faculty of Engineering and Information Sciences, University of Wollongong, Wollongong, NSW 2522, Australia; dominic@uow.edu.au
7 Department of Mechanical and Materials Engineering, Queen's University, Kingston, ON K7L 3N6, Canada; levente.balogh@queensu.ca
* Correspondence: spalanisamy@swin.edu.au

Abstract: Wedge-shaped samples were manufactured by four different Additive Manufacturing (AM) processes, namely selective laser melting (SLM), electron beam melting (EBM), direct metal deposition (DMD), and wire and arc additive manufacturing (WAAM), using Ti-6Al-4V as the feed material. A high-resolution powder diffractometer was used to measure the diffraction patterns of the samples whilst rotated about two axes to collect detected neutrons from all possible lattice planes. The diffraction pattern of a LaB_6 standard powder sample was also measured to characterize the instrumental broadening and peak shapes necessary for the Diffraction Line Profile Analysis. The line profile analysis was conducted using the extended Convolution Multiple Whole Profile (eCMWP) procedure. Once analyzed, it was found that there was significant variation in the dislocation densities between the SLMed and the EBMed samples, although having a similar manufacturing technique. While the samples fabricated via WAAM and the DMD processes showed almost similar dislocation densities, they were, however, different in comparison to the other two AM processes, as expected. The hexagonal (HCP) crystal structure of the predominant α-Ti phase allowed a breakdown of the percentage of the Burgers' vectors possible for this crystal structure. All four techniques exhibited different combinations of the three possible Burgers' vectors, and these differences were attributed to the variation in the cooling rates experienced by the parts fabricated using these AM processes.

Keywords: Ti-6Al-4V; additive manufacturing; selective laser melting (SLM); electron beam melting (EBM); direct metal deposition (DMD); wire and arc additive manufacturing (WAAM); diffraction line profile analysis; extended convolution multiple whole profile (eCMWP)

---

1. Introduction

Additive Manufacturing (AM) of metallic materials is receiving increasing attention worldwide [1–3]. There are two main AM approaches, and they are the powder bed approach, and the direct deposition approach. In the powder bed approach, a layer of powder is swept over a platform and the powder is melted together using either a laser or an electron beam, known as selective laser melting (SLM) and electron beam melting (EBM), respectively. The platform is then lowered and a new layer of powder is swept over. The melting process is then performed again according to the G-code given by the pre-processing software. This is repeated until the desired part is formed [4,5]. The direct deposition approach is used to melt either powder blown onto the substrate or wire fed into the melt pool of the substrate, using a heat source that is usually either a high powered laser or an electric arc, known as direct metal deposition (DMD) and wire arc additive manufacturing (WAAM), respectively. Tracks of material are placed side-by-side, layer-upon-layer, until the desired shape is formed. The powder bed approach easily produces complex shapes, with the aid of support structures, while the direct deposition approach produces more basic shapes but can produce much larger sized components because it is not limited by the size of the powder bed chamber [6].

In this study, the titanium alloy Ti-6Al-4V has been investigated, which has been widely used in aerospace and medical applications. Four leading metallic AM processes, namely SLM, EBM, DMD, and WAAM, were employed to manufacture wedge-shaped Ti-6Al-4V samples. While all four of these techniques are relatively mature technologies and have proven capable to build 3D shapes, the metallurgical character of the deposits, in particular differences between the different technologies, is not very well known and may play a role in identifying which of the technologies should be employed for the manufacturing of a particular component. One aspect of the metallurgical character of the deposits is the dislocation content, which has a significant influence on the mechanical properties of the deposits, in particular the strength and ductility. This knowledge aids in understanding why the mechanical properties of these AM processes varies, as that reported by Sames et al. [7] and Frazier [8]. Therefore, diffraction line profile analysis was employed to measure the dislocation contents produced by these AM technologies.

Diffraction line profile analysis (DLPA) is a diffraction analysis technique where the number and type of dislocations present in a structure can be determined quantitatively, together with other microstructural features, such as average sub-grain size, planar fault frequency, and the breakdown of the Burgers' vectors of the different dislocations and their relative percentages [9,10]. Moreover, DLPA is an indirect method used to derive average microstructural characteristics from neutron or X-ray diffraction patterns. In the case of neutron diffraction, the volume of the diffracting material is in the range of cubic centimetres, which allows a non-destructive and bulk characterization of the microstructure of the investigated material. It is most commonly used during plastic deformation, in parallel with plasticity models to understand the various slip system activities [11,12]. Using neutron DLPA, it has been shown that the initial dislocation density of as-built stainless steel samples depends on the type and the parameters of the applied AM process and can be altered by subsequent heat treatments or plastic deformation [13,14]. Results strongly suggest that the flow strength of the as-built AM stainless steel is primarily controlled by the dislocation density present in the material, making DLPA a useful characterization tool for such materials [13]. This type of analysis, which provides quantitative characteristics on the dislocation structure, deepens the understanding of the deformation mechanisms operating in metals and their correlation with the mechanical properties of the bulk polycrystal. This capability can be particularly useful for materials having hexagonal close-packed (HCP) crystal structures, as that of titanium, which possess anisotropic properties, and their dislocation structures vary with deformation temperature and grain orientation. The main source of dislocations in the metal AM structures form during the martensitic phase transformation occurring due to rapid solidification of the molten metal. As the crystal lattice changes from one crystal structure to another—in the case of Ti-6Al-4V, from β-Ti (HCP crystal structure) to α'-Ti (BCC crystal structure) due to displacive transformation—dislocations are formed at the transformation interface to allow for the misfit between the two crystal lattices, HCP and BCC [15]. Ahmed and Rack [16] have shown that the crystallography of the two types of martensite that forms for Ti-6Al-4V is different and is dependent on the cooling rate. Therefore, the dislocation content is an indication as to the nature and extent of the martensitic phase transformation, which is why the diffraction line profile analysis is vital in understanding the effect of the various AM processes on the microstructure and mechanical properties of Ti-6Al-4V. Moreover, this type of analysis may elucidate this change in the martensite formed by the change in the percentage of dislocations formed, as different orientation relationships at the transforming interface will change the type and number of dislocations formed, as that reported by Carroll et al. [17].

The primary objective of this study is to measure the different dislocation densities that form during four prominent metal AM processes, namely SLM, EBM, DMD, and WAAM, when fabricating Ti-6Al-4V wedge-shaped samples. This titanium alloy has been widely used for both aerospace and medical applications. Hence, in this study, the samples were irradiated with neutrons and the diffraction data was collected and analyzed using DLPA technique.

2. Materials and Methods

2.1. Sample Preparation

A wedge-shaped sample, with dimensions shown in Figure 1, was fabricated using each of the four AM processes, namely SLM, EBM, DMD, and WAAM. The wedge geometry was chosen to manifest changes in the character of the builds as a function of section size (e.g., thinner sections may cool faster than the thicker sections), which may influence microstructure formation and hence the mechanical properties. The processing parameters for the four different techniques were different and were optimized in a separate study by the various research providers.

An EOSINT M280 machine (Electro Optical Systems EOS GmbH, Krailling, Germany) was used to fabricate the titanium wedge-shape sample in the horizontal orientation using the SLM process. Gas atomized Ti-6Al-4V powder with particle size up to 63 μm was used. The process parameters used were: laser power 280 W, scan speed 1200 mm/s, layer thickness 30 μm, and hatch spacing 140 μm.

An Arcam A1 machine (Arcam AB, Mölndal, Sweden) was used for the EBM process. The titanium sample was fabricated using Ti-6Al-4V ASTM Grade 23 powder (average particle size of 73.52 μm). The standard Arcam theme 3.2.121 (Arcam AB, Mölndal, Sweden) was employed, which had an acceleration voltage of 60 kV, beam current of 1–10 mA, beam spot size of 200 μm, speed factor of 98, scanning line offset of 0.1 mm, layer thickness of 50 μm, and preheating temperature of 730 °C.

A 5 kW Trumpf-POM machine was used for the DMD process. A 210 mm × 115 mm × 6 mm titanium base plate in the annealed condition and Ti-6Al-4V powder (average particle size of 60 μm) supplied by TLS Technik were used for deposition. The powder was delivered to the deposition area. The fabrication of the wedge-shaped sample was conducted in an argon and helium gas atmosphere to minimize oxygen contamination. A laser power of 1600 W, laser spot size of about 2.2 mm, laser head traverse speed of 60 mm/min, and powder feed rate of 4.3 g/min were employed for processing.

A gas tungsten arc welding (WAAM) process was used for fabricating the Ti-6Al-4V sample. A 250 mm × 100 mm × 12 mm titanium base plate and a 1 mm diameter hard-drawn Ti-6Al-4V wire were used for deposition. However, the thick track dimensions of the WAAM process made it impossible to produce the wedge sample to the required dimensional tolerances by using only the deposition process. Instead, a rectangular block of material of dimensions 25 mm × 55 mm × 12 mm was deposited and the wedge shape was subsequently produced by wire cutting and machining. A current of 140 Amps, travel speed of 152 mm/min, arc length of 3.5 mm, and wire feed speed of 1.34 m/min (0.28 kg/h) were used as processing parameters. Welding grade argon (99.995% purity) was used as shielding gas at a flow rate of 25 L/min.

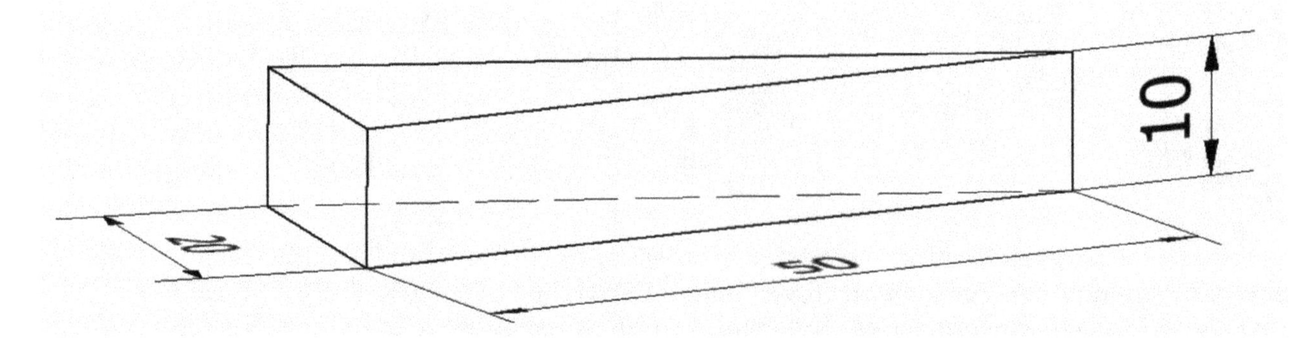

Figure 1. Dimensions of the wedge-shaped sample (in mm) used for the diffraction line profile analysis.

2.2. Neutron Diffraction Set Parameters and Extended Convolution Multiple Whole Profile (eCMWP) Analysis

Neutron diffraction data for the Ti-6Al-4V samples was collected on the high-resolution diffractometer Echidna at the OPAL facility (ANSTO, Lucas Heights, Australia), using neutrons of wavelength 1.6215 Å. During data collection, each sample was rotated around its two-fold axis to reduce the effect of preferred orientation, if any. To determine instrumental contribution to the peak width, calibration data was collected for a powder LaB_6 sample (NIST SRM 660c). The entire sample was simultaneously exposed to the neutron beam during the rotations in both orientations.

The analysis in this study was performed using the peak line broadening analysis software package eCMWP ("extended Convolution Multiple Whole Profile" procedure, 2017, G. Ribárik, et al., Budapest, Germany) [9,18]. The eCMWP software constructs a theoretical diffraction pattern based on well-established physical models of the microstructure, e.g., sub-grain size distributions, dislocation structures, and planar faults. The final shape of the various peaks of a diffraction pattern, $I^{PM}(2\theta)$, is a convolution of the contribution of various lattice defects and the contribution of the diffraction instrument itself, which is calculated using Equation (1) [9,18].

$$I^{PM}(2\theta) = \sum_{hkl} I^{S}_{hkl} * I^{D}_{hkl} * I^{PD}_{hkl} * I^{INST}_{hkl} + I_{BG} \quad (1)$$

where the defect related profile functions are the size, I^{S}_{hkl}, represents the dislocation cell or sub-grain size distribution [19], I^{D}_{hkl}, represents the contribution of the dislocations [20,21], I^{PD}_{hkl}, represents the contribution of the planar defects such as twin boundaries or stacking faults [22,23], I^{INST}_{hkl}, is the instrumental peak broadening and shape [24], and I_{BG} is the background of the diffraction pattern usually represented by a cubic spline. The instrumental peak shapes, I^{INST}_{hkl}, were determined by measuring a LaB_6 standard powder sample, which has no detectable microstructure (i.e., it is coarse grained, strain and dislocation free), thereby generating a result which is indicative of the peak broadening caused by the diffraction instrument itself. The resulting theoretical diffraction pattern presented in Equation (1) is fitted to the experimental data using a least-squares algorithm [18]. The fitting variables of the theoretical diffraction pattern are the quantitative features of the microstructure, such as the median and width of the sub-grain size distribution, density, type, and arrangement of the dislocations, and frequency of planar faults. The eCMWP software determines these quantitative characteristics by forward modelling the microstructure until a match is found between the theoretical and measured whole diffraction patterns [9,18].

Figure 2 shows the eCMWP refinement for the SLM sample. The open circles represent the measured data, the continuous line represents the modeled pattern, and the difference between the two is also shown. The 004, 202, and 104 reflections were not included in the analysis due to their low signal-to-noise ratio. Even though 200 and 210 also have low signal-to-noise ratios they were included, because ignoring them would have decreased the quality of the modeling for the overlapping high-intensity reflections. The eCMWP software matches the measured and the modeled full pattern by refining the microstructural parameters. In the present case, the refined microstructural parameters were: area weighted average sub-grain size $<X>_A$, total dislocation density ρ, and ratio of sub-densities having $<a>$, $<c+a>$, and $<c>$ Burgers' vector type, as described by Máthis et al. [10] and Ungár et al. [25]. It is important to note that the area weighted average sub-grain size $<X>_A$ represents a domain size which is defined by low-angle grain boundaries or dislocation walls, thus it will be referred to as sub-grain size or dislocation cell size; as reported by Ungár et al. [19]; hence, it is not the grain size defined by high angle grain boundaries visible in an optical microscope or in a low resolution TEM.

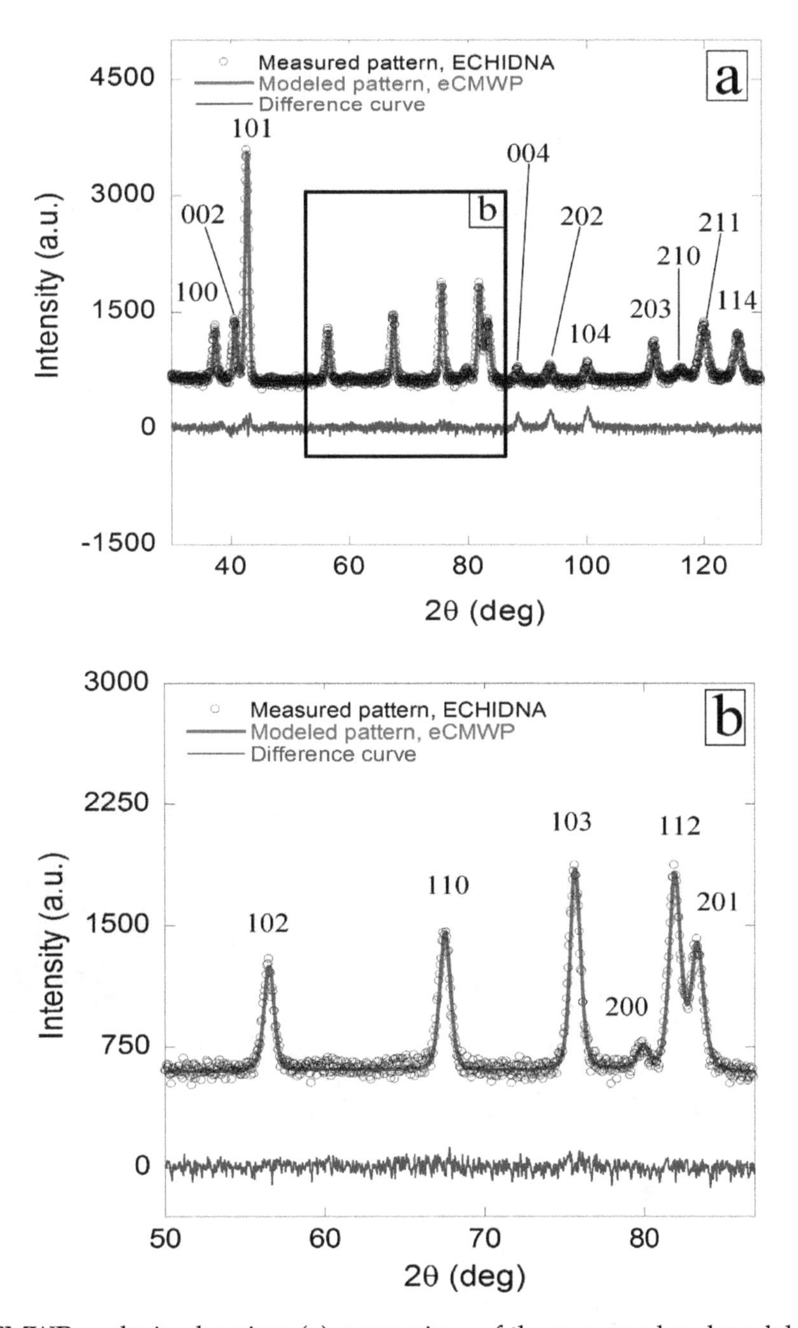

Figure 2. The eCMWP analysis, showing: (**a**) comparison of the measured and modeled pattern for the SLMed Ti-6Al-4V sample; and (**b**) diffraction plots between 50–85° (2θ) consisting of 102, 110, 103, 200, 112, and 201 phase reflections.

In order to qualitatively assess the neutron diffraction measurements, a Williamson-Hall (WH) plot was constructed. The WH plot presents the full width at half maximum (FWHM) of the peaks as a function of peak position. The FWHM values have been corrected for the instrumental broadening, thus, they represent only the broadening induced by the microstructural features found in the samples.

2.3. Residual Stress Analysis

Residual stress measurements were carried out using the contour method. The samples were wire-cut and surface profiles were measured on a Brown and Sharpe coordinate measuring machine (TESA USA, North Kingstown, RI, USA) equipped with a low force touch probe and 1 mm diameter ruby-tipped stylus. Each cut surface was measured with a 0.1 mm × 0.1 mm grid spacing, producing approximately 20,000 data points. The residual stresses were calculated from the raw contour data using MATLAB (Version 8.4, The Mathworks Inc., Natick, MA, USA) scripts and ABAQUS (Version 6.13, Dassault Systèmes Simulia Corp., Johnston, RI, USA) Finite Element code.

2.4. Microstructural Analysis

For microstructural characterization, the samples were cut from the mid-section of the wedge and prepared for metallographic examination. The samples were polished and etched with Kroll's reagent. The microstructures were examined under the Olympus BX-61 optical microscope (Olympus Corporation, Shinjuku, Japan).

3. Results and Discussion

The representative microstructures of the 3D printed titanium samples, fabricated using four different AM processes, are shown in Figure 3. It was noticed that all the samples consisted of α'-Ti martensitic phase along with $\alpha+\beta$-Ti matrix and prior β-Ti grain boundaries. However, the morphology of the α'-Ti martensite phase was slightly different, depending upon the AM process. The SLMed sample consisted of fine martensitic laths resembling a needle-like shape. The EBMed sample consisted of a similar type of fine martensitic laths, as that of the SLMed sample, but these laths were short and less in quantity. On the other hand, the DMDed sample consisted of thick and short martensitic laths with distinct prior β-Ti grain boundaries. The WAAMed sample consisted of a longer prior β-Ti grain boundary.

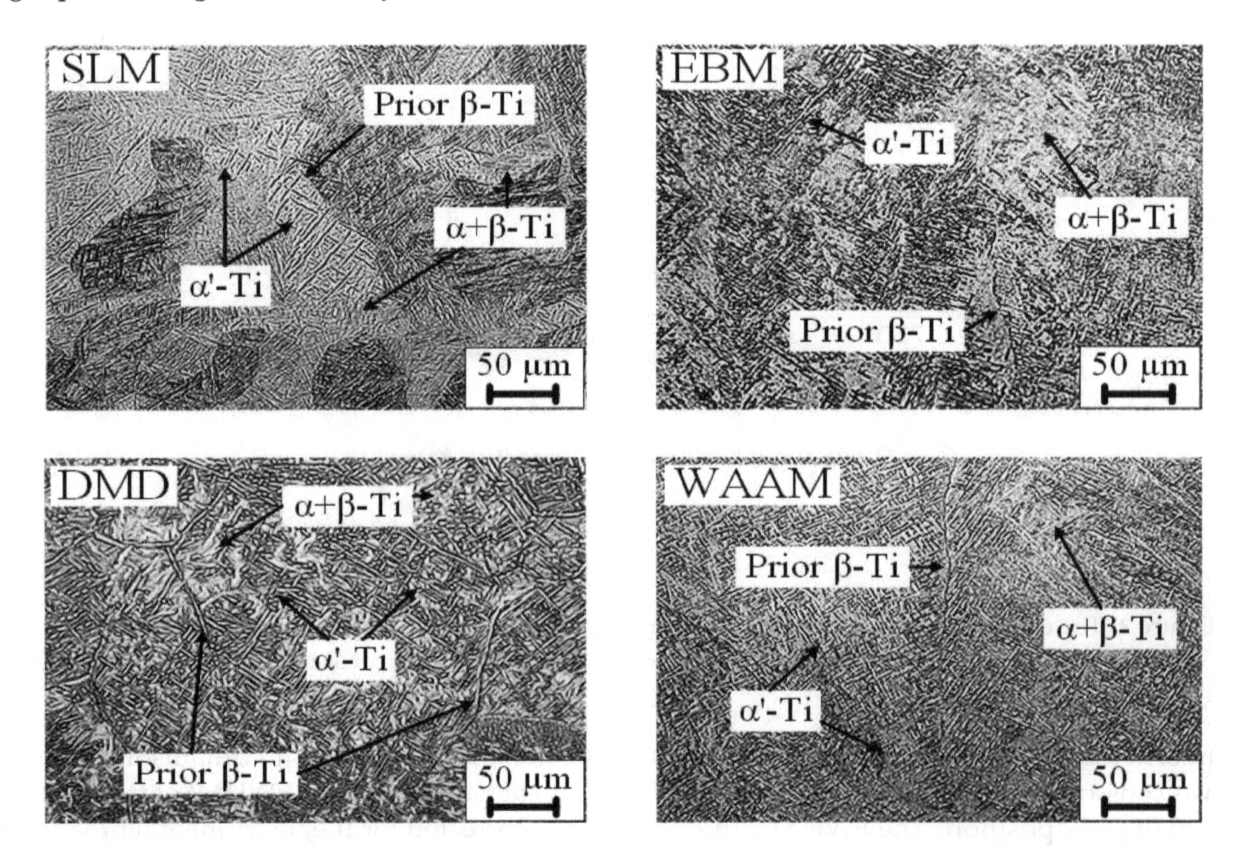

Figure 3. Microstructures of the 3D printed Ti-6Al-4V samples using four different AM processes.

The Williamson-Hall (WH) plot is shown in Figure 4, and the quantitative results obtained using eCMWP are presented in Table 1. From both the WH plot and the eCMWP data, it can be observed that the SLMed sample consisted of the highest dislocation density, whereas the EBMed sample had the lowest dislocation density in comparison to the DMDed and WAAMed samples, which exhibited similar dislocation densities. In contrast, the FWHM of the different diffraction peaks does not increase monotonously with increasing 'K', the reciprocal of the lattice spacing. This is due to the well-known effect of strain anisotropy [26], which is an indication of a significant dislocation density present in the material. The mathematical description of strain anisotropy is provided by the dislocation contrast factors, which describe the broadening of a given diffraction peak as a function of the hkl Miller indices and the different dislocation types. The evaluation of the contrast factors is handled internally by the eCMWP line profile analysis software, and the results can be used to determine the ratio of the dislocations densities having <*a*>, <*c*+*a*>, and <*c*> type Burgers' vectors [10].

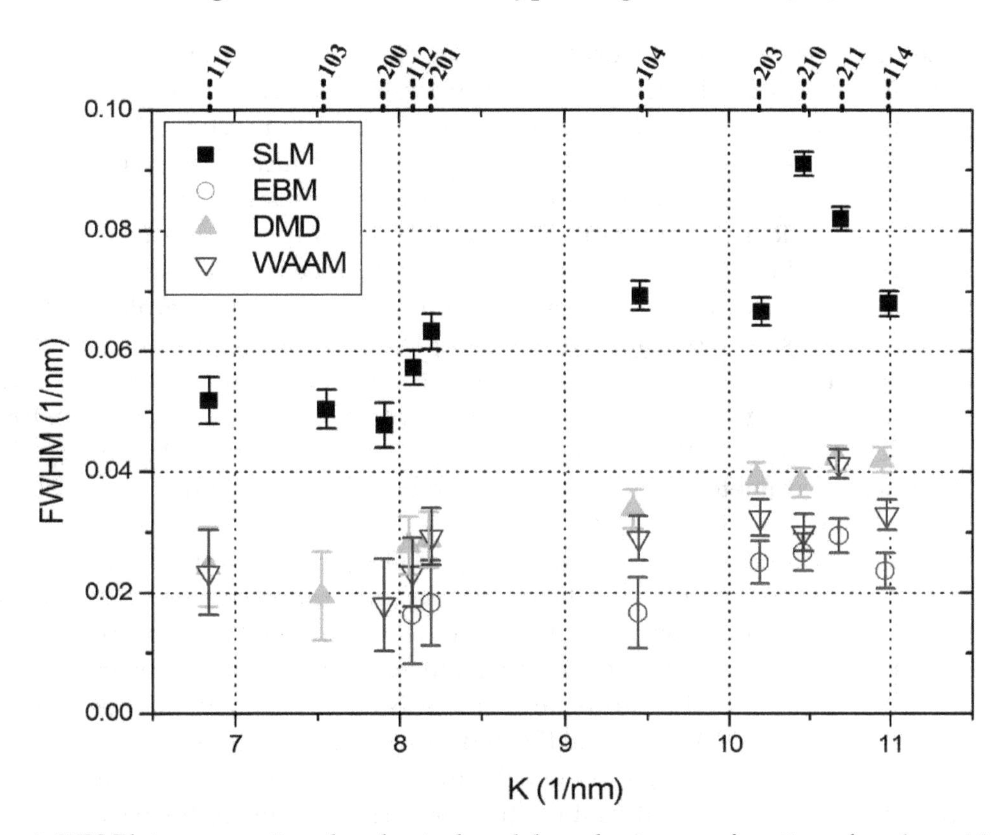

Figure 4. WH Plot representing the physical peak broadening as a function of peak position 'K'.

Table 1. Dislocation sub-cell sizes $<X>_A$, total dislocation densities ρ_{TOTAL}, and the ratios of sub-densities having <*a*>, <*c*+*a*>, and <*c*> Burgers' vector types for the Ti-6Al-4V specimen fabricated using four metal AM processes.

AM Process	$<X>_A$ (nm)	ρ_{TOTAL} (m^{-2})	<*a*> %	<*c*+*a*> %	<*c*> %
SLM	100 ± 15	$(24 \pm 3) \times 10^{14}$	85 ± 10	10 ± 10	5 ± 10
EBM	>500	$(1.4 \pm 0.5) \times 10^{14}$	70 ± 15	30 ± 15	0 ± 15
DMD	120 ± 20	$(4.1 \pm 0.5) \times 10^{14}$	60 ± 15	30 ± 15	10 ± 15
WAAM	>500	$(4.1 \pm 0.5) \times 10^{14}$	80 ± 10	20 ± 10	0 ± 10

From the eCMWP results presented in Table 1, it is quite evident that there is significant variation in the dislocation sizes and densities, which is process-dependent. It should be noted here that the entire sample was irradiated with neutrons simultaneously, and the values presented in Table 1 are an average across the sample. The WAAMed sample exhibited a large sub-grain size that can be attributed to the high heat input of the gas tungsten arc-welding (GTAW) torch used for its fabrication.

While not measured in this study, if the temperature is below the martensite finish temperature but still elevated during processing, it is possible that recovery can occur, thus increasing the sub-grain size [27]. This is further supported by the evidence that the WAAMed sample consisted of long prior β-Ti grain boundaries, which is possible when the cooling rates are low [28]. Likewise, the EBMed sample also has a high sub-grain size, which can be attributed to the high bed temperature, and subsequent slow cooling after deposition, allowing recovery processes to occur (i.e., the temperature allows the dislocations to annihilate each other). Therefore, the EBMed sample consisted of fewer α'-Ti martensitic laths (shown in Figure 3) compared to other samples [29]. On the other hand, the DMDed and SLMed samples exhibit smaller sub-cell sizes that indicates that the cooling rate during the process was higher than WAAM process, as well as the recovery rate being significantly reduced. However, the heat input in the DMD process was considerably higher than the SLM process, which was retained in the sample longer, resulting in thicker martensitic laths, as observed in the microstructures of these samples [15].

The EBMed sample has very low dislocation density, which can be the result of the high powder bed temperature (730 °C) employed during the EBM process. This provides sufficiently large amount of energy to drive recovery of any dislocations that form due the martensitic transformation [27]. However, the dislocation density in the SLMed sample is ~6× larger than in the WAAMed and DMDed samples and ~20× larger than the EBMed sample. This large increase in the dislocation content for the SLMed sample is not evident, given that these samples experience martensitic phase transformation during rapid cooling irrespective of the AM process used. There are three potential sources of the formation of dislocations during fabrication of these four samples, as follows; plastic deformation due to residual stress formation as the sample cools after deposition [30]; the dislocation formation of the displacive martensitic phase transformation [31]; and the plastic strain that the existing martensite laths undergo as new martensite laths form during the phase transformation [32]. As reported by Vasinonta et al. [30], the formation of residual stress in the 3D printing of Ti-6Al-4V is dependent on the process parameters. Therefore, this can be the first potential source of an increase in the dislocation content in the SLMed sample.

The residual stresses present in the Ti-6Al-4V samples along the central cross-section are shown in Figure 5. The SLMed sample contained a significant amount of compressive residual stress in the lower central portion of the wedge sample and tensile residual stress in the top and bottom edges. Such a steep residual stress gradient can result in the increased formation of dislocations in the sample [33,34]. In contrast, the EBMed sample had a more uniform residual stress state across the cross-section, ranging between −200 MPa and 200 MPa. The EBMed sample showed a much lower level of residual stress state compared to the SLMed sample, primarily due to the high powder bed temperature and the vacuum atmosphere maintained during the fabrication process [29]. The DMDed sample also consisted of uniform residual stresses, apart from a couple of pockets of compressive residual stresses and tensile residual stresses closer to the edges of the wedge sample, similar to that reported by Cottam et al. [35]. The WAAMed sample consisted of patches of mild tensile residual stress in the range of 100 to 300 MPa, dispersed throughout the cross-section of the wedge-shaped specimen owing to the much higher heat input than the laser-based DMD process [29,36].

The SLMed and WAAMed samples have approximately the same breakdown of dislocation types, from Table 1. The DMDed sample has a Burgers' vectors breakdown with more pyramidal, <*c+a*>, and prismatic (<*c*>) dislocations, whereas the EBMed sample consists of higher proportion of pyramidal, <*c+a*> type dislocations. Perhaps, this can be due to easier recovery of <*a*> type dislocations than the other two types. This is reasonable as the <*a*> type dislocations make up most of the dislocation content for Ti-6Al-4V samples fabricated using all four processes. Therefore, the probability that two <*a*> dislocations will meet and annihilate is higher than two <*c+a*> dislocations, resulting in an increase in the proportion of <*c+a*> type dislocations with increasing levels of recovery [27].

The crystallography of the strain associated with the martensitic phase transformation is invariant and as such the level of dislocations it introduces will be relatively consistent for the four 3D printing processes. The plastic strain of the existing martensitic laths as new martensitic laths form is dependent on the stress state of the material, as the martensitic transformation proceeds to a level which is a combination of the residual stress state and the local stress in the grain. Therefore, since the residual stress state of the SLMed sample is higher than for the WAAMed and DMDed samples, the plastic strain during the transformation will increase, and as a result, the amount of dislocations that will form in the SLMed sample will increase. Ali et al. [37] reported a decomposition of the α'-Ti martensite structure into a more homogeneous $\alpha+\beta$ phase upon preheating the SLM powder bed to about 570 °C. Furthermore, each of these AM processes experience different cooling rates, as graphically illustrated in Figure 6. Although not experimentally analyzed, it is expected that the cooling rate of the powder-bed metal AM processes is quite high, and for Ti-6Al-4V could be more than 525 °C/s, whereas for the other two AM processes, DMD and WAAM, the cooling rates are likely to be lower than SLM and EBM. Moreover, due to the high heat input in the WAAM process, the cooling rate can even be below 410 °C/s. Therefore, this explains the resultant microstructures, dislocation densities, and residual stresses of the titanium samples fabricated using these four different AM processes. It should be noted here that there will be process variations associated with each of the above AM processes which might yield a different result to that reported in this work. Therefore, further in-depth investigation is required to comprehend the dependency of the properties of the printed parts in terms of microstructure, phases, and dislocation densities on the process variables.

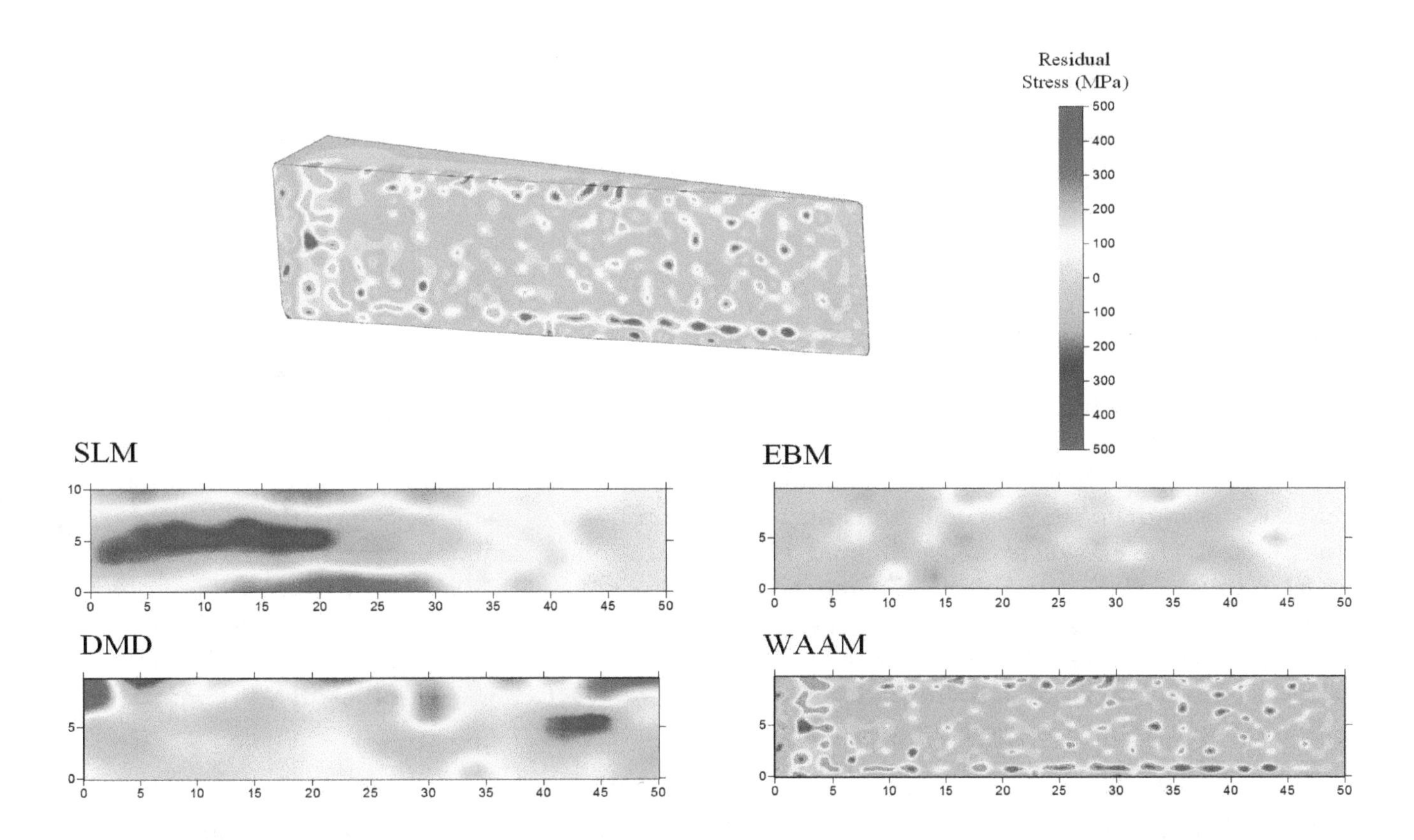

Figure 5. Residual stresses in the Ti-6Al-4V samples fabricated using different AM processes.

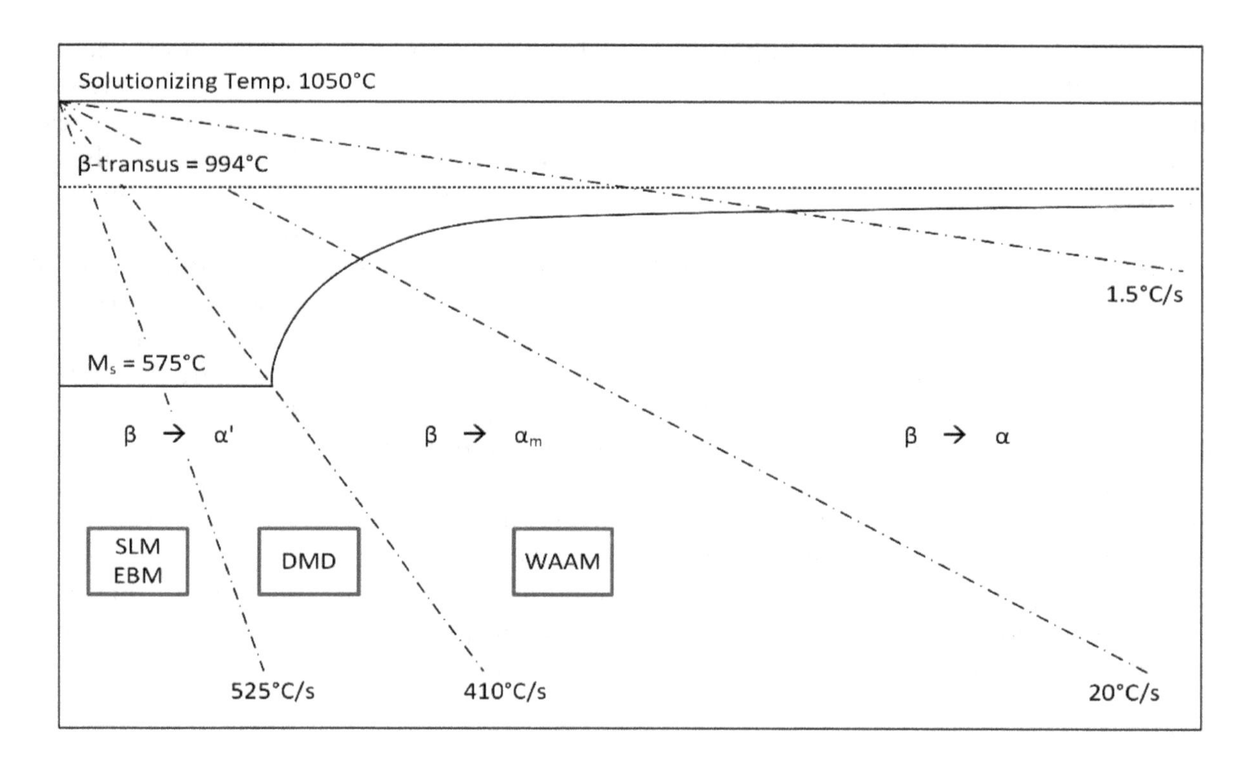

Figure 6. Schematic continuous cooling diagram for Ti-6Al-4V solution treated at 1050 °C for 30 min and quenched using the Jominey end quench test, showing the fit in terms of the cooling rates experienced during the four AM processes. (Reproduced with permission from Ahmed and Rack, Phase transformations during cooling in α+β titanium alloys, published by Elsevier, 1998 [16]).

4. Conclusions

Titanium alloys, such as Ti-6Al-4V, are widely used in aerospace and medical applications. In this work, the variation in the dislocation content of Ti-6Al-4V samples produced by the four metal AM processes, SLM, EBM, DMD, and WAAM, was investigated, and it was attributed to the different process characteristics. The SLMed sample contained a high dislocation content, the source of which was attributed to the volume of the martensitic phase transformation as well as the residual stresses in the sample. The DMDed sample had intermediate dislocation content and a significant amount of <*c*> and <*c*+*a*> dislocations due to the intermediate cooling rate experienced by the sample during this process. The EBMed sample exhibited a low dislocation content and a large dislocation cell size, which was attributed to the high temperature of the powder bed of 730 °C during the printing process, which facilitated recovery of the <*a*> type dislocations, thereby resulting in a decrease in the α'-Ti martensite phase. The WAAMed sample produced a dislocation content similar to that of the DMDed sample, but the breakdown of the dislocation of Burgers' vectors was significantly different, which was attributed to the low cooling rate during the process.

Author Contributions: Conceptualization, R.C. and S.P.; data curation, R.C. and M.A.; formal analysis, M.A.; funding acquisition, S.P.; investigation, R.C. and M.A.; methodology, R.C.; project administration, S.P.; resources, S.P., M.A., T.J., C.H., and D.C.; supervision, S.P.; validation, M.A. and L.B.; writing—review and editing, R.A.R.R.

Acknowledgments: This paper includes research that was supported by DMTC Limited (Australia). The authors have prepared this paper in accordance with the intellectual property rights granted to partners from the original DMTC project. Swinburne University of Technology would also like to thank Girish Thipperudrappa for operating the DMD during the manufacturing of the wedge-shaped sample.

Notations

3D	Three Dimensional
AM	Additive Manufacturing
BCC	Body Centered-Cubic
EBM	Electron Beam Melting
eCMWP	extended Convolution Multiple Whole Profile
DLPA	Diffraction Line Profile Analysis
DMD	Direct Metal Deposition
FWHM	Full Width at Half Maximum
HCP	Hexagonal Close-Packed
$I^{PM}(2\theta)$	Convolution of diffraction peaks at 2θ diffraction angle
I^{S}_{hkl}	Dislocation cell/sub-grain size distribution at [hkl] crystal plane
I^{D}_{hkl}	Contribution of the dislocations at [hkl] crystal plane
I^{PD}_{hkl}	Contribution of the planar defects at [hkl] crystal plane
I^{INST}_{hkl}	Instrumental peak broadening and shape at [hkl] crystal plane
I_{BG}	Background of the diffraction pattern
K	Reciprocal of the lattice spacing
SLM	Selective Laser Melting
TEM	Transmission Electron Microscope
WAAM	Wire Arc Additive Manufacturing
WH	Williamson-Hall
$<X>_A$	Area weighted average sub-grain size
ρ_{Total}	Total dislocation density

References

1. Kruth, J.P.; Leu, M.C.; Nakagawa, T. Progress in additive manufacturing and rapid prototyping. *CIRP Ann. Manuf. Technol.* **1998**, *47*, 525–540. [CrossRef]
2. Attar, H.; Ehtemam-Haghighi, S.; Kent, D.; Dargusch, M.S. Recent developments and opportunities in additive manufacturing of titanium-based matrix composites: A review. *Int. J. Mach. Tools Manuf.* **2018**, *133*, 85–102. [CrossRef]
3. Rashid, R.; Masood, S.H.; Ruan, D.; Palanisamy, S.; Rahman Rashid, R.A.; Elambasseril, J.; Brandt, M. Effect of energy per layer on the anisotropy of selective laser melted AlSi12 aluminium alloy. *Addit. Manuf.* **2018**, *22*, 426–439. [CrossRef]
4. Ponnusamy, P.; Masood, S.H.; Palanisamy, S.; Rahman Rashid, R.A.; Ruan, D. Characterization of 17-4PH alloy processed by selective laser melting. *Mater. Today* **2017**, *4*, 8498–8506. [CrossRef]
5. Agius, D.; Kourousis, K.; Wallbrink, C. A Review of the As-Built SLM Ti-6Al-4V Mechanical Properties towards Achieving Fatigue Resistant Designs. *Metals* **2018**, *8*, 75. [CrossRef]
6. Hoye, N.; Cuiuri, D.; Rahman Rashid, R.A.; Palanisamy, S. Machining of GTAW additively manufactured Ti-6Al-4V structures. *Int. J. Adv. Manuf. Technol.* **2018**, *99*, 313–326. [CrossRef]
7. Sames, W.J.; List, F.A.; Pannala, S.; Dehoff, R.R.; Babu, S.S. The metallurgy and processing science of metal additive manufacturing. *Int. Mater. Rev.* **2016**, *61*, 315–360. [CrossRef]
8. Frazier, W.E. Metal additive manufacturing: A review. *J. Mater. Eng. Perform.* **2014**, *23*, 1917–1928. [CrossRef]
9. Ribárik, G.; Ungár, T. Characterization of the microstructure in random and textured polycrystals and single crystals by diffraction line profile analysis. *Mater. Sci. Eng. A* **2010**, *528*, 112–121. [CrossRef]
10. Máthis, K.; Nyilas, K.; Axt, A.; Dragomir-Cernatescu, I.; Ungár, T.; Lukáč, P. The evolution of non-basal dislocations as a function of deformation temperature in pure magnesium determined by X-ray diffraction. *Acta Mater.* **2004**, *52*, 2889–2894. [CrossRef]
11. Glavicic, M.G.; Salem, A.A.; Semiatin, S.L. X-ray line-broadening analysis of deformation mechanisms during rolling of commercial-purity titanium. *Acta Mater.* **2004**, *52*, 647–655. [CrossRef]
12. Glavicic, M.G.; Semiatin, S.L. X-ray line-broadening investigation of deformation during hot rolling of Ti–6Al–4V with a colony-alpha microstructure. *Acta Mater.* **2006**, *54*, 5337–5347. [CrossRef]

13. Brown, D.W.; Adams, D.P.; Balogh, L.; Carpenter, J.S.; Clausen, B.; King, G.; Reedlunn, B.; Palmer, T.A.; Maguire, M.C.; Vogel, S.C. In Situ Neutron Diffraction Study of the Influence of Microstructure on the Mechanical Response of Additively Manufactured 304L Stainless Steel. *Metall. Mater. Trans. A Phys. Metall. Mater. Sci.* **2017**, *48*, 6055–6069. [CrossRef]
14. Pokharel, R.; Balogh, L.; Brown, D.W.; Clausen, B.; Gray, G.T., III; Livescu, V.; Vogel, S.C.; Takajo, S. Signatures of the unique microstructure of additively manufactured steel observed via diffraction. *Scr. Mater.* **2018**, *155*, 16–20. [CrossRef]
15. Rahman Rashid, R.A.; Palanisamy, S.; Attar, H.; Bermingham, M.; Dargusch, M.S. Metallurgical features of direct laser-deposited Ti6Al4V with trace boron. *J. Manuf. Process.* **2018**, *35*, 651–656. [CrossRef]
16. Ahmed, T.; Rack, H.J. Phase transformations during cooling in α+β titanium alloys. *Mater. Sci. Eng. A* **1998**, *243*, 206–211. [CrossRef]
17. Carroll, B.E.; Palmer, T.A.; Beese, A.M. Anisotropic tensile behavior of Ti–6Al–4V components fabricated with directed energy deposition additive manufacturing. *Acta Mater.* **2015**, *87*, 309–320. [CrossRef]
18. Ribarik, G.; Ungar, T.; Gubicza, J. MWP-fit: a program for multiple whole-profile fitting of diffraction peak profiles by ab initio theoretical functions. *J. Appl. Crystallogr.* **2001**, *34*, 669–676. [CrossRef]
19. Ungár, T.; Tichy, G.; Gubicza, J.; Hellmig, R. Correlation between subgrains and coherently scattering domains. *Powder Diffr.* **2005**, *20*, 366–375. [CrossRef]
20. Ungár, T.; Tichy, G. The Effect of Dislocation Contrast on X-Ray Line Profiles in Untextured Polycrystals. *Phys. Status Solidi* **1999**, *171*, 425–434. [CrossRef]
21. Borbély, A.; Ungár, T. X-ray line profiles analysis of plastically deformed metals. *C. R. Phys.* **2012**, *13*, 293–306. [CrossRef]
22. Balogh, L.; Ribárik, G.; Ungár, T. Stacking faults and twin boundaries in fcc crystals determined by x-ray diffraction profile analysis. *J. Appl. Phys.* **2006**, *100*, 023512. [CrossRef]
23. Balogh, L.; Tichy, G.; Ungár, T. Twinning on pyramidal planes in hexagonal close packed crystals determined along with other defects by X-ray line profile analysis. *J. Appl. Crystallogr.* **2009**, *42*, 580–591. [CrossRef]
24. Stokes, A.R. A numerical Fourier-analysis method for the correction of widths and shapes of lines on X-ray powder photographs. *Proc. Phys. Soc.* **1948**, *61*, 382. [CrossRef]
25. Ungár, T.; Balogh, L.; Ribárik, G. Defect-Related Physical-Profile-Based X-Ray and Neutron Line Profile Analysis. *Metall. Mater. Trans. A* **2010**, *41*, 1202–1209. [CrossRef]
26. Ungár, T. Dislocation model of strain anisotropy. *Powder Diffr.* **2008**, *23*, 125–132. [CrossRef]
27. Humphreys, F.J.; Hatherly, M. Chapter 6 - Recovery After Deformation. In *Recrystallization and Related Annealing Phenomena*, 2nd ed.; Humphreys, F.J., Hatherly, M., Eds.; Elsevier: Oxford, UK, 2004; pp. 169–213.
28. Dąbrowski, R. The Kinetics of Phase Transformations During Continuous Cooling of the Ti6Al4V Alloy from the Single-Phase β Range. *Arch. Metall. Mater.* **2011**, *56*, 703. [CrossRef]
29. Li, C.; Liu, Z.Y.; Fang, X.Y.; Guo, Y.B. Residual Stress in Metal Additive Manufacturing. *Proc. CIRP* **2018**, *71*, 348–353. [CrossRef]
30. Vasinonta, A.; Beuth, J.L.; Griffith, M. Process Maps for Predicting Residual Stress and Melt Pool Size in the Laser-Based Fabrication of Thin-Walled Structures. *J. Manuf. Sci. Eng.* **2006**, *129*, 101–109. [CrossRef]
31. Bhadeshia, H.K.D.H. Developments in martensitic and bainitic steels: Role of the shape deformation. *Mater. Sci. Eng. A* **2004**, *378*, 34–39. [CrossRef]
32. Christien, F.; Telling, M.T.F.; Knight, K.S. Neutron diffraction in situ monitoring of the dislocation density during martensitic transformation in a stainless steel. *Scr. Mater.* **2013**, *68*, 506–509. [CrossRef]
33. Mishurova, T.; Cabeza, S.; Artzt, K.; Haubrich, J.; Klaus, M.; Genzel, C.; Requena, G.; Bruno, G. An Assessment of Subsurface Residual Stress Analysis in SLM Ti-6Al-4V. *Materials* **2017**, *10*, 348. [CrossRef] [PubMed]
34. Yadroitsev, I.; Yadroitsava, I. Evaluation of residual stress in stainless steel 316L and Ti6Al4V samples produced by selective laser melting. *Virtual Phys. Prototyp.* **2015**, *10*, 67–76. [CrossRef]
35. Cottam, R.; Thorogood, K.; Lui, Q.; Wong, Y.C.; Brandt, M. The Effect of Laser Cladding Deposition Rate on Residual Stress Formation in Ti-6Al-4V Clad Layers. *Key Eng. Mater.* **2012**, *520*, 309–313. [CrossRef]

36. Szost, B.A.; Terzi, S.; Martina, F.; Boisselier, D.; Prytuliak, A.; Pirling, T.; Hofmann, M.; Jarvis, D.J. A comparative study of additive manufacturing techniques: Residual stress and microstructural analysis of CLAD and WAAM printed Ti–6Al–4V components. *Mater. Des.* **2016**, *89*, 559–567. [CrossRef]
37. Ali, H.; Ma, L.; Ghadbeigi, H.; Mumtaz, K. In-situ residual stress reduction, martensitic decomposition and mechanical properties enhancement through high temperature powder bed pre-heating of Selective Laser Melted Ti6Al4V. *Mater. Sci. Eng. A* **2017**, *695*, 211–220. [CrossRef]

12

Self-Propagating High Temperature Synthesis of TiB_2–$MgAl_2O_4$ Composites

Nina Radishevskaya [1], Olga Lepakova [1], Natalia Karakchieva [2,*], Anastasiya Nazarova [1], Nikolai Afanasiev [1], Anna Godymchuk [3,4] and Alexander Gusev [4,5]

[1] Tomsk Scientific Centre SB RAS, Tomsk 634055, Russia; osm.ninaradi@yandex.ru (N.R.); klavdievna.k@yandex.ru (O.L.); osm.nazarova@yandex.ru (A.N.); Af42@yandex.ru (N.A.)
[2] Physical-Technical Institute, Tomsk State University, Tomsk 634050, Russia
[3] Department of Nanomaterials and Nanotechnologies, National Research Tomsk Polytechnic University, Tomsk 634050, Russia; godymchuk@mail.ru
[4] Department of Functional Nanosystems and High-Temperature Materials, National University of Science and Technology MISIS, Moscow 119991, Russia; nanosecurity@mail.ru
[5] Research Institute of Environmental Science and Biotechnology, G.R. Derzhavin Tambov State University, Tambov 392000, Russia
* Correspondence: kosovanatalia@yandex.ru

Abstract: Metal borides are widely used as heat-insulating materials, however, the range of their application in high-temperature conditions with oxidative medium is significantly restricted. To improve the thermal stability of structural materials based on titanium boride, and to prevent the growth of TiB_2 crystals, additives based on alumina-magnesia spinel with chemical resistant and refractory properties have been used. The aim of this work is to study the structure of TiB_2 with alumina-magnesia spinel additives obtained by self-propagating high-temperature synthesis (SHS). TiB_2 structure with uniform fine-grained distribution was obtained in an $MgAl_2O_4$ matrix. The material composition was confirmed by X-ray diffraction analysis (DRON-3M, filtered Co kα-emission), FTIR spectroscopy (Thermo Electron Nicolet 5700, within the range of 1300–400 cm^{-1}), and scanning electron microscopy (Philips SEM 515). The obtained material represents a composite, where the particles of TiB_2 with a size of 5 μm are uniformly distributed in the alloy of alumina-magnesia spinel.

Keywords: titanium diboride; alumina-magnesia spinel; self-propagating high-temperature synthesis; composites

---

1. Introduction

Self-propagating high-temperature synthesis (SHS) is used to develop new technologies for the production of refractory nonmetallic composite materials with defined properties. In spite of the fact that metal carbides and borides are widely used as insulation materials, the range of their application in oxidative mediums at high temperatures is very restricted. To increase the refractory properties of metal carbides and borides, alumina-magnesia spinel $MgAl_2O_4$ with the melting temperature of 2105 °C, which corresponds to the high level of refractoriness [1], is used as an additive.

Magnesium and aluminothermic synthesis is widely used for the production of refractory ceramic materials, e.g., with the use of metallothermic reduction in a TiO_2–MgO–Al_2O_3–Al system, the refractory materials based on $MgAl_2O_4$ and titanium carbonitrides are obtained [2]. High-strength porous ceramic material, containing in its composition $MgAl_2O_4$, TiB_2, TiO_2, $Al_4B_2O_6$, and $Mg_2B_2O_5$ was obtained in a TiO_2–B_2O_3–Al system with MgO additives. This material can be used as a catalyst at temperatures of 600 °C–700 °C in an open atmosphere [3]. Moreover, aluminum is widely used in the synthesis of composite materials. In the structure of composites, the intermetallic matrices from both $TiAl/Ti_3Al$ and $MgAl_2O_4$ are incorporated [4]. In all of the abovementioned works, $MgAl_2O_4$ is synthesized in the form of particles.

Another method of heat-resistant composite production is through titanium diboride synthesis from its elements with the use of chemical-resistant and refractory alumina-magnesia spinel ($MgAl_2O_4$). This method allows decelerating high-temperature solid-phase oxidative reactions in the process of material exploitation.

The aim of this work is to study the phase composition and microstructure of a TiB_2 + $MgAl_2O_4$ heat-resistant composite obtained by self-propagating high-temperature synthesis with $MgAl_2O_4$ additives of different concentrations.

At high temperatures (~3000 °C), spinel melts and spreads along the surface of TiB_2 grains, forming the matrix that protects the TiB_2 grain surface with the spinel.

2. Materials and Methods

To prepare reaction mixtures, dried in a vacuum at temperature of 200 °C for 2 h, titanium powders (TPP-8, JSC "Avisma"; titanium composition ~96 wt %; particle size < 160 μm), amorphous boron (B-99A-TU-6-02-585-75), and alumina-magnesia spinel (TU 6-09-01-136) were used. Four mixtures of different compositions were prepared: (1) 90% (Ti + 2B) + 10% $MgAl_2O_4$; (2) 75% (Ti + 2B) + 25% $MgAl_2O_4$; (3) 60% (Ti + 2B) + 40% $MgAl_2O_4$; (4) 50% (Ti + 2B) + 50% $MgAl_2O_4$. Powders were thoroughly mixed to obtain homogenous blends. Then, from the obtained mixtures, porous (40–45%) cylindrical particles were formed with a diameter of 20 mm and a length of 30–32 mm by using a hydraulic press. Self-propagating high-temperature synthesis was conducted in a constant pressure setup in argon atmosphere at a pressure of ~6 atm. Samples ignition was carried out using an ignition mixture of powders (Ti + 2B) with the help of a tungsten filament, which was supplied with a short-term electrical impulse. The maximal combustion temperature was detected by the tungsten-rhenium thermocouple BP5-BP20 with a diameter of 100 μm. Temperature registration was conducted with the use of an analog-to-digital converter LA-20USB connected with a personal computer.

The compositions of the obtained materials were proved by X-ray phase analyses (Dron-3M, filtered Co kα-emission, Saint Petersburg, Russia), IR spectroscopy (FTIR spectrometer Nicolet-5700, Thermo Electron Corporation, Atkinson, USA). Measurements were carried out using an add-in device of scattering reflection in KBr at a frequency interval of 1300–400 cm^{-1}. To study the microstructure, an optical microscope (Axiovert 200M, OM, Karl Zeiss, Germany) and a scanning electron microscope (SEM-515, Philips, Amsterdam, The Netherlands) were used.

3. Results and Discussion

Among gas-free systems, the Ti-B system is characterized by the highest exothermicity. For a powder mixture with the ratio of components Ti:B = 1:2 the adiabatic temperature of combustion is Tad = 3190 K [5]. Alumina-magnesia spinel $MgAl_2O_4$ is inert in relation to the mixture Ti-2B. In Table 1, the physicochemical properties of spinel are presented [6,7].

Table 1. Physicochemical properties of compounds.

Compound	Melting Temperature, °C	Density, g/cm^3	$-\Delta H^{\circ}_{form}$, kJ/mol
$MgAl_2O_4$	2135	3.8	2307.8
TiB_2	2850	4.45–4.50	293.3
$MgTiO_3$	1680	3.91	1573.6
α–Al_2O_3	2045	3.99	1675.0

Figure 1 shows the combustion thermogram of the TiB_2 (75 wt %) + $MgAl_2O_4$ (25 wt %) system. The maximal combustion temperature is 2300 °C, which is higher than the spinel melting temperature. Synthesis was conducted layer-by-layer in the steady state combustion conditions. Similar combusting conditions were observed for Ti + 2B + xCu and Ti + 2B + xFe systems. Depending on their content, different metal alloys partially or fully surround particles of titanium borides [8,9].

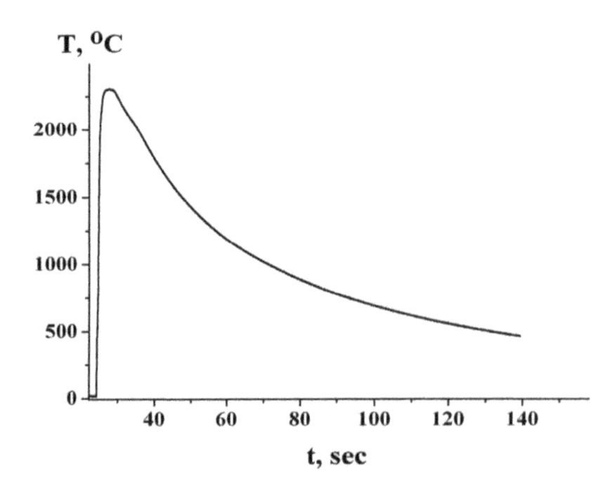

Figure 1. Combustion thermogram of the TiB_2 (75 wt %) + $MgAl_2O_4$ (25 wt %) system.

Studies on the microstructure of the composite blends based on TiB_2 with different $MgAl_2O_4$ compositions showed that, depending on the amount of added spinel, the composite structure change (Figure 2). If the amount of added $MgAl_2O_4$ is <10%, the grains of titanium diboride in the microstructure of the composite are partially surrounded by a solidified alloy of $MgAl_2O_4$ (Figure 2a). The best results were obtained at a spinel composition of 25%. The fine-grain microstructure from TiB_2 grains (light crystals) was observed, which is fully surrounded by spinel (dark areas). When 40% $MgAl_2O_4$ was added to the blend during the synthesis, the formation of a non-homogeneous structure was observed. The structure contains areas with the fine-grained titanium diboride and adjusting areas from alumina-magnesia spinel (Figure 2c).

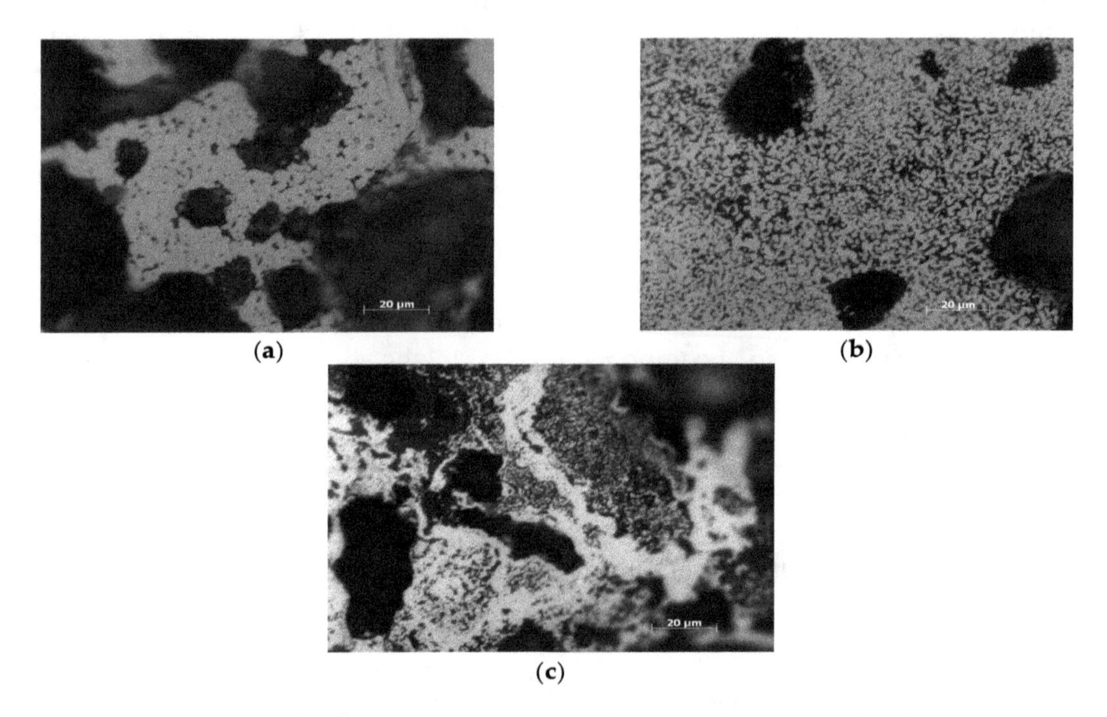

Figure 2. Microstructure of SHS composites based on titanium diboride with additions of $MgAl_2O_4$: (**a**) 90% (Ti + 2B) + 10% $MgAl_2O_4$; (**b**) 75% (Ti + 2B) + 25% $MgAl_2O_4$; (**c**) 60% (Ti + 2B) + 40% $MgAl_2O_4$.

When 45% $MgAl_2O_4$ is added to the composite, the mixture does not burn in this case, because $MgAl_2O_4$ is inert.

Complete information on the structure of the product formed during SHS can be obtained by analyses of fracture surfaces, studied with scanning electron microscopy. Figure 3 shows the microstructure of fractures of SHS ceramic samples based on titanium diboride with the addition of 25% $MgAl_2O_4$ (Figure 3a,b), and 0% $MgAl_2O_4$ (Figure 3c,d).

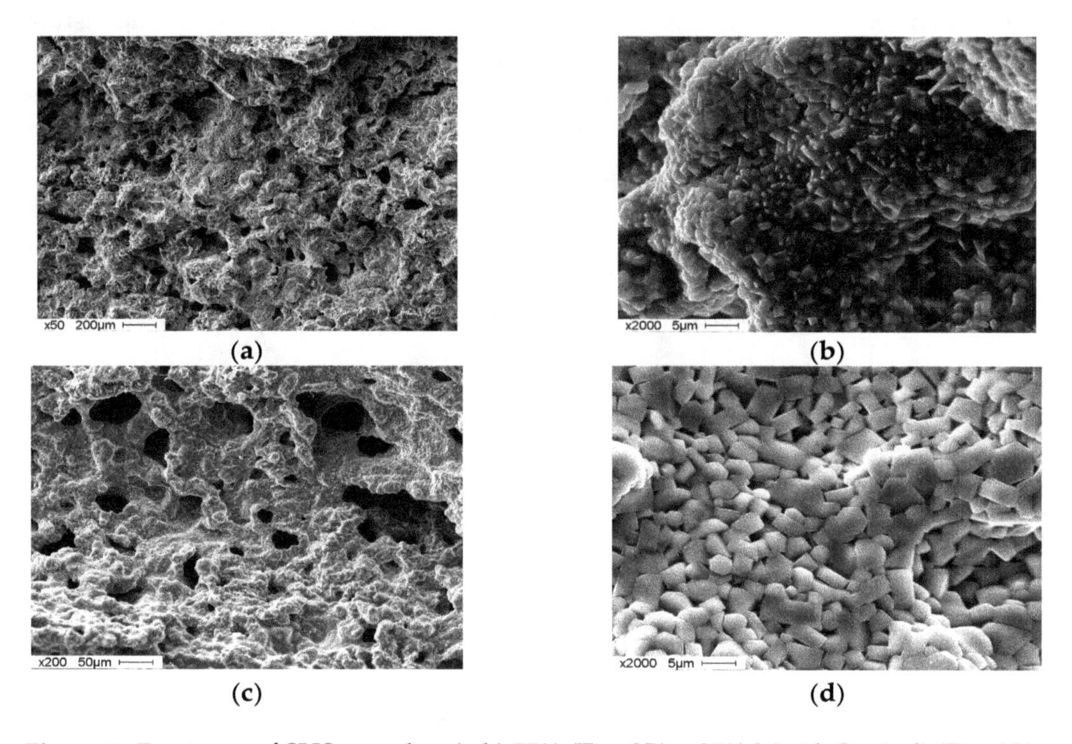

Figure 3. Fractures of SHS samples: (**a**,**b**) 75% (Ti + 2B) + 25% $MgAl_2O_4$; (**c**,**d**) (Ti + 2B).

As can be seen from Figure 3, the addition of 25% $MgAl_2O_4$ leads to the decreasing of TiB_2 crystals (~2 µm), which are surrounded by a solidified alloy of alumina-magnesia spinel. The microstructure of the SHS sample with Ti + 2B composition is formed by large TiB_2 faceted crystals.

Figure 4 shows the diffraction patterns of TiB_2 composites with different amounts of spinel. X-ray diffraction analyses showed that in the composition of alumina-magnesia spinel, there is 12 wt % of $MgAl_2O_4$. Figure 4 shows that spinel is identified in the composite containing 25 wt % of $MgAl_2O_4$, though, metallographically the spinel is identified at 10 wt % of $MgAl_2O_4$.

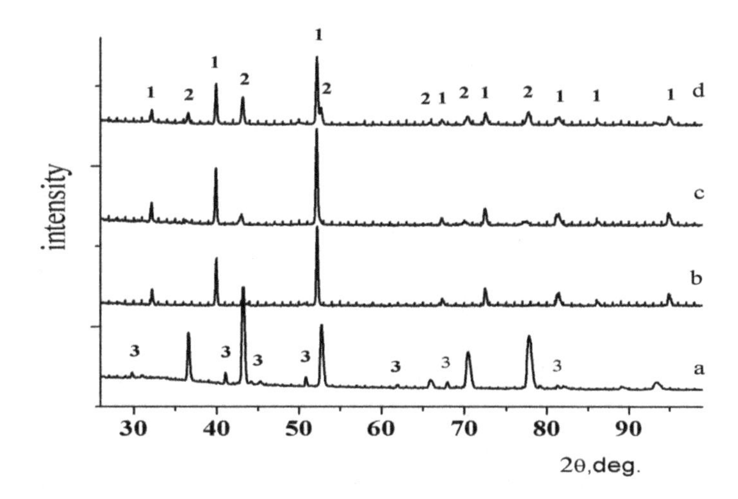

Figure 4. X-ray diffraction patterns of TiB_2 composites with different contents of alumina-magnesia spinel: (**a**) $MgAl_2O_4$; (**b**) TiB_2 + 10% $MgAl_2O_4$; (**c**) TiB_2 + 25% $MgAl_2O_4$; (**d**) TiB_2 + 40% $MgAl_2O_4$. 1-TiB_2, 2-$MgAl_2O_4$, 3-Al_2O_3.

The composite with the fine-grained microstructure containing 25 wt % of $MgAl_2O_4$ was studied by FTIR spectroscopy. Figure 5 shows the FTIR spectrum of $MgAl_2O_4$, TiB_2–$MgAl_2O_4$ composite, corundum, and TiB_2.

Figure 5 (pattern 1) shows that alumina-magnesia spinel has two different absorption bands with maximums at 692.0 cm^{-1} and 540.0 cm^{-1}, related to the tetrahedral coordinated magnesium MgO_4 and octahedral coordinated aluminum of AlO_6. The small peak in the frequency range of 800–900 cm^{-1} proves the presence of Al_2O_3 in spinel content. Irregularity of the spinel structure leading to a change of binding force in the cation sub-lattice is identified by the emergence of an absorption band at 558.7 cm^{-1} [10].

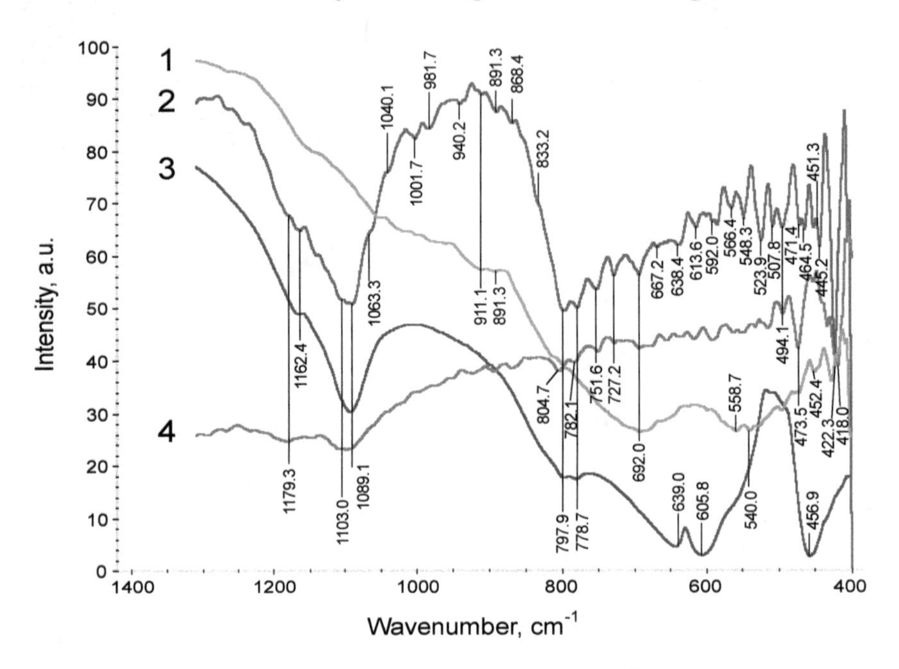

Figure 5. FTIR spectrum in the frequency range of 400–1300 cm^{-1}: (**1**) $MgAl_2O_4$; (**2**) TiB_2 composite 25 wt % of $MgAl_2O_4$; (**3**) gray corundum; (**4**) TiB_2.

FTIR spectrum of composite (TiB_2 + 25 wt % of $MgAl_2O_4$) consists of numerous absorption bands typical for titanium diboride, spinel, and corundum (pattern 2).

According to the burning thermogram for the 75 wt % TiB_2 + 25 wt % $MgAl_2O_4$ system, the burning temperature is 2300 °C. Therefore, $MgAl_2O_4$ is partially decomposed with corundum formation.

$$MgAl_2O_4 \quad \overset{T}{\rightarrow} \quad Al_2O_3 + MgO$$

Pattern 3 shows the FTIR spectrum of gray corundum. Along with absorption bands at 639.0 cm^{-1}, 605.8 cm^{-1}, and 456.9 cm^{-1} typical for octahedral coordinated aluminum AlO_6 in α–Al_2O_3, there are absorption bands at 1089.1 cm^{-1}, 797.9 cm^{-1}, and 778.7 cm^{-1}, related to the tetrahedral coordinated aluminum AlO_4 [11]. The same absorption bands are observed in the composite spectrum.

It is well known that α–Al_2O_3 contains aluminum atoms which are octahedrally coordinated by oxygen [10,12]. According to the literature data [13], the gray color of corundum is caused by the presence of aluminous spinel $AlOAl_2O_3$. This spinel was identified during the electrocorundum synthesis in reducing medium [13]. The melting temperature of spinel is 1980 °C [1].

At the interference level, the absorption bands at 940.2 cm^{-1}, 727.2 cm^{-1}, and 507.8 cm^{-1} are observed. They can be referred to $MgTiO_3$ [12]. The formation of $MgTiO_3$ is possible during the synthesis at the phase boundary between TiB_2 and $MgAl_2O_4$.

Oxygen and MgO can be borrowed during the thermal decomposition of spinel. In this case, aluminum is moved from an octahedral coordination to a tetrahedral one with the formation of both $MgTiO_3$ and aluminous spinel with an intensive absorption band at 1089.1 cm^{-1}. It is well-known [14] that at high temperatures over Al_2O_3, the gas phase is formed as a result of thermal dissociation. The gas phase contains aluminum sub-oxides Al_2O and AlO^-.

$$Al_2O_3 \xrightarrow{T} Al_2O \uparrow + O_2 \uparrow$$

Aluminum sub-oxides can also participate in the formation of aluminous spinel $AlOAl_2O_3$.

$$2Al_2O_3 + Al_2O + O \rightarrow 2AlOAl_2O_3$$

The FTIR spectrum of this composite (pattern 2) represents the envelope line along the spectrum of alumina-magnesia spinel. The overlap of numerous bond oscillation frequencies, related to the TiB_2, corundum, aluminous spinel, and $MgTiO_3$, is observed.

Studies showed that the obtained composite consists of TiB_2 fine grains, which are homogeneously distributed in the alumina-magnesia matrix containing α–Al_2O_3. Traces of $MgTiO_3$ and aluminous spinel are also present in the composite.

According to the literature data [5], 12 mol % of MgO and 85.5 mol % of Al_2O_3 can be dissolved in alumina-magnesia spinel. In Table 2, the eutectic melting temperatures in the MgO–Al_2O_3 system are presented.

Table 2. Eutectic melting temperatures in the MgO–Al_2O_3 system.

Chemical Compounds in Eutectics	Al_2O_3 Composition, wt %	Melting Temperature, °C
MgO, $MgAl_2O_4$	55	1995
$MgAl_2O_4$, Al_2O_3	98	1920

Melting temperatures of TiB_2, α–Al_2O_3, $MgTiO_3$, and $MgAl_2O_4$ as well as their eutectics are presented in Tables 1 and 2. As can be seen from Table 2, all values of the melting temperatures are very high, which proves that the obtained ceramic material is refractory.

4. Conclusions

It was shown that structure with a homogeneous fine-grained distribution of TiB_2 grains was obtained by using 25 wt % of $MgAl_2O_4$.

The formed surface layer of $MgAl_2O_4$ on the grains boundary of TiB_2 serves as a blocking protection from titanium diboride oxidation and prevents the growth of TiB_2 crystals.

A partial decomposition of spinel occurred during the composite synthesis. This is proved by the presence of $MgTiO_3$ and corundum traces in the composite, which were identified by FTIR spectroscopy.

Acknowledgments: The work was carried out with financial support from the Ministry of Education and Science of the Russian Federation in the framework of Increase Competitiveness Program of MISIS.

Author Contributions: Nina Radishevskaya performed FTIR-spectroscopy experiments and analyzed the data; Olga Lepakova conducted the microstructure research of samples; Natalia Karakchieva conducted the X-ray phase analyses of samples; Anastasiya Nazarova conducted the synthesis of samples; Nikolai Afanasiev wrote the paper; Anna Godymchuk and Alexander Gusev studied SHS characteristics, such as combustion temperature and combustion wave propagation mode and velocity.

References

1. Li, Y.B.; Li, N.; Ruan, G.Z.; Li, X.H. Reaction in the aluminotheric reduction nitridation reaction to synthesize $MgAl_2O_4$/TiN. *Ceram. Int.* **2005**, *31*, 825–829. [CrossRef]

2. Omid, E.K.; Naghizadeh, R.; Rezaie, H.R. Synthesis and comparison of $MgAl_2O_4$–Ti (C,N) composites using aluminothermic-carbothermal reduction and molten salts routes. *J. Ceram. Process. Res.* **2013**, *14*, 445–447.
3. Bae, Y.; Jun, B. Preparation of ultrafine TiC, $MgAl_2O_4$ and AlON composite powder using chemical furnace. *J. Ceram. Process. Res.* **2008**, *9*, 661–665.
4. Zaki, Z.I.; Ahmed, Y.M.Z.; Abdel Gawad, S.R. In situ synthesis of porous magnesia spinel/TiB_2 composite by combustion technique. *J. Ceram. Soc. Jpn.* **2009**, *117*, 719–723. [CrossRef]
5. Horvitz, D.; Gotman, I. Pressure-assisted DHD synthesis of $MgAl_2O_4$–TiAl in Situ composites with interpenetrating networks. *Acta Mater.* **2002**, *50*, 1961–1971. [CrossRef]
6. Horoshavin, L.B. *Spinel Nanorefractory Materials*; UB RAS: Ekaterinburg, Russia, 2009; p. 600.
7. Merzhanov, A.G. *Processes of Burning and Materials Synthesis*; ISMAN: Chernogolovka, Russia, 1998; p. 511.
8. Lepakova, O.K.; Raskolenko, L.G.; Maksimov, Y.M. Self-propagating high-temperature synthesis of composite material TiB_2–Fe. *J. Mater. Sci.* **2004**, *39*, 3723–3732. [CrossRef]
9. Vadchenko, S.G.; Filimonov, I.A. Burning modes of diluted system Ti + 2B. *Phys. Burn. Combust.* **2003**, *39*, 48–55.
10. Barabanov, V.F.; Goncharov, G.N.; Zorina, M.L. *Modern Physical Methods on Geochemistry*; Leningrad University: Pushkin, Russia, 1990; p. 390.
11. Chernyakova, K.V.; Vrubelevskii, I.A.; Ivanovskaya, M.I.; Kotikov, D.A. Defective structure of anode alumina oxide, formed by method of bilateral anodic oxidation. *J. Appl. Spectrosc.* **2012**, *79*, 83–89.
12. Nakamoto, K. *IK-Spektry i Spektry KR Neorganicheskikh i Koordinatsionnykh Soedinenii (Infrared and Raman Spectra of Inorganic and Coordination Compounds)*; Mir: Moscow, Russia, 1991; p. 536.
13. Solodkii, E.N.; Solodkii, N.F. Reasons for coloring corundum ceramics. *Glass Ceram.* **2000**, *11*, 24–26.
14. Kulikov, I.S. *Metals Deoxidation*; Metallurgy: Moscow, Russia, 1975; p. 504.

The Effects of Prestrain and Subsequent Annealing on Tensile Properties of CP-Ti

Le Chang, Chang-Yu Zhou * and Xiao-Hua He

School of Mechanical and Power Engineering, Nanjing Tech University, Nanjing 211816, China; chellechang@163.com (L.C.); xh_he@njtech.edu.cn (X.-H.H.)

* Correspondence: changyu_zhou@163.com

Academic Editor: Mark T. Whittaker

Abstract: The aim of the present work is to investigate the effects of prestrain and subsequent annealing on tensile properties of commercial pure titanium (CP-Ti). According to tensile test results, yield strength and ultimate tensile strength increase with the increase of prestrain. Elongation and uniform strain decrease linearly with prestrain. In the case of prestrain that is higher than 3.5%, the macro-yield of specimens changes from gradual yielding to discontinuous yielding. It is supposed that considerable numbers of dislocations introduced into the material lead to the appearance of yield plateau. The quantitative analysis of the contribution of dislocation hardening to the strain hardening shows that dislocation-associated mechanisms play an important role in strain hardening. Moreover, a modified Fields-Backofen model is proposed to predict the flow stress of prestrained CP-Ti at different strain rates. Both strain rate sensitivity and strain hardening exponent decrease with prestrain. Fracture surfaces of the specimens show that fracture mechanism of all tested specimens is dimple fracture. The more ductile deformation in prestrained CP-Ti after annealing indicates that its ductility is improved by annealing.

Keywords: CP-Ti; prestrain; yield plateau; tensile properties; flow stress; fracture

1. Introduction

During the manufacture of materials (stamping, cold rolling, equal channel angular pressing, bending) and the installation and service history of the equipment components (such as creep, overload), different degrees of plastic deformation will happen in the materials. The extent of prior plastic strain introduced into the material can significantly change the mechanical properties and consequently affect the plastic deformation behavior. At present, the investigation of the influence of prestrain on the materials has been focused on steels [1–7], titanium alloys [8–10], aluminum alloys [11,12], magnesium alloys [13–15], Zr-based alloys [16] and so on. Usually, prior heavy cold work leads to a considerable increase in strength by creating dislocation barriers to inhibit subsequent dislocation movement during plastic deformation at room temperature. For instance, Zhang et al. [2] found that by pre-straining and bake hardening, the strength of C–Mn–Si TRIP steel was enhanced. Further, they supposed that the unlocking from weak carbon atmospheres of dislocations newly formed during prestraining led to the appearance of yield point on the stress–strain curve [3]. Lee et al. [6,7] found that work hardening rate and strain rate sensitivity (SRS) of 304 L stainless steel were dependent on the variation of prestrain. Sarker et al. [13,14] suggested that the variation of strain hardening rate of AM30 magnesium alloy with pre-straining was related to deformation twinning and detwinning. Thus, it can be seen that prestrain has important effects on the subsequent plastic deformation behavior for many materials. Also, it provides that the effects of prestrain on the performance of hexagonal close-packed materials are associated with twinning structures produced in the prior deformation [13–15]. Many researchers have studied the effects of prior severe plastic deformation, such as cold rolling and equal channel angular pressing (ECAP) on mechanical properties of CP-Ti [17–19]. The activation of twinning in CP-Ti is affected by many factors, such as deformation temperature, strain rate, deformation value, deformation direction [20,21]. Prior deformation (cold rolling and ECAP) produces many twins in the initial microstructure of CP-Ti, which is different from tensile prestrain [18,19,22–24]. The effects of stretching prestrain on the mechanical properties of CP-Ti have not been examined. Thus, in this paper, different amounts of prestrain were applied to CP-Ti and a uniaxial tensile experiment was conducted to investigate the effects of prestrain on mechanical properties, work hardening rate, flow stress and fracture mechanism.

2. Experimental Materials and Procedures

The materials used in the present study were CP-Ti in the form of a cold-rolled and annealed plate. The chemical composition was given in Table 1. The specimens with 3 mm thickness (see Figure 1) were machined from the as-received plate by wire electrical discharge machining. In order to exclude the effect of surface roughness, the specimens were ground and polished by metallographic abrasive paper. The tensile testing was carried out by Instron tensile experimental equipment (INSTRON, Norwood, MA, USA). The strain was automatically measured by extensometer (INSTRON, Norwood, MA, USA). Different levels of pre-deformation along rolling direction were applied at the same strain rate (0.0005 s^{-1}) with the control of extensometer. Then, the specimens were unloaded and reloaded to fracture. The strain rate during reloading ranged from 0.00005 to 0.004 s^{-1}. In addition, to investigate the effects of subsequent annealing on tensile properties of prestrained samples, isothermal heat-treatment was conducted using a vacuum furnace equipped with an auto tune temperature controller (HUAHONG, Suzhou, China). The annealing temperatures were 500 °C and 600 °C with dwell times ranging from 30 min to 1 h. The detailed annealing experimental scheme was listed in Table 2. The microstructure was characterized using optical metallography. The specimens were mechanically ground and chemically etched in a solution consisting of 10 mL nitric acid, 2 mL hydrofluoric acid and 88 mL distilled water. Study on the fracture surface of tensile specimens was carried out using a JSM-6360LV scanning electron microscope (JEOL, Tokyo, Japan).

Table 1. Chemical composition of used commercial pure (CP)-titanium (wt. %).

Element	Ti	Fe	C	N	H	O
Composition	Balance	0.06	0.01	<0.01	0.001	0.12

Table 2. Annealing experimental scheme.

Prestrain (ε_{pre})/%	Annealing			
	500 °C, 30 min	500 °C, 40 min	500 °C, 60 min	600 °C, 30 min
1	√	√	-	√
2	√	√	√	-
3.5	√	√	√	√
5	√	√	√	-
6.5	√	√	-	√

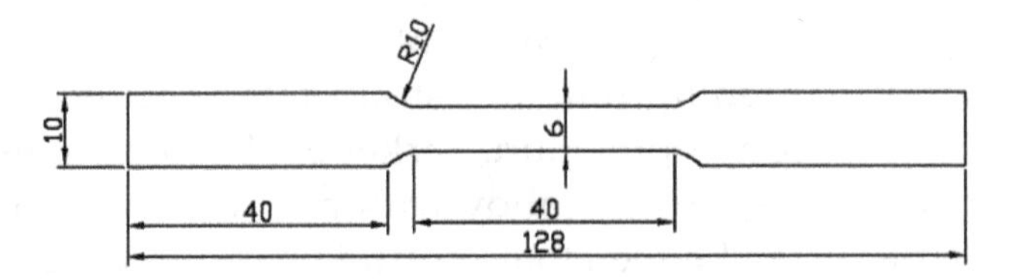

Figure 1. The geometry of the test specimens.

3. Results and Discussion

3.1. Tensile Properties of As-Received Commercial Pure (CP)-Ti

According to uniaxial tensile experiment, mechanical properties of CP-Ti were determined. Engineering stress–strain curves of the as-received material were given in Figure 2. Plastic deformation process can be divided into three stages by the three characteristic points (A, B, C) on the tensile curve. The first feature point is yield strength point (point A), deformation before this point is the elastic deformation stage. Obviously, in the range of 0.00005–0.004 s^{-1}, the yield of CP-Ti is continuous as no yield plateau appears. The second feature point is the highest point (point B). Generally, when maximum tensile force reaches, sample necking starts, the corresponding point B is the demarcation between uniform plastic deformation and non-uniform plastic deformation. When plastic deformation is concentrated, load-bearing capacity decreases quickly, the corresponding feature point is point C. The strain of diffuse necking and localized necking were expressed as follows:

$$\varepsilon_d = n \tag{1}$$

$$\varepsilon_l = 2n \tag{2}$$

where ε_d is initial strain of diffuse necking, ε_l is initial strain of localized necking and n is strain hardening index. The stress–strain curve in plastic stage is divided into three stages, via uniform deformation, diffuse necking stage and localized necking stage by the two perpendicular lines. As the work hardening exponent is different under different strain rates, distinguish area of stress–strain curve was only given at strain rate of 0.0005 s^{-1}, as shown in Figure 2. It can be seen that CP-Ti has obvious strain rate sensitivity. With the increase of strain rate, tensile stress–strain curve of CP-Ti ascends. Moreover, the yield strength and tensile strength increase and the elongation decreases.

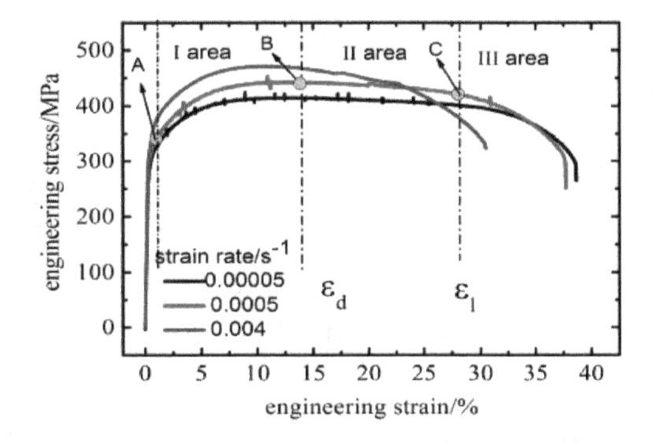

Figure 2. Engineering stress–strain curves of as-received CP-Ti.

3.2. Tensile Properties of Prestrained Specimens

The prestrain was taken from the uniform plastic deformation zone and the dispersion instability region (1%, 2%, 3.5%, 5%, 6.5%, 10%, 15%, respectively). According to tensile stress–strain curves of different prestrained specimens, as shown in Figure 3a, it can be seen that, after pre-stretching, tensile stress–strain curve ascends. Both yield strength and ultimate tensile strength increase. The strength of CP-Ti is enhanced at the cost of ductility, as fracture strain decreases a lot with prestrain. By comparing the engineering stress–strain curves of 3.5%, 5% and 6.5% prestrained specimens, it can be seen that engineering stress–strain curves of these specimens are close, that means when prestrain is ranging from 3.5% to 6.5%, flow stress increases slightly with prestrain. When prestrain reaches 10% and above, the ultimate strength is rapidly reached and yield strength is very close to tensile strength. It is worth noting that as prestrain reaches 3.5%, the macro-yield changes from gradual yielding to discontinuous yielding, as shown in Figure 3b. The discontinuous increase of flow stress shows the appearance of yield plateau.

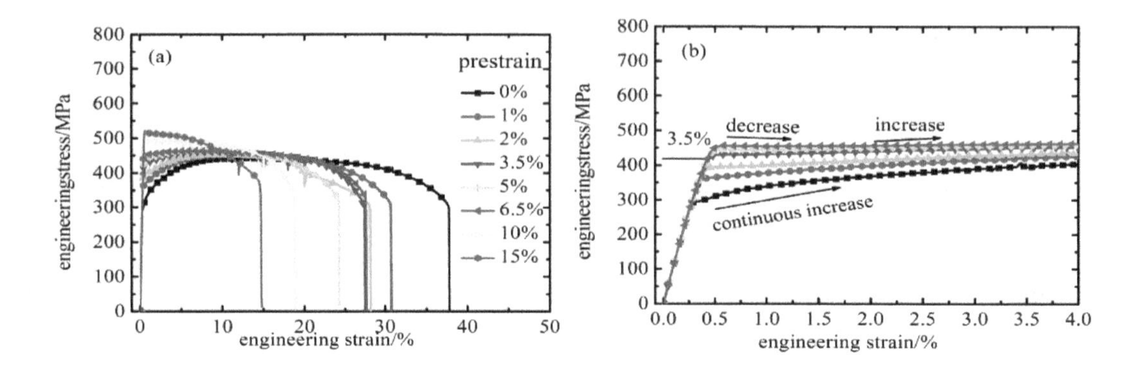

Figure 3. Engineering stress–strain curves of different prestrained specimens. (**a**) 0%–15% prestrained CP-Ti and (**b**) the appearance of yield plateau.

In order to evaluate the variation of flow stress with true strain, work hardening rate θ (dσ/dε) is derived from the true stress–strain curve. Figure 4 displays variations of θ with true strain for as-received and prestrained CP-Ti. An often reported feature for pure titanium, in particular in compression conditions, consists in a three-stage character of deformation curve [25,26]. Work hardening rate of as-received CP-Ti shows three decrease stages (A, B and C stage), as shown in Figure 4a. Work hardening rate decreases sharply at the beginning of plastic deformation (stage A) and then decreases gradually at moderate strain (stage B). In stage C, work hardening rate decreases slowly with strain, as the slopes flatten with the increase of true strain. Between stages A and B the transition occurs at a true plastic strain of 0.018, between stages B and C at a true plastic strain of 0.096. The transition points between stages A and B and that between stages B and C of as-received CP-Ti are consist with results of Becker et al. [27] and Hama et al. [28]. For prestrained CP-Ti in Figure 4b, its θ is dependent on true strain and prestrain. With the increase of prestrain ($\varepsilon_{pre} \geq 2\%$), a rise stage B of work hardening rate occurs. Stage B is characterized by an increasing strain hardening rate with true strain at the range of 0.5%–2%. With increasing prestrain, the rise tendency in stage B is more remarkable. In the case of $\varepsilon_{pre} \geq 3.5\%$, strain hardening rate at the end of stage A is negative, also indicates that yield plateau appears in tensile stress–strain curves, which is consist with Figure 3b. Additionally, compared with as-received CP-Ti, θ of prestrained CP-Ti decreases, indicating the decrease of work hardening ability. Li et al. [29] and Ghaderi et al. [30] investigated the effect of grain size on the tensile deformation mechanisms of CP-Ti and found that as grain size decreased to about 26 μm, yield plateau occurred. In this paper, the grain size of CP-Ti is about 30 μm, as shown in Figure 5a, determined by linear intercept method. Li et al. [29] supposed there were two possible reasons for the observed variation of the macro-yield. First is the high number of dislocation tangles and the mobile dislocation density. The second reason is the presence of deformation twinning. However, as grain size decreases to 30 μm, the activation of twinning in CP-Ti during tensile deformation is non-significant [29]. Thus, ascribing the initial yield plateau of CP-Ti during tensile deformation to the presence of twinning is unreasonable. It is no doubt that dislocation structure exists in the prestrained samples and its density increases with prestrain. However, the volume fraction of twinning is tremendously small, as shown in Figure 5b–f, though its density increases with prestrain. Almost no twinning structure can be observed in 2%–3.5% prestrained samples and few twins exists in 5%–15% prestrained samples indicated by white arrows. According to quantitative statistical analysis of microstructure in the process of tensile deformation along rolling direction [31–34], the twin volume fraction is always considerably small and the plastic deformation can be attributed to dislocation slip. Therefore, as prestrain reaches 3.5%, high dislocation density in the initial microstructure leads to the presence of yield plateau.

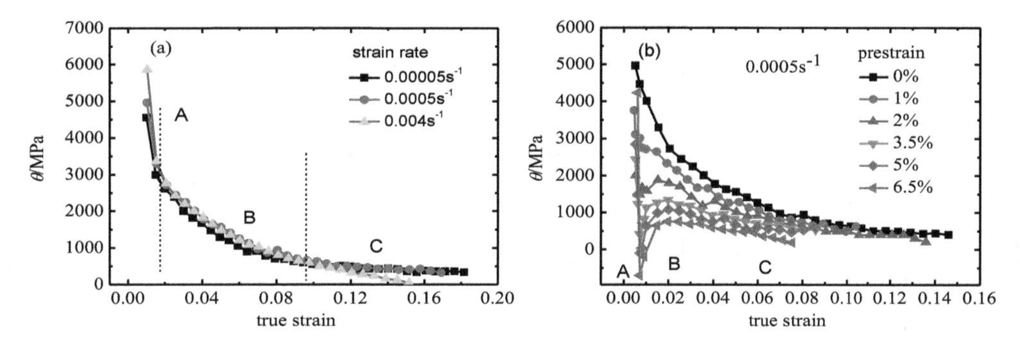

Figure 4. The variation of work hardening rate with true strain. (**a**) As-received CP-Ti at different strain rates and (**b**) prestrained CP-Ti at 0.0005 s^{-1}.

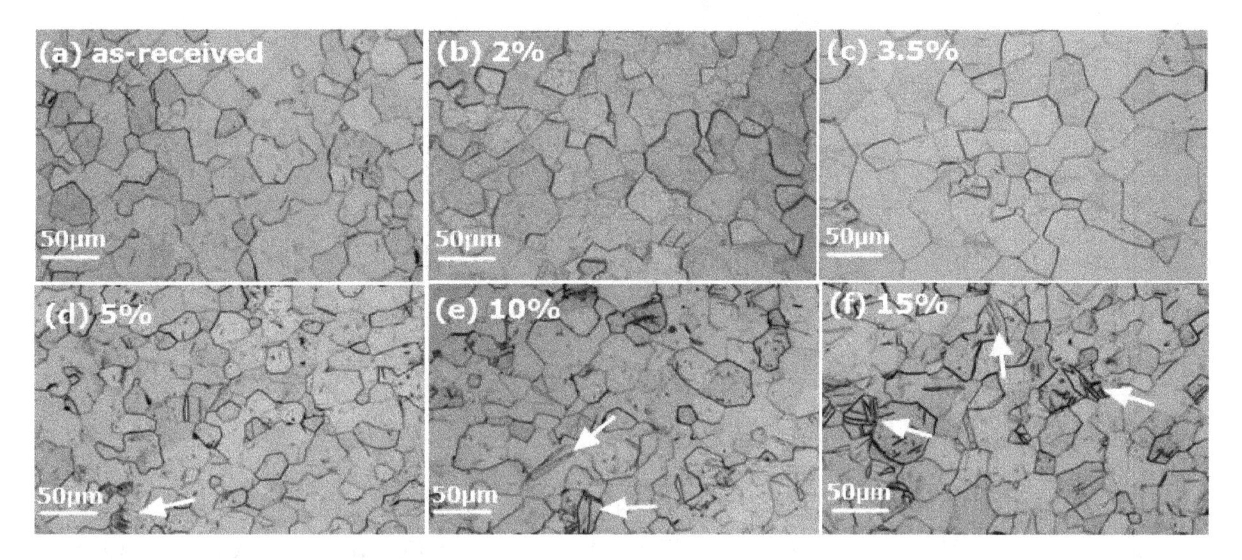

Figure 5. Metallographic figures of different prestrained samples: (**a**) as-received; (**b**) 2%; (**c**) 3.5%; (**d**) 5%; (**e**) 10% and (**f**) 15% prestrained specimens. Twins are indicated by white arrows in Figure 5**d**–**f**.

The variations of strength and ductility of CP-Ti as functions of tensile prestrain are shown in Figures 6 and 7 respectively. It is observed that with the increase of prestrain, both yield strength and tensile strength increase. The increment in yield strength with the amount of prestrain is gradually slow, which can be seen from the slope of variation of yield strength with the increase of prestrain. The increase of yield strength with prestrain can be rationalised by following two reasons. One is that dislocation density increases quickly with increasing prestrain. As a result, the resistance to start plastic deformation on reloading increases sharply with increasing prestrain. The other is that as prestrain increases, few twins also contribute to the increase of yield strength. The yield strength increases more and more slowly with prestrain. The logarithmic curve of the yield strength versus the prestrain is linear. Thus, a power law can be used to describe the relationship between prestrain and yield strength, expressed as follow:

$$\sigma = 68.08{\varepsilon_{pre}}^{0.43} + 305.34 \tag{3}$$

It is shown that the curve of the Equation (3) is consistent with the original data point, and the correlation coefficient R is 0.999. When prestrain is less than 6.5%, ultimate tensile strength increases slowly with prestrain. As prestrain increases up to 10%, ultimate tensile strength increases rapidly and yield strength is close to tensile strength. According to relationships between uniform strain, elongation (ε), total elongation ($\varepsilon_{total} = \varepsilon_{pre} + \varepsilon$) and prestrain, both uniform strain and elongation decrease linearly with prestrain. As prestrain increases up to 10%, plastic deformation of CP-Ti quickly develops into the diffuse necking stage. However, total elongation of CP-Ti remains constant about 30.5%, slightly less than the as-received material, that means the total ductility keeps in constant. During the initial prestrain process the higher the prestrain is, the lower elongation on reloading is.

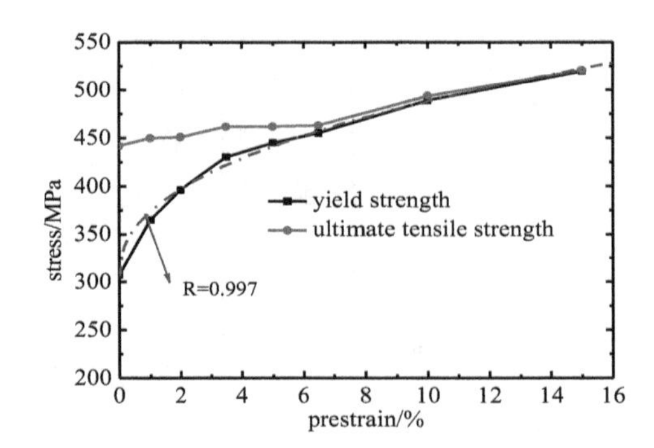

Figure 6. Variation of strength parameters.

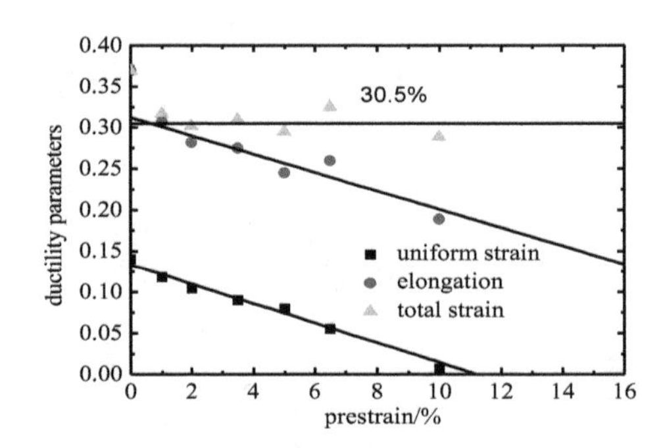

Figure 7. Variation of ductility parameters with amount of prestrain.

3.3. Tensile Properties of Prestrained Specimens after Annealing

In order to further analyze the phenomenon of discontinuous yielding occurs in higher prestrained specimens, subsequent annealing was conducted to remove dislocations. According to National military standard of the People's Republic of China [35], stress relieving temperature is in the range of 445–595 °C and heating holding time is from 15 to 360 min. In order to establish proper heating parameters, the prestrained specimens were conducted heat-treatment at the temperature of 500 °C with 30–60 min and 600 °C with 30 min respectively. The results show that when prestrained specimens are heated at 500 °C, the yield and tensile strength no longer decrease as holding time up to 40 min. Additionally, compared with prestrained specimens without annealing, yield and tensile strength decrease. That means when holding time reaches to 40 min at 500 °C, most of dislocations can be eliminated. At the same time, after heat treatment at 600 °C and 30 min holding time, yield and tensile strength of prestrained specimens are equal to those of the as-received material. That indicates that recrystallization process of grain could be realized and both dislocations, twinning structures and other defects have been removed through heat-treatment at 600 °C and 30 min holding time. According to metallographic figures of different prestrained samples after annealing at 500 °C and 40 min, grain size and shape do not change, as shown in Figure 8. No twinning is observed in the two percent prestrained specimen and twinning structures in the five percent prestrained specimen keep same as these without annealing. This also implies that twinning structures in prestrained specimens couldn't be annealing twins. Therefore, the heat treatment temperature and holding time were established as 500 °C and 40 min, which can eliminate the dislocations generated by prestrain and remain the twins. Engineering stress–strain cures of prestrained specimens after annealing are shown in Figure 9. It is observed that yield plateau of relative higher prestrain ($\varepsilon_{pre} \geq 3.5\%$) specimens disappears after annealing. Also, work hardening rate of prestrained CP-Ti after annealing is calculated according to true stress–strain curve. As most of dislocations is eliminated after annealing, it is expected that the variation of θ for prestrained specimens with annealing is consist with that of as-received CP-Ti, as shown in Figure 10. Work hardening rates of all specimens are positive and no rising stage can be observed. Additionally, the transitions between different stages of θ are consistent with those in Figure 4. Therefore, it is obviously demonstrated that the high dislocation density in prestrained specimens results in the appearance of yield plateau on reloading and rising stage of work hardening rate.

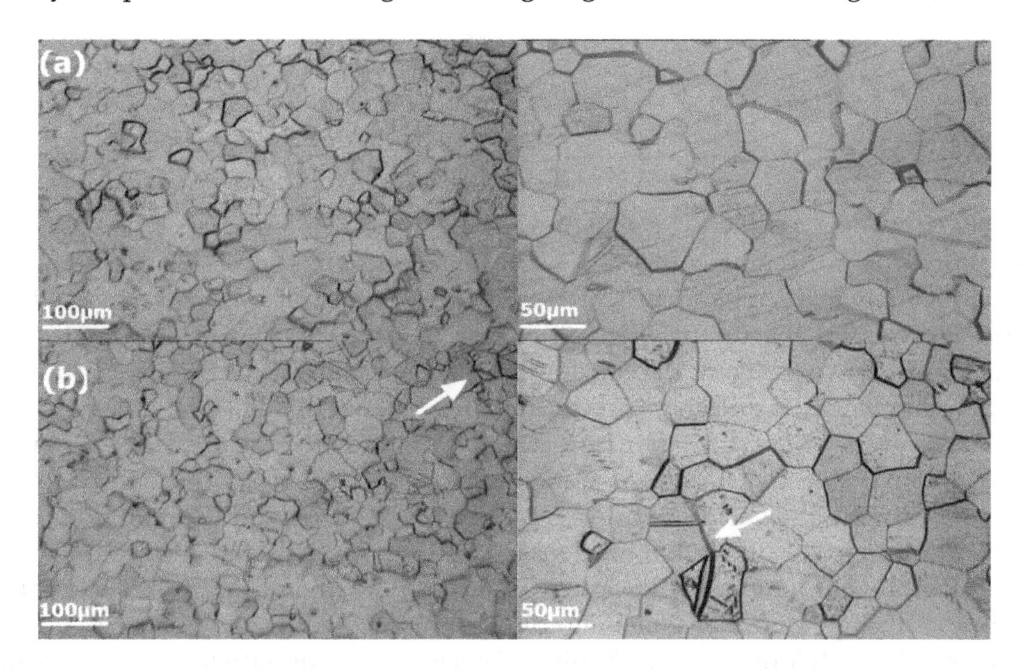

Figure 8. Metallographic figures of different prestrained samples after heat-treatment (500 °C, 40 min): (**a**) two percent prestrain with annealing; (**b**) five percent prestrain with annealing. Twins are indicated by white arrows in Figure 8**b**.

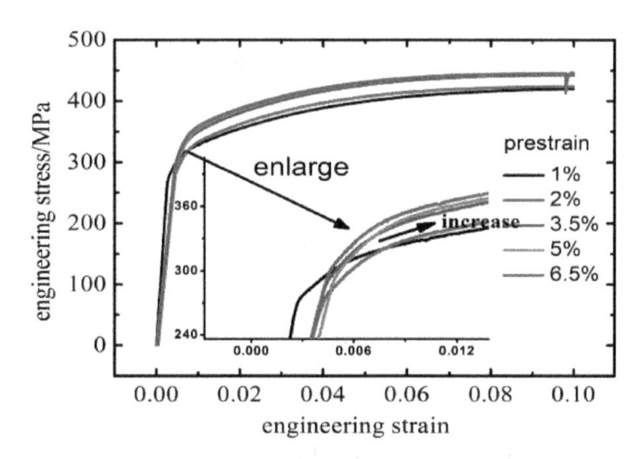

Figure 9. Engineering stress–strain cures of prestrained specimens after annealing (500 °C, 40 min).

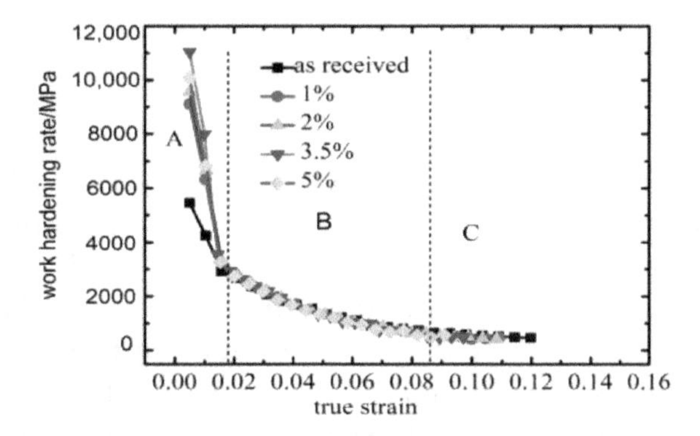

Figure 10. Three stages of work hardening rate in prestrained specimens after annealing: Sharp decrease in stage A, moderate decrease in stage B and slow decrease in stage C.

Further quantitative analysis of the contribution of dislocation hardening to the strain hardening of the material was made. In this paper, two parameters ($\Delta\sigma_1$, $\Delta\sigma_2$) were introduced, expressed as follows:

$$\Delta\sigma_1 = \sigma_R - \sigma_{RA} \tag{4}$$

$$\Delta\sigma_2 = \sigma_{RA} - \sigma_S \tag{5}$$

$\Delta\sigma_1$, $\Delta\sigma_2$ represents hardening contributions from dislocations associated mechanisms and other mechanisms, respectively. σ_S, σ_R and σ_{RA} refer to yield strength of as-received CP-Ti, prestrained specimens and prestrained specimens with annealing, respectively. From Figure 11, it can be seen that below the strain of two percent, the hardening mainly comes from the dislocations associated mechanisms and other hardening mechanisms are not important at this stage (about 3–4 MPa). As strain increases to 3.5%, $\Delta\sigma_2$ (about 21–31 MPa) increases fast. With the increase of strain, the contribution on strain hardening from both the dislocations associated mechanisms and other mechanisms increases. Nevertheless, according to the variation of $\Delta\sigma_1/(\Delta\sigma_1 + \Delta\sigma_2)$ with prestrain in Figure 11, despite the decrease of contribution from dislocation hardening, dislocations associated mechanisms still occupy 80% of the overall strain hardening. Therefore, dislocations play an important role in strain hardening of CP-Ti during room temperature tensile along rolling direction. Also, it indicates that dislocation slip is the predominant plastic deformation mechanism. This is consistent with the results of Amouzou [31] and Roth et al. [33]. Based on Taylor-type relations between dislocation density and local flow stress, the quantified contribution of dislocation density to the increase of strength is expressed as follows:

$$\sigma_p = \sigma_0 + M\alpha Gb\rho^{1/2} \tag{6}$$

where M, α, G and b are the Taylor factor, Taylor constant, shear modulus and Burgers vector, respectively; and σ_0 is the lattice friction stress. According to the literature [31], the relation between dislocation density (ρ) and tensile strain (ε) is linear:

$$\rho = a_1 + a_2\varepsilon \tag{7}$$

where a_1 and a_2 are constants. Thus, the variation of dislocation hardening stress ($\Delta\sigma_1$) with prestrain (ε_{pre}) should satisfy the following relationship:

$$\Delta\sigma_1 = c_1 + c_2\left(a_1 + a_2\varepsilon_{pre}\right)^{1/2} \tag{8}$$

where c_1 and c_2 are constants. From Figure 11, it is obviously that the relationship between $\Delta\sigma_1$ and ε_{pre} satisfies Equation (8) and the correlation coefficient R is 0.97. After annealing treatment, ultimate tensile strength decreases, compared with prestrained specimens. The decrease of ultimate tensile strength in the 3.5% prestrained specimen is maximum, as shown in Figure 12. The elongation of prestrained specimens increases after annealing, as shown in Figure 13. When prestrain is less than 3.5%, the elongation of prestrained specimens after annealing is higher than the as-received CP-Ti. However, when prestrain is above five percent, the elongation of prestrained specimens after annealing starts to decrease compared with the as-received CP-Ti. This indicates that by proper heat treatment and certain prior plastic deformation, both the strength and ductility of CP-Ti can be improved.

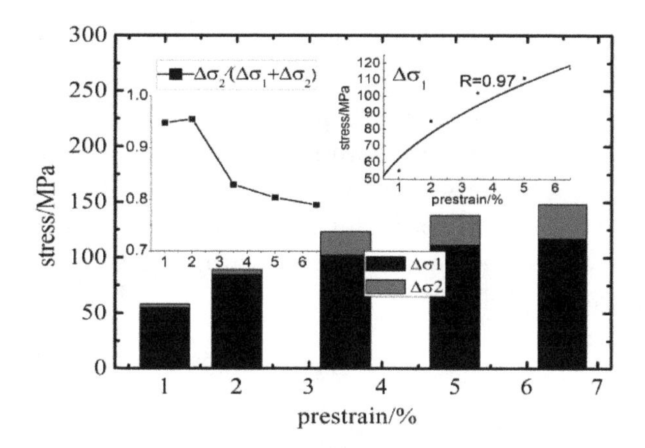

Figure 11. The quantified analysis of effects of dislocations associated mechanisms ($\Delta\sigma_1$) and other mechanisms ($\Delta\sigma_2$) on strain hardening.

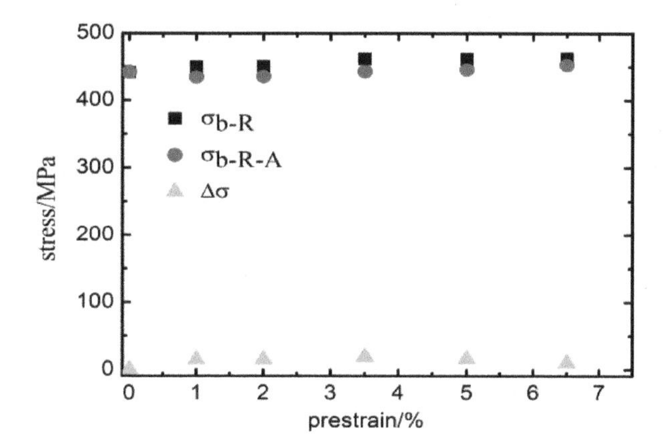

Figure 12. Decline of ultimate tensile strength after annealing: $\sigma_{\text{b-R}}$, $\sigma_{\text{b-R-A}}$, Δ_σ represent tensile strength of prestrained specimens, prestrained specimens with annealing and the difference between prestrained and prestrained specimens with annealing.

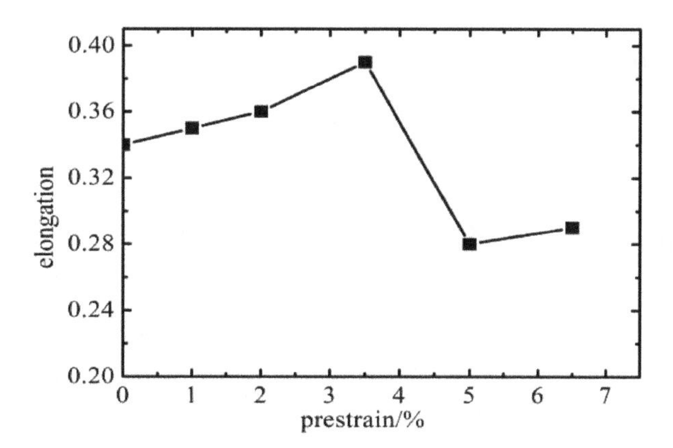

Figure 13. Elongation of prestrained specimens with annealing.

3.4. Constitutive Model of Prestrained CP-Ti

The Fields-Backofen model [36], expressed as follows:

$$\sigma = k\varepsilon^n \dot{\varepsilon}^m \quad (9)$$

is commonly used to quantitatively describe the effects of strain rate sensitivity and strain strengthening on stress–strain curves of the homogeneous strain strengthening region, where k is the strength coefficient, n is the strain hardening exponent, and m is the strain rate sensitivity exponent. According to true stress–strain curves of prestrained CP-Ti at different strain rates, the variation of strain rate sensitivity index $m(\log\sigma/\log\dot{\varepsilon})$ with prestrain is presented in Figure 14a. SRS of prestrained CP-Ti decreases linearly with prestrain. Strain hardening exponent $n(\log\sigma/\log\varepsilon)$ and strength coefficient $k(\sigma/\varepsilon^n\dot{\varepsilon}^m)$ of prestrained CP-Ti at different strain rates are calculated in Figure 14b,c. The values of strain hardening exponent and strength coefficient almost keep constant at different strain rates, as shown in Figure 14b,c. Both n and k decrease with prestrain. Based on experiment data, m, n and k as functions of prestrain were fitted as follows:

$$m = -0.17\varepsilon_{pre} + 0.03 \tag{10}$$

$$n = -1.19{\varepsilon_{pre}}^{0.87} + 0.14 \tag{11}$$

$$k = -4440.70\varepsilon_{pre} + 859.49 \tag{12}$$

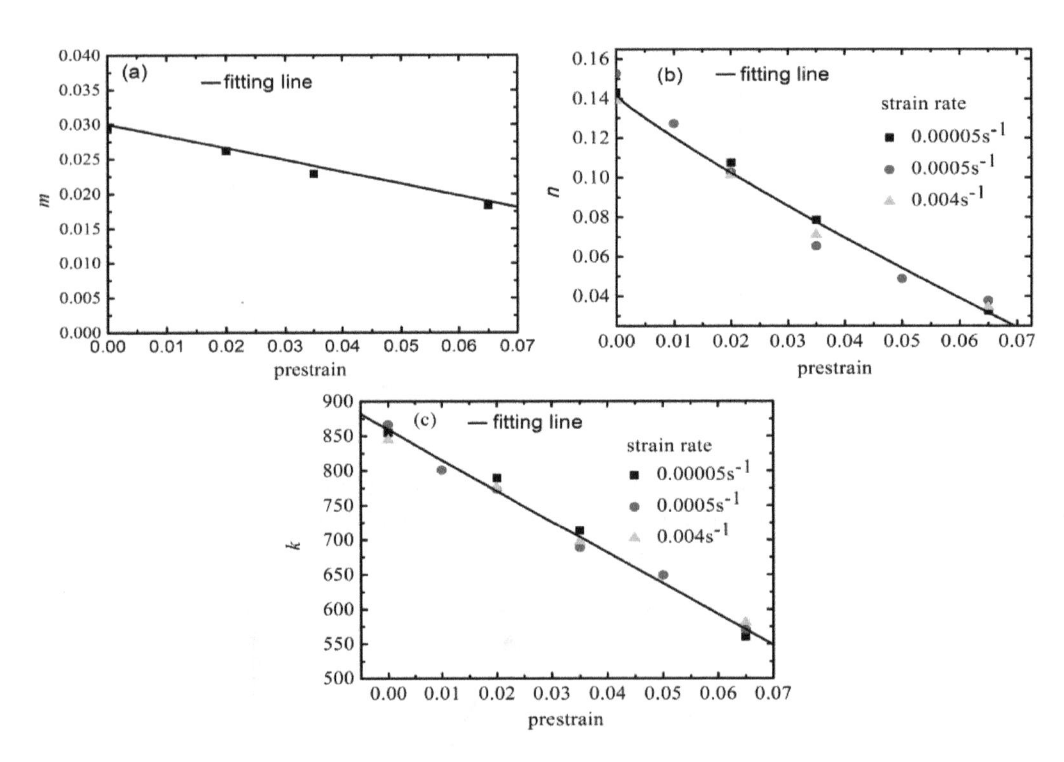

Figure 14. Variation of parameters with prestrain: (**a**) Strain rate sensitivity; (**b**) strain hardening exponent and (**c**) strength coefficient.

Thus, the modified Fields–Backofen model containing prestrain is finally obtained as follows:

$$\sigma = \left(-4440.70\varepsilon_{pre} + 859.49\right)\varepsilon^{-1.19{\varepsilon_{pre}}^{0.87}+0.14}\dot{\varepsilon}^{-0.17\varepsilon_{pre}+0.03} \tag{13}$$

In order to investigate the prominence of the modified constitutive model, comparisons between the experimental data and the flow stress predicted by the modified constitutive model at different strain rates are shown in Figure 15. In addition, the predictability of the constitutive equation is verified via employing standard statistical parameters, such as absolute deviation (ΔA) and correlation coefficient (R), as shown in Figure 15a–d. At different strain rates, the values of correlation coefficient R in Figure 15a–d are above 0.97, hence the modified Fields-Backofen model shows a very high degree of goodness of fit. It is found that the absolute deviation of flow stress obtained from the modified constitutive model varies from 1.87 to 6.88 MPa. The constitutive equations gives the least absolute deviation of 1.87 at 0.004 s^{-1} of as-received CP-Ti and the largest absolute deviation of 6.88

at 0.00005 s^{-1} of 3.5% prestrained sample. Thus, the proposed constitutive equation presents a good estimate of the plastic flow stress for prestrained CP-Ti at different strain rates.

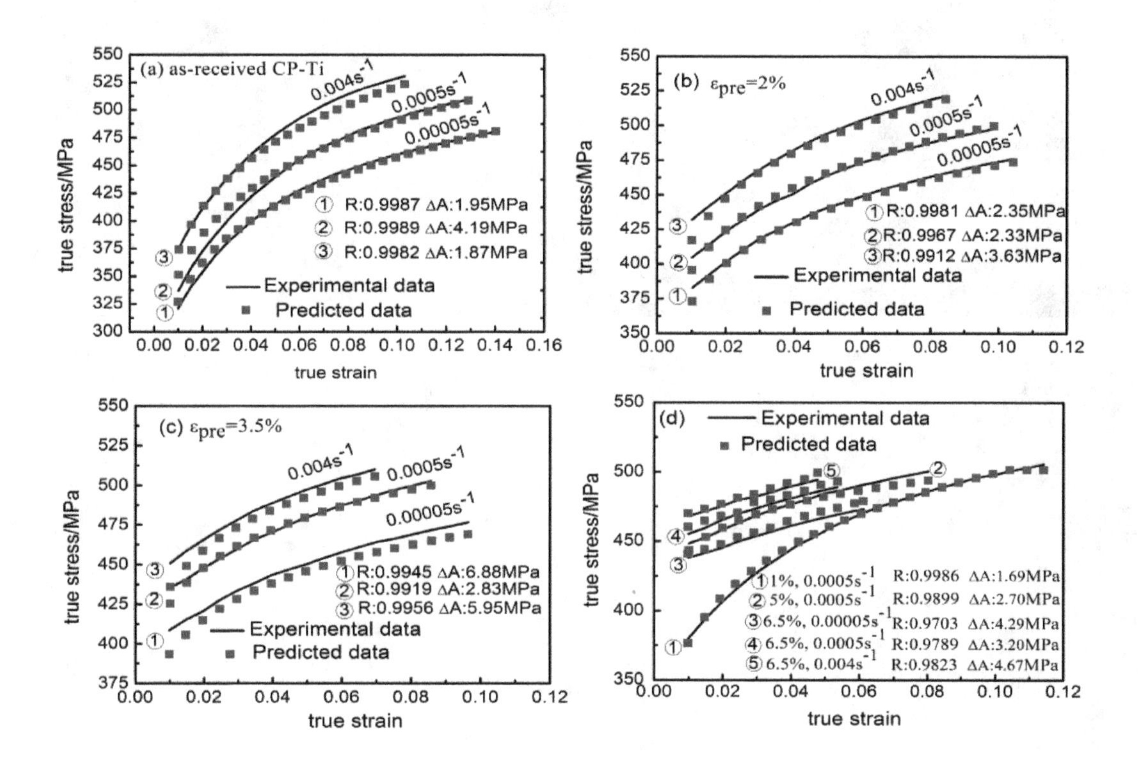

Figure 15. Comparison between the experimental and predicted flow stress data from the modified Fields-Backofen model: (**a**) as-received CP-Ti; (**b**) 2% prestrained CP-Ti; (**c**) 3.5% prestrained CP-Ti; (**d**) 1%, 5%, 6.5% prestrained CP-Ti.

3.5. Fracture Behavior

Scanning electron microscopy (SEM) images of the fracture surfaces of as-received CP-Ti were presented in Figure 16. SEM images of the fracture surfaces show that the fracture mode of as-received CP-Ti is mostly ductile at room temperature. In Figure 16a, the central region is a crack propagation area with massive strong nests, and the outer region is a shear lip zone with shallow dimples and micro-pores. The central region shows more ductile deformation features than the outer shear-lip region, which has a flat surface with shallow dimples, as shown in Figure 16c,d. The mechanism for ductile crack growth can be characterized by micro-void nucleation, growth and coalescence. As the specimen is loaded, local strains and stresses at the crack tip become sufficient to nucleate void. These voids grow and link with the main crack [37].

The failure locations of prestrained and prestrained samples with annealing present pronounced necking feature, as shown in Figure 17 from macro view. The failure locations of all tested samples show that ragged fracture surface with some macro voids across the whole cross section of fracture surface, and relatively flat shear lip zone was detected in marginal area. Magnified images of central region of fracture surfaces were presented in Figure 18. It is obvious that the size of dimples shown in Figure 18d,f is larger than that in Figure 18a–c, the depth of dimples is much deeper and its distribution is more uniform. The more ductile deformation in prestrained CP-Ti after annealing indicates that the ductility of prestrained CP-Ti is improved by annealing. The increase in ductility is a result of high dislocation mobility, where the crack tip is proceeded by a plastic deformation mechanism that forms dense arrays of dimples without cleavage steps and facets [38].

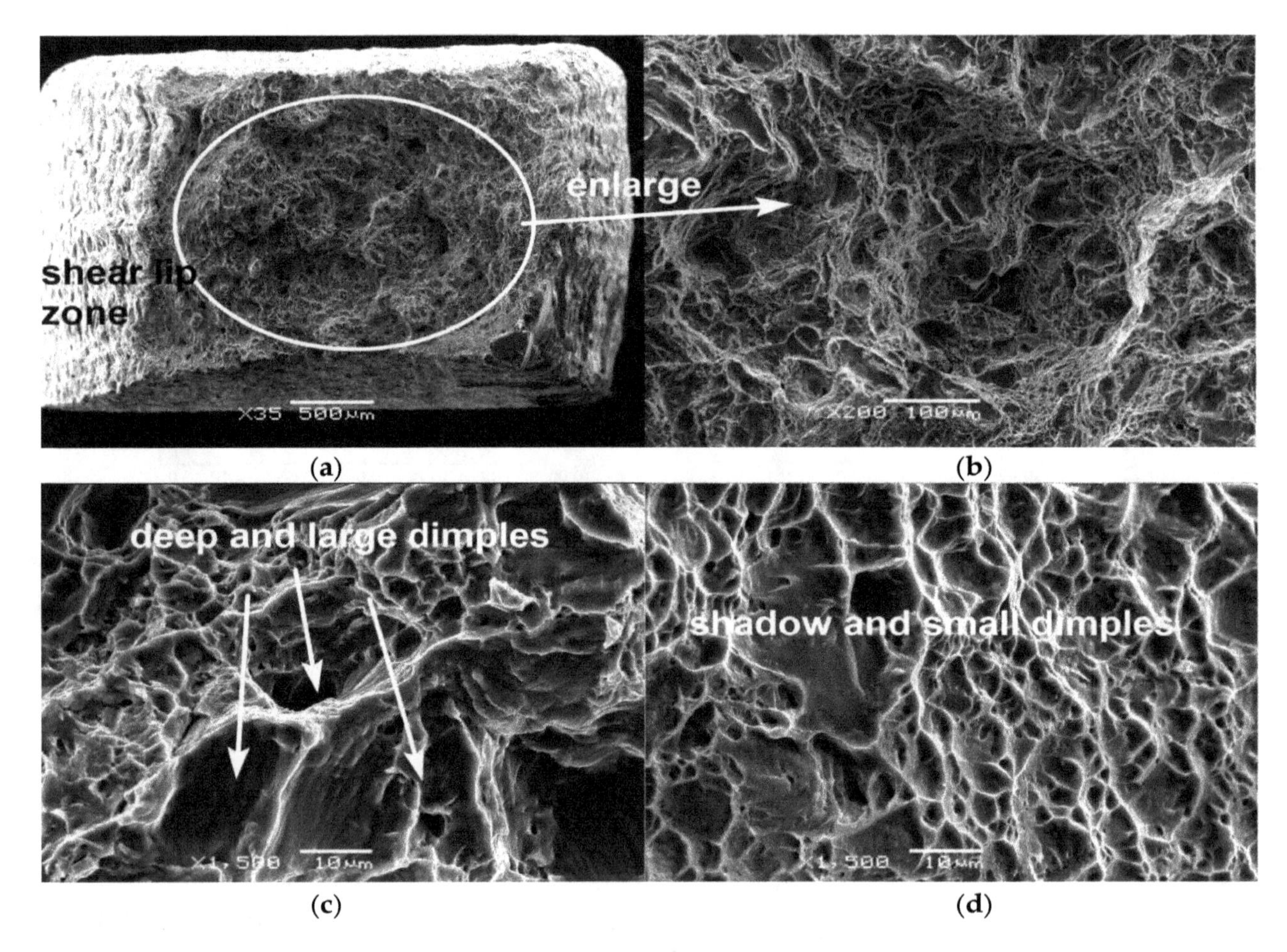

Figure 16. Micrographs of fractured surfaces of as-received CP-Ti with no prestrain: (**a**) macro view showing necking with microcracks and shear lip zone at the outer area; (**b**) magnified view of crack propagation region from central region in Figure 16a; (**c**) magnified image showing deep and large dimples from central region; (**d**) magnified image showing the shallow dimples at the shear-lip region.

Figure 17. Macro view of fractured surfaces of prestrained CP-Ti: (**a**) 1% prestrained; (**b**) 3.5% prestrained; (**c**) 6.5% prestrained; (**d**) 1% prestrained with annealing; (**e**) 3.5% prestrained with annealing; (**f**) 6.5% prestrained with annealing.

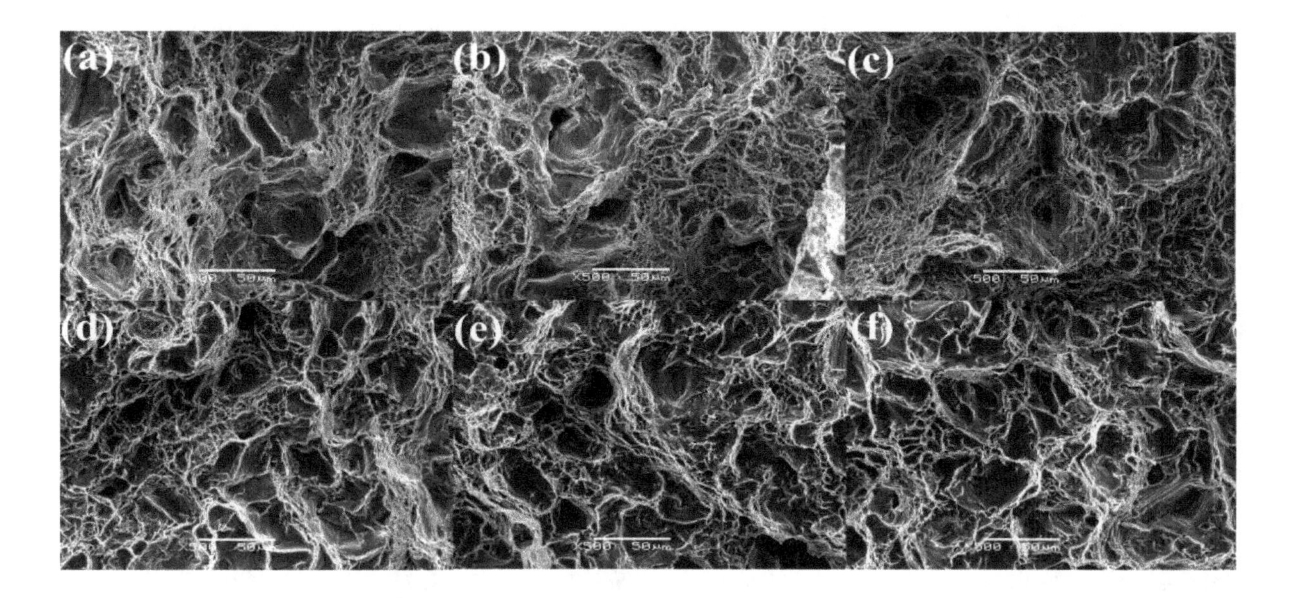

Figure 18. Micrographs of fractured surfaces of prestrained CP-Ti: (**a**) 1% prestrained; (**b**) 3.5% prestrained; (**c**) 6.5% prestrained; (**d**) 1% prestrained with annealing; (**e**) 3.5% prestrained with annealing; (**f**) 6.5% prestrained with annealing.

4. Conclusions

In the present paper, the effect of prestrain and subsequent annealing on the mechanical behaviors of CP-Ti along cold rolling direction was investigated. The main results presented in this paper are as follows:

- Engineering yield strength and tensile strength increase with the tensile prestrain. The relationship between prestrain and yield strength can be described by power law. Elongation and uniform strain decrease linearly with increasing the tensile prestrain, and total strain is almost constant, which is slightly less than as-received CP-Ti.
- At the condition of prestrain higher than 3.5%, yield plateau is observed in the engineering stress–strain curves. Through metallographic observations and heat treatment, considerable number of dislocations produced in prestrain is determined as the factor account for the appearance of yield plateau during reloading.
- Dislocations play an important role in strain hardening along cold rolling direction tension. With the increase of strain, the contribution on strain hardening from both the dislocation-associated mechanisms and other mechanisms increases.
- The strain rate sensitivity exponent, strain hardening exponent and strength coefficient decrease with the increase of prestrain. Flow stress of prestrained CP-Ti is predicted accurately by the modified Fields-Backofen model.
- With annealing at 500 °C and 40 min, the 3.5% prestrained specimen shows the highest ductility, and its yield strength is enhanced. The more ductile deformation features on the fracture surfaces after annealing confirms that the ductility of prestrained CP-Ti can be improved by annealing.

Acknowledgments: The authors gratefully acknowledge the financial supports of the National Natural Science Foundation of China (51475223) and the Graduate Student Scientific Innovative Project of Jiangsu Province (KYLX15_0801).

Author Contributions: All contributed to the design of the experimental plan. Le Chang performed the experiments; Chang-Yu Zhou and Xiao-Hua He analyzed the data; Le Chang wrote the paper.

References

1. Ahmed, I.I.; Grant, B.; Sherry, A.H.; Quinta, J. Deformation path effects on the internal stress development in cold worked austenitic steel deformed in tension. *Mater. Sci. Eng. A* **2014**, *614*, 326–337. [CrossRef]
2. Zhang, L.C.; Timokhina, I.B.; La Fontaine, A.; Ringer, S.P.; Hodgson, P.D.; Pereloma, E.V. Effect of pre-straining and bake hardening on the microstructure and mechanical properties of CMnSi TRIP steels. *Metall. Ital.* **2009**, *101*, 49–55.
3. Pereloma, E.; Beladi, H.; Zhang, L.C.; Timokhina, I. Understanding the behavior of advanced high-strength steels using atom probe tomography. *Metall. Mater. Trans. A* **2012**, *43*, 3958–3971. [CrossRef]
4. Baeka, J.; Kim, Y.; Kim, C.; Kim, W.; Seok, C. Effects of pre-strain on the mechanical properties of API 5L X65 pipe. *Mater. Sci. Eng. A* **2010**, *527*, 1473–1479. [CrossRef]
5. Robertson, L.T.; Hilditch, T.B.; Hodgson, P.D. The effect of prestrain and bake hardening on the low-cycle fatigue properties of TRIP steel. *Int. J. Fatigue* **2008**, *30*, 587–594. [CrossRef]
6. Lee, W.S.; Lin, C.F. Effects of prestrain and strain rate on dynamic deformation characteristics of 304L stainless steel: Part 1—Mechanical behavior. *Mater. Sci. Technol.* **2002**, *18*, 869–876. [CrossRef]
7. Lee, W.S.; Lin, C.F. Effects of prestrain and strain rate on dynamic deformation characteristics of 304L stainless steel: Part 2—Microstructural study. *Mater. Sci. Technol.* **2002**, *18*, 877–884. [CrossRef]
8. Whittaker, M.T.; Evans, W.J. Effect of prestrain on the fatigue properties of Ti834. *Int. J. Fatigue* **2009**, *31*, 1751–1757. [CrossRef]
9. Whittaker, M.; Jones, P.; Pleydell-Pearce, C.; Rugg, D.; Williams, S. The effect of prestrain on low and high temperature creep in Ti834. *Mater. Sci. Eng. A* **2010**, *527*, 6683–6689. [CrossRef]
10. Song, Z.Y.; Sun, Q.Y.; Xiao, L.; Liu, L.; Sun, J. Effect of prestrain and aging treatment on microstructures and tensile properties of Ti–10Mo–8V–1Fe–3.5Al alloy. *Mater. Sci. Eng. A* **2010**, *527*, 691–698. [CrossRef]
11. Werber, A.; Liewald, M. Measurement and analysis of differential work hardening behavior of pure titanium sheet using spline function. *Int. J. Mater. Form.* **2011**, *4*, 193–204.
12. Wowk, D.; Pilkey, K. Effect of prestrain with a path change on the strain rate sensitivity of AA5754 sheet. *Mater. Sci. Eng. A* **2009**, *520*, 174–178. [CrossRef]
13. Sarker, D.; Friedman, J.; Chen, D.L. Influence of pre-strain on de-twinning activity in an extruded AM30 magnesium alloy. *Mater. Sci. Eng. A* **2014**, *605*, 73–79. [CrossRef]
14. Sarker, D.; Chen, D.L. Dependence of compressive deformation on pre-strain and loading direction in an extruded magnesium alloy: Texture, twinning and de-twinning. *Mater. Sci. Eng. A* **2014**, *596*, 134–144. [CrossRef]
15. Hama, T.; Nagao, H.; Kuchinomachi, Y.; Takuda, H. Effect of pre-strain on work-hardening behavior of magnesium alloy sheets upon cyclic loading. *Mater. Sci. Eng. A* **2014**, *591*, 69–77. [CrossRef]
16. Chen, L.; Li, J.; Zhang, Y.; Lu, W.; Zhang, L.C.; Wang, L.; Zhang, D. Effect of low-temperature pre-deformation on precipitation behavior and microstructure of a Zr-Sn-Nb-Fe-Cu-O alloy during fabrication. *J. Nucl. Sci. Technol.* **2016**, *53*, 496–507. [CrossRef]
17. Zherebtsov, S.V.; Dyakonov, G.S.; Salem, A.A.; Malysheva, S.P.; Salishchev, G.; Semiatin, S.L. Evolution of grain and subgrain structure during cold rolling of commercial-purity titanium. *Mater. Sci. Eng. A* **2011**, *528*, 3474–3479. [CrossRef]
18. Chun, Y.B.; Yu, S.H.; Semiatin, S.L.; Hwang, S.K. Effect of deformation twinning on microstructure and texture evolution during cold rolling of CP-titanium. *Mater. Sci. Eng. A* **2005**, *398*, 209–219. [CrossRef]
19. Nasiri-Abarbekoh, H.; Ekrami, A.; Ziaei-Moayyed, A.A.; Shohani, M. Effects of rolling reduction on mechanical properties anisotropy of commercially pure titanium. *Mater. Des.* **2012**, *34*, 268–274. [CrossRef]
20. Chichili, D.R.; Ramseh, K.T.; Hemker, K.J. The high-strain-rate response of alpha-titanium: Experiments, deformation mechanisms and modeling. *Acta Mater.* **1998**, *46*, 1025–1043. [CrossRef]
21. Peng, J.; Zhou, C.Y.; Dai, Q.; He, X.H. The temperature and stress dependent primary creep of CP-Ti at low and intermediate temperature. *Mater. Sci. Eng. A* **2014**, *611*, 123–135. [CrossRef]
22. Zhao, X.C.; Yang, X.R.; Liu, X.Y.; Wang, C.T.; Huang, Y.; Langdon, T.G. Processing of commercial purity titanium by ECAP using a 90 degrees die at room temperature. *Mater. Sci. Eng. A* **2014**, *607*, 482–489. [CrossRef]
23. Sordi, V.L.; Ferrante, M.; Kawasaki, M.; Langdon, T.G. Microstructure and tensile strength of grade 2 titanium processed by equal-channel angular pressing. *J. Mater. Sci.* **2012**, *47*, 7870–7876. [CrossRef]

24. Roodposhit, P.S.; Farahbakhsh, N.; Sarkar, A.; Murty, K.L. Microstructural approach to equal channel angular processing of commercially pure titanium—A review. *T. Nonferr. Metal. Soc.* **2015**, *25*, 1353–1366. [CrossRef]
25. Salem, A.A.; Kalidindi, S.R.; Doherty, R.D. Strain hardening of titanium: Role of deformation twinning. *Acta Mater.* **2003**, *51*, 4225–4237. [CrossRef]
26. Salem, A.A.; Kalidindi, S.R.; Doherty, R.D. Strain hardening regimes and microstructure evolution during large strain compression of high purity titanium. *Scr. Mater.* **2002**, *46*, 419–423. [CrossRef]
27. Becker, H.; Pantleon, W. Work-hardening stages and deformation mechanism maps during tensile deformation of commercially pure titanium. *Comp. Mater. Sci.* **2013**, *76*, 52–59. [CrossRef]
28. Hama, T.; Nagao, H.; Kobuki, A.; Fujimoto, H.; Takuda, H. Work-hardening and twinning behaviors in a commercially pure titanium sheet under various loading paths. *Mater. Sci. Eng. A* **2015**, *620*, 390–398. [CrossRef]
29. Li, L.; Zhang, Z.; Shen, G. Effect of Grain Size on the Tensile Deformation Mechanisms of Commercial Pure Titanium as Revealed by Acoustic Emission. *J. Mater. Eng. Perform.* **2015**, *24*, 1975–1986. [CrossRef]
30. Ghaderi, A.; Barnett, M. Sensitivity of deformation twinning to grain size in titanium and magnesium. *Acta Mater.* **2011**, *59*, 7824–7839. [CrossRef]
31. Amouzou, K.E.A.; Richeton, T.; Roth, A.; Lebyodkin, M.A.; Lebedkina, T.A. Micromechanical modeling of hardening mechanisms in commercially pure α-titanium in tensile condition. *Int. J. Plast.* **2016**, *80*, 222–240. [CrossRef]
32. Won, W.J.; Park, K.T.; Hong, S.G.; Lee, C.S. Anisotropic yielding behavior of rolling textured high purity titanium. *Mater. Sci. Eng. A* **2015**, *637*, 215–221. [CrossRef]
33. Roth, A.; Lebyodkin, M.A.; Lebedkina, T.A.; Lecomte, J.S.; Richeton, T.; Amouzou, K.E.A. Mechanisms of anisotropy of mechanical properties of α-titanium in tension conditions. *Mater. Sci. Eng. A* **2014**, *596*, 236–243. [CrossRef]
34. Barkia, B.; Doquet, V.; Couzini, J.; Guillot, I.; Hripre, E. In situ monitoring of the deformation mechanisms in titanium with different oxygen contents. *Mater. Sci. Eng. A* **2015**, *636*, 91–102. [CrossRef]
35. Shi, J.; Sha, A.X. GJB 3763A. In *Heat Treatment for Titanium and Titanium Alloys*; Defense Science and Technology Industry Committee: Beijing, China, 2004.
36. Fields, D.S.; Backofen, W.A. Determination of strain hardening characteristics by torsion testing. *Proc. Am. Soc. Test. Mater.* **1957**, *57*, 1259–1272.
37. Chuluunbat, T.; Lu, C.; Kostryzhev, A.; Tieu, K. Investigation of X70 line pipe steel fracture during single edge-notched tensile testing using acoustic emission monitoring. *Mater. Sci. Eng. A* **2015**, *640*, 471–479. [CrossRef]
38. Chauhan, A.; Litvinov, D.; Aktaa, J. High temperature tensile properties and fracture characteristics of bimodal 12Cr-ODS steel. *J. Nucl. Mater.* **2016**, *468*, 1–8. [CrossRef]

14

Study on Hot Deformation Behavior and Microstructure Evolution of Ti-55 High-Temperature Titanium Alloy

Fengyong Wu [1,2], Wenchen Xu [1,2,*], Xueze Jin [1,2,*], Xunmao Zhong [1], Xingjie Wan [1], Debin Shan [1,2] and Bin Guo [1,2]

1 School of Materials Science and Engineering & National Key Laboratory for Precision Hot Processing of Metals, Harbin Institute of Technology, Harbin 150001, China; wfy2000cn@163.com (F.W.); zxmhit@126.com (X.Z.); goqa@foxmail.com (X.W.); shandebin@hit.edu.cn (D.S.); bguo@hit.edu.cn (B.G.)

2 National Key Laboratory for Precision Hot Processing of Metals, Harbin Institute of Technology, Harbin 150001, China

* Correspondence: xuwc_76@hit.edu.cn (W.X.); jinxzabc@163.com (X.J.)

Abstract: The isothermal compression experiment of as-rolled Ti-55 alloy was carried out on a Gleeble-3800 thermal simulation test machine at the deformation temperature range of 700–1050 °C and strain rate range of 0.001–1 s^{-1}. The hot deformation behavior and the microstructure evolution were analyzed during thermal compression. The results show that the apparent activation energy Q in $\alpha + \beta$ dual-phase region and β single-phase region were calculated to be 453.00 KJ/mol and 279.88 KJ/mol, respectively. The deformation softening mechanism was mainly controlled by dynamic recrystallization of α phase and dynamic recovery of β phase. Discontinuous yielding behavior mainly occurred in β phase region, which weakened gradually with the increase of deformation temperature (>990 °C) and strain rate (0.01–1 s^{-1}) in β phase region. The processing map derived from Murty's criterion was more accurate in predicting the hot workability than that derived from Prasad's criterion. The optimized hot working window was 850–975°C/0.001–1 s^{-1}, in which sufficient dynamic recrystallization occurred and $\alpha + \beta$-transus microstructure was obtained. When deformed at higher temperature ($\geq$1000 °C), coarsened lath-shape β-transus microstructure was formed, while deformed at lower temperature ($\leq$825 °C) and higher strain rate ($\geq$0.1 s^{-1}), the dynamic recrystallization was not sufficient, thus flow instability appeared because of shear cracking.

Keywords: Ti-55 titanium alloy; hot deformation behavior; dynamic recrystallization; processing map

---

1. Introduction

Ti-55 alloy is a near α titanium alloy with the nominal chemical composition Ti-(5.0–6.0)Al-(3.0–4.0) Sn-(2.5–3.3)Zr-(0.3–1.5)Mo-(0.2–0.7)Ta-(0.2–0.7)Nb-(0.1–0.5)Si (wt %), which shows high strength and excellent corrosion resistance. Due to the addition of Si, Ta and Nb elements, the thermal stability and oxidation resistance are obviously improved, so the alloy can meet the requirement of long-term service with the temperature no less than 550 °C [1]. As a potential structural material for engine compressor, blade, sheet components in aviation and aerospace industries, Ti-55 alloy has gained increasing attention in China in recent years [2,3].

Similar to other high temperature titanium alloys, such as Ti-1100 and IMI834, this alloy is quite difficult to be formed into a complex shape because of its poor workability and high strength. Moreover, the accurate control of microstructure morphologies and properties of final components is very difficult for those high temperature titanium alloys with high alloying elements in forming process [4,5]. In order to develop plastic forming methods of high temperature titanium alloys, such as forging, extrusion and sheet hot forming, it is necessary to characterize the deformation behavior, including the flow stress behavior, deformation mechanism and microstructure evolution of the materials. Currently, quite a few investigations have been conducted to analyze the hot deformation behavior of high temperature titanium alloys. For instance, Liu and Bake [6] analyzed the deformation characteristics of IMI685 alloy and revealed its dynamic softening mechanism in β phase field, which was helpful for widening its forging temperature range. Wanjara et al. [4,5] explored the flow stress behavior and the microstructure evolution of near-α IMI834 alloy in β and α + β phase regions, respectively, through isothermal compression experiment. Niu et al. [7] investigated the high temperature behavior of a near-α Ti-600 alloy and determined its optimized superplastic forming window.

Although there is some research on the thermal stability and tensile formability of Ti-55 alloy [1,3,8], the hot compression behavior has been rarely reported, which limits the application of hot plastic forming processes, such as forging and extrusion, to this kind of titanium alloy. In this study, the isothermal compression of Ti-55 titanium alloy has been conducted at different strain rates and temperatures both in the α + β and β regions. The flow stress behavior has been analyzed and the microstructure evolution has been observed. Subsequently, the processing map has been established to understand the deformation mechanism during hot compression, and the optimum hot working window has been determined for hot processing of Ti-55 alloy.

2. Materials and Methods

The starting material used in the study was as-rolled sheet of Ti-55 alloy with the thickness of 12 mm, supplied by Baoji Titanium Industry Company Limited in China. The actual chemical composition (wt %) of the present Ti-Al-Sn alloy was determined by an inductively coupled plasma (ICP) test as follows: Al-5.2, Sn-3.3, Zr-2.9, Mo-1.0, Ta-0.4, Nb-0.4, Si-0.2 and the rest Ti. The β-transus temperature was calculated to be about 990 °C by a differential thermal analysis (DTA) test and metallographic method. The experimental specimens with dimension of Φ6 mm × 8 mm, whose axial direction was aligned with the sheet thickness direction (i.e., normal direction), were electro-discharge wire cut from the as-rolled sheet for hot compression, as shown in Figure 1. The isothermal compression test was carried out in the temperature range of 700–1050 °C with 50 °C intervals and strain rate range of 0.001–1 s^{-1} on a Gleeble-3800 simulator manufactured by DSI Company, which locate at Sao Paulo, Minnesota, America, with the low vacuum atmosphere of 1×10^{-3} Torr. All the specimens were heated to the deformation temperature at a heating rate of 10 °C/s and held for 2 min before hot compression, and the deformation temperature was recorded by Pt–Rh thermocouple wires spot-welded in the surface of the hot deformation specimens. In order to maintain the hot deformation microstructure, the specimens were quenched by water immediately after hot compression with 60% height reduction in low vacuum atmosphere. The tantalum chip of 0.1 mm was placed between crosshead and specimen to provide lubrication and prevent cementation during hot compression. The specimens were sectioned parallel to the compression axis after hot compression for microstructure observation. The specimens for optical microscopy (OM) observation were etched in a solution of 2 mL hydrofluoric acid, 4 mL nitric acid and 94 mL H_2O. Electropolishing was carried out in a solution of 12.5% perchloric acid and 87.5% acetic acid with the voltage of 25 V and temperature of −20 °C to prepare the samples for scanning electron microscopy (SEM) and electron back scattering diffraction (EBSD) analysis. The specimens for transmission electron microscopy (TEM) analysis were mechanically ground to about 100 μm, followed by two-jet electro polished at an ion voltage of 20 V under the temperature of about −20 °C. The SEM, EBSD and TEM testing were conducted on a Quanta 200FEG scanning electron microscope and TecnaiG2F30 transmission electron microscope, respectively, manufactured by FEI Company located at Hillsborough, Oregon, America.

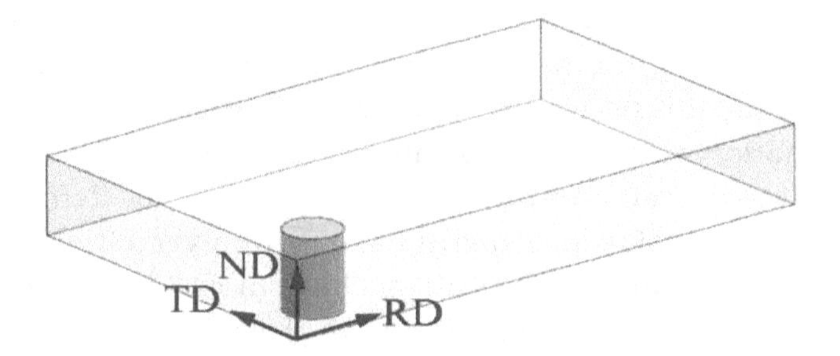

Figure 1. Schematic diagram of specimens cutting from the as-rolled Ti-55 sheet (ND: Normal Direction; RD: Rolling Direction; TD: Transverse Direction).

3. Results

3.1. Flow Stress Behavior

3.1.1. Flow Stress-Strain Curves

Figure 2 shows the stress-strain curves of as-rolled Ti-55 alloy during isothermal compression at different temperatures and strain rates. Clearly, the flow stress increased with the increase of true strain at the initial deformation stage to the peak stress because of work hardening, and then the flow stress decreased with further increase of true strain due to flow softening occurring during hot compression. Besides, the flow stress decreased with increasing deformation temperature and decreasing strain rate, indicating the flow stress was sensitive to deformation temperature and strain rate. When deformed at higher temperature and lower strain rate ($\geq$900 °C and $\leq$0.01 s^{-1}), the flow stress-strain curves reached to steady values with the increase of true strain, indicating the softening mechanism, including dynamic recrystallization (DRX) of α phase or dynamic recovery (DRV) of β phase, proceeded quickly to balance the rate of work hardening [9,10]. Besides this, the other secondary phase particles rich in Sn, Mo, Ta and Nb elements at the grain boundaries may have pinning effect on dislocation motion (see Figure 7a), which partly balanced dynamic softening during hot compression of Ti-55 alloy. However, when deformed in other conditions, such as lower temperature and higher strain rate, the curves exhibited continuous softening behavior after peak stress without the steady-state condition occurring due to insufficient softening behavior.

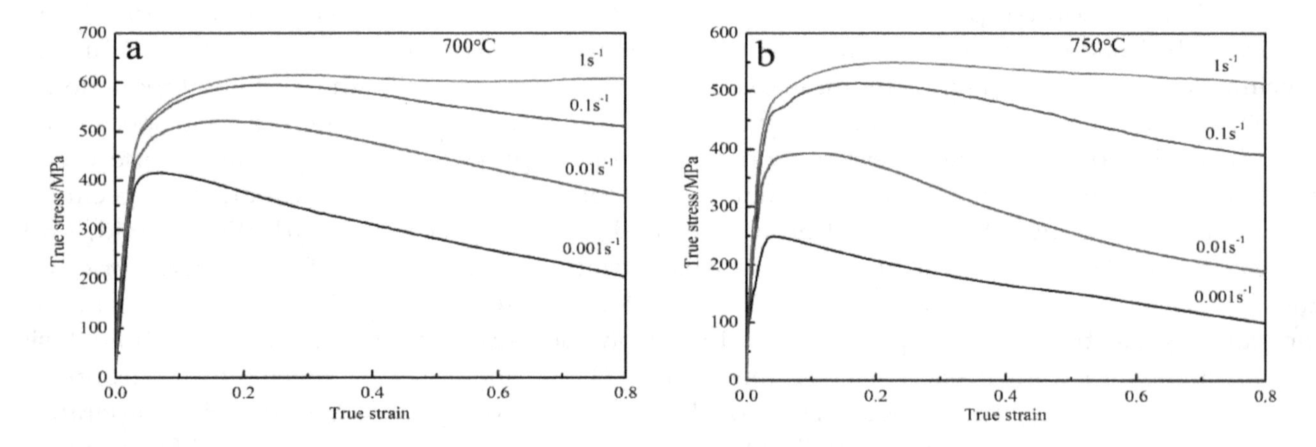

Figure 2. *Cont.*

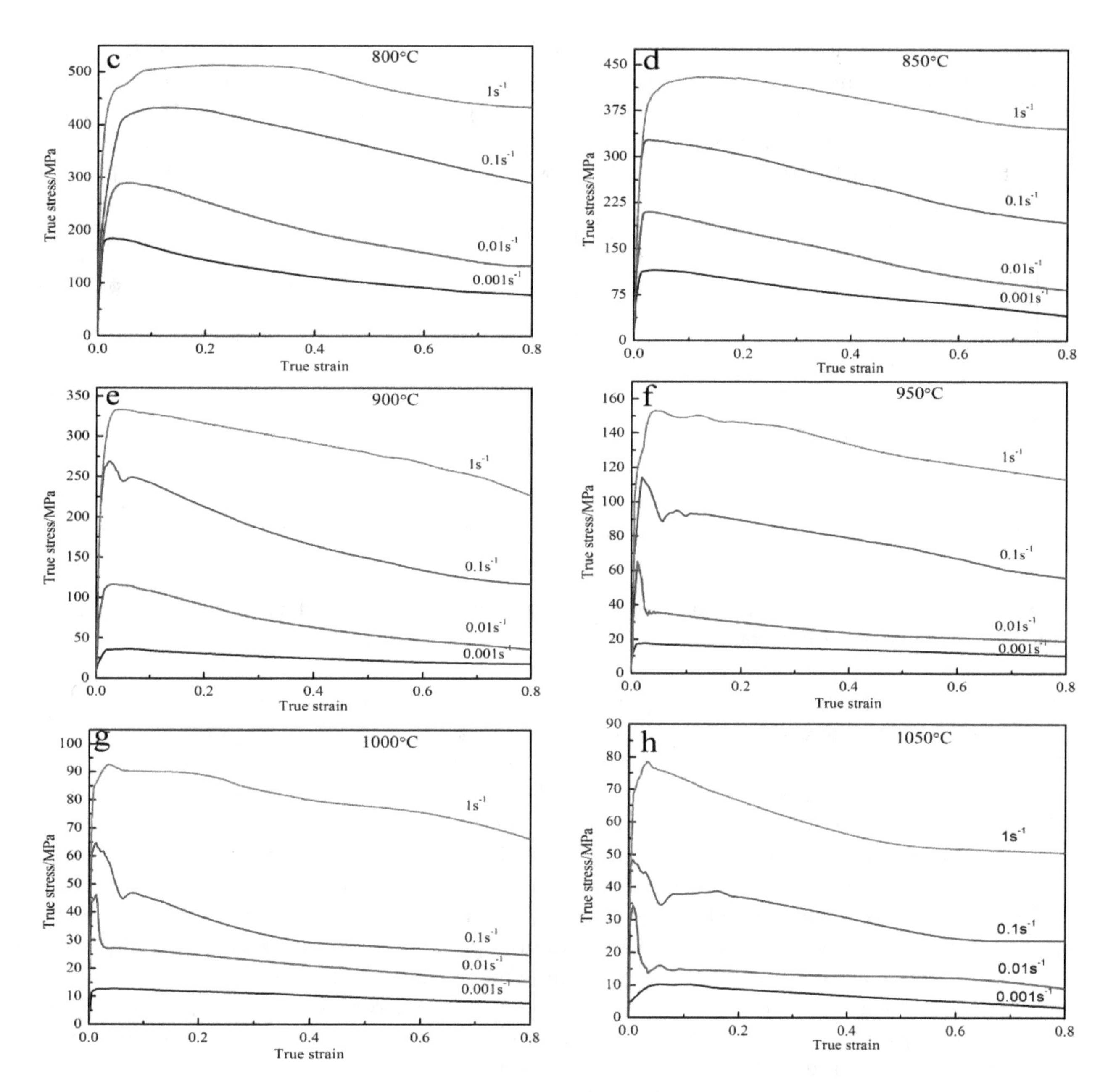

Figure 2. Typical stress-strain curves at different temperatures and strain rates of as-rolled Ti-55 alloy: (**a**) 700 °C; (**b**) 750 °C; (**c**) 800 °C; (**d**) 850 °C; (**e**) 900 °C; (**f**) 950 °C; (**g**) 1000 °C; (**h**) 1050 °C.

3.1.2. Discontinuous Yielding Behavior

Discontinuous yielding behavior, i.e., the sudden drop of flow stress beyond peak stress, was observed for the deformation conditions performed above 900 °C at strain rates of 0.01–1.0 s^{-1}, while there was no obvious discontinuous yielding behavior appearing at lower strain rate of 0.001 s^{-1}. Figure 3 shows the magnitude of the yield drop (i.e., $\sigma_{UY} - \sigma_{LY}$) at various deformation temperatures and strain rates. It can be seen that the yield drop reduced as the temperature increased over 950 °C, and the yield drop occurred only at the middle strain rate of 0.1 s^{-1} at 900 °C. The discontinuous yielding behavior has been found in many β titanium alloys, such as Ti-10V-4.5Fe-1.5Al, Ti-6.8Mo-4.5Fe-1.5Al and Ti40 [11–13]. Recently, this phenomenon has been reported in some α + β and near α titanium alloys. For instance, Li et al. and Wang et al. [14,15] discovered the discontinuous flow stress drop in α + β alloy Ti-3Al-5V-5Mo and TC8, respectively. In addition, Jia et al. [10] demonstrated the behavior in near α alloy of Ti60. Generally, the discontinuous yield phenomenon could be explained mainly by two theories: static theory and dynamic theory. The first theory involved the dislocation locking and unlocking, and the second one associated discontinuous yielding with the abrupt formation of large quantities of new mobile dislocations originated from the grain boundary sources [14]. For titanium alloys, more researchers indicated that the discontinuous yielding should be attributed to dynamic theory, rather than static theory during hot compression at elevated temperatures [12,16,17].

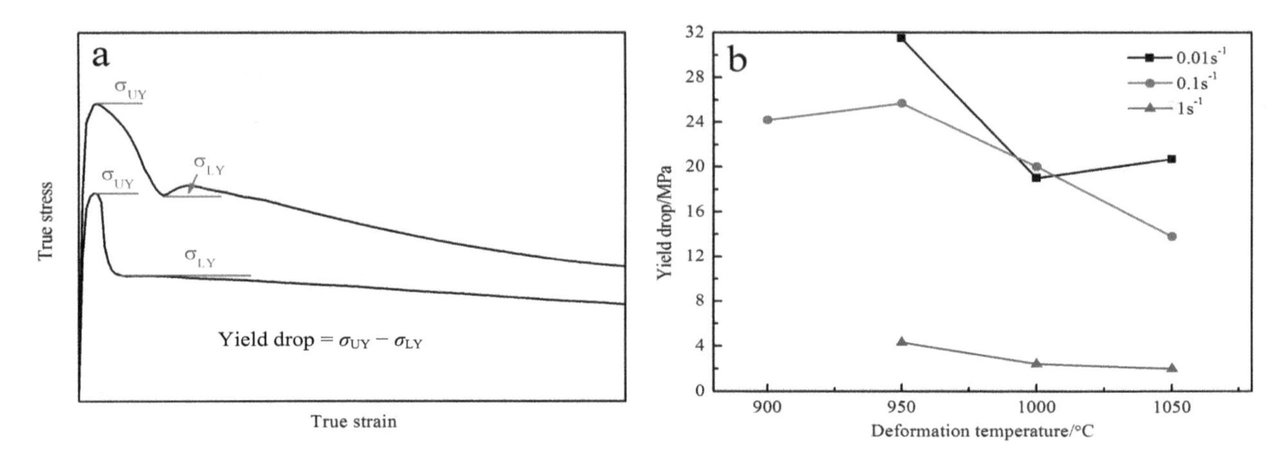

Figure 3. Schematic diagram of yield stress drop (**a**) and magnitude of yield drop in Ti-55 alloy as a function of temperature and strain rate (**b**).

However, the influence of processing parameters, such as deformation temperature and strain rate, on the discontinuous yielding was not clearly understood. For the present titanium alloy, the yield drop appeared most pronouncedly at 950 °C and reduced gradually with further increase of deformation temperature, as shown in Figure 3b. Usually, the increase of deformation temperature may induce two opposite results. On the one hand, the increase of deformation temperature could enhance the thermal activation to promote the generation of new mobile dislocations. On the other hand, the temperature rising may reduce the dislocation density and weaken the stress concentration to restrain the operation of new mobile dislocation. At lower temperature less than 900 °C, the increase of dislocation density and stress concentration could not operate new mobile dislocations significantly due to lower thermal activation and strong pinning effect of high content of alloy elements of Ti-55 alloy. With the increase of deformation temperature in β phase region (>990 °C), the reducing of dislocation density and stress concentration played a main role in restraining the generation of new mobile dislocations, leading to the decrease of the yielding drop.

Besides this, the discontinuous yield behavior weakened progressively with the increase of strain rate from 0.01 s^{-1} to 1 s^{-1} in this study. The magnitude of the yield drop changed slightly as the strain rate increased from 0.01 s^{-1} to 0.1 s^{-1}, and pronouncedly reduced to the minimum when the strain rate increased from 0.1 s^{-1} to 1 s^{-1} for all the tests at higher temperatures (>900 °C). A similar evolution tendency was found in the hot compression of Ti60 alloy at the strain rate of 1 s^{-1} [10]. The possible reason was that high strain rate induced intense work hardening, which may conceal the discontinuous yielding behavior. When the strain rate was too low (less than 0.001 s^{-1}), the dislocations were easily propagated and hard to be accumulated, which would lessen the stress concentration and lead to the disappearance of yield drop. Moreover, dynamic recovery rather than dynamic recrystallization was more prone to take place during hot compression in β phase region, which reduced dynamic softening of titanium alloy. Therefore, the yield drop at low strain rate of 0.001 s^{-1} vanished during hot deformation both in β and α + β phase regions.

3.2. Kinetic Analysis

During hot plastic deformation, the relationship among the flow stress, strain rate and deformation temperature can be described by a hyperbolic sine law [18]:

$$\dot{\varepsilon} = A[\sinh(\alpha\sigma_p)]^n \exp(-\frac{Q}{RT}). \tag{1}$$

This also can be given by the Zener–Hollomon parameter as follows [19]:

$$Z = \dot{\varepsilon}\exp(\frac{Q}{RT}) = \mathrm{A}[\sinh(\alpha\cdot\sigma_p)]^n, \tag{2}$$

where Z is the Zener–Hollomon parameter; A, α and n is materials constants, and $\alpha = \beta/n_1$, $n_1 = \partial \ln\dot{\varepsilon}/\partial \ln\sigma$, $\beta = \partial \ln\dot{\varepsilon}/\partial\sigma$; R is the gas constant; $\dot{\varepsilon}$ is the strain rate; σ_p is the peak stress; T is the deformation temperature; Q is the activation energy.

From Equation (1), the activation energy Q can be calculated as:

$$Q = R\{\frac{\partial \ln\dot{\varepsilon}}{\partial \ln[\sinh(\alpha\sigma_p)]}\}_T\{\frac{\partial \ln[\sinh(\alpha\sigma_p)]}{\partial(\frac{1}{T})}\}_{\dot{\varepsilon}}. \tag{3}$$

By linear regression of the relations of $\sigma_\mathrm{p} - \ln\dot{\varepsilon}$ and $\ln\sigma_\mathrm{p} - \ln\dot{\varepsilon}$ at different deformation conditions, the value of α was calculated as 0.00573 and 0.0269 in the temperature range of 700–950 °C and 1000–1050 °C, respectively, as shown in Figure 4a,b. Through linear fitting of $\ln[\sinh(\alpha\sigma_p)]$ vs. $\ln\dot{\varepsilon}$ and $\ln[\sinh(\alpha\sigma_p)]$ vs. $1/T$ in the temperature ranges of 700–950 °C and 1000–1050 °C shown in Figure 4c–e, the average activation energy Q of as-rolled Ti-55 alloy were calculated to be 453.00 KJ/mol and 279.88 KJ/mol, respectively. Based on the linear relationship of $\ln[\sinh(\alpha\sigma_p)]$ vs. $\ln Z$, the stress constant n of as-rolled Ti-55 alloy in the temperature range of 700–950 °C and 1000–1050 °C was calculated as 3.3851 and 2.6125, respectively, as shown in Figure 4f. Besides this, the correlation coefficient of as-rolled Ti-55 alloy in the temperature range of 700–950 °C and 1000–1050 °C for the linear relationship of $\ln Z$ vs. $\ln[\sinh(\alpha\sigma_p)]$ were 0.9150 and 0.9500, indicating it was reliable to describe the hot deformation behavior of as-rolled Ti-55 alloy by using hyperbolic sine law. Therefore, the dependence of peak stress on the strain rate and deformation temperature of as-rolled Ti-55 alloy in the temperature range of 700–900 °C and 950–1050 °C, respectively, could be expressed as:

$$\dot{\varepsilon} = 5.0129 \times 10^{18}[\sinh(5.73 \times 10^{-3}\sigma_p)]^{3.3851} \exp(-\frac{453000}{8.314T}), \tag{4}$$

$$\dot{\varepsilon} = 2.2101 \times 10^{9}[\sinh(2.69 \times 10^{-2}\sigma_p)]^{2.6125} \exp(-\frac{279880}{8.314T}). \tag{5}$$

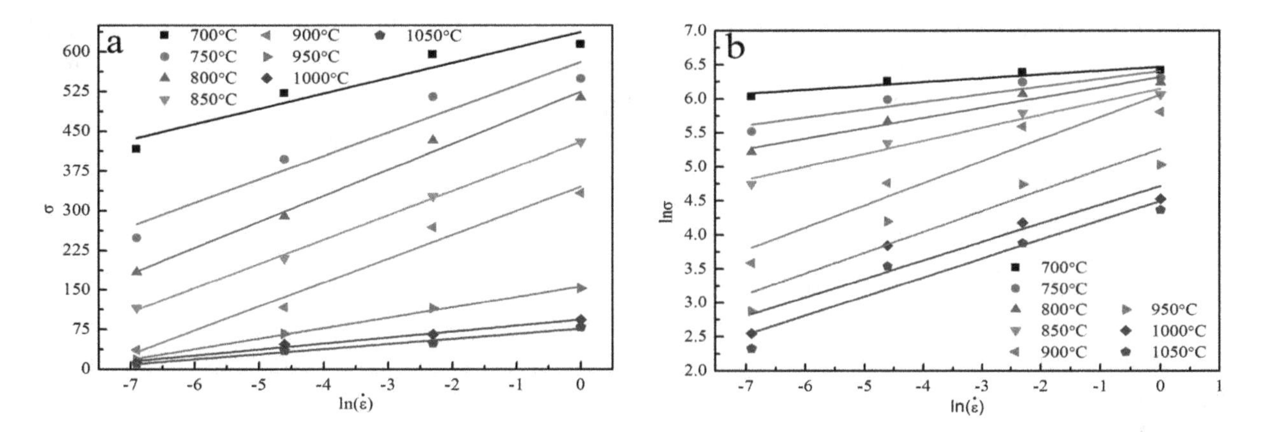

Figure 4. *Cont.*

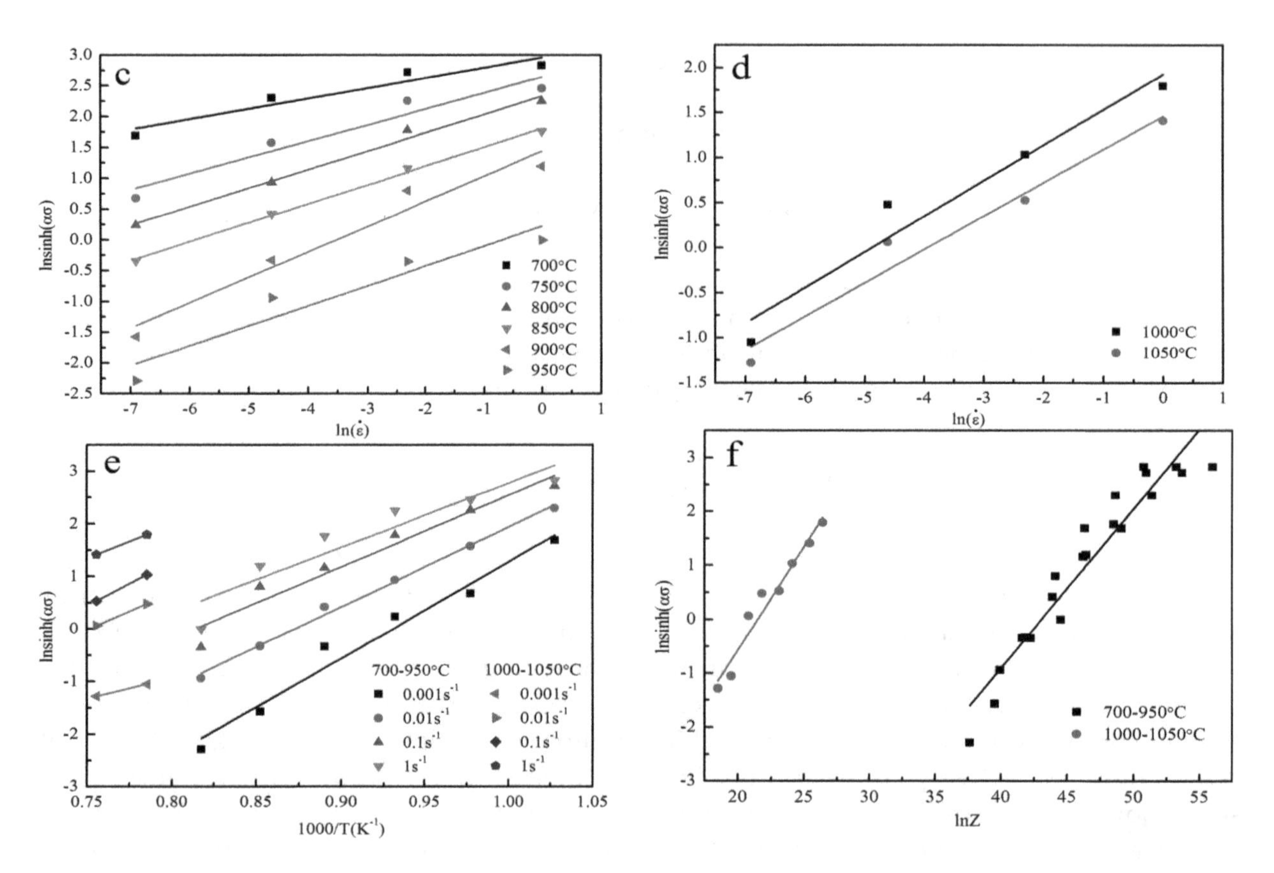

Figure 4. Linear relationships fitting of peak stress with deformation temperature and strain rate for the as-rolled Ti-55 alloy: (**a**) $\sigma_p - \ln\dot{\varepsilon}$; (**b**) $\ln\sigma_p - \ln\dot{\varepsilon}$; (**c**) and (**d**) $\ln[\sinh(\alpha\sigma_p)] - \ln\dot{\varepsilon}$; (**e**) $\ln[\sinh(\alpha\sigma_p)]$-$1/T$; (**f**) $\ln[\sinh(\alpha\sigma_p)]$-$\ln Z$.

The deformation activation energy of as-rolled Ti-55 alloy in α + β dual-phase region and β single-phase region was calculated to be 453.00 KJ/mol and 279.88 KJ/mol, respectively, both of which were greater than the lattice self-diffusion energy of α-Ti (150 KJ/mol) and β-Ti (153 KJ/mol) [4,9]. The results indicated that the main softening mechanism should be dynamic recrystallization in α + β dual-phase region and dynamic recovery in β single-phase region, respectively [20,21]. It should be noted that the activation energy in β phase region was mostly reported in the range of 180–220 kJ/mol during hot deformation of some titanium alloys [10,21–23], while the activation energy in β region of the Ti-55 alloy reached 279.88 kJ, which was higher than other titanium alloys. The possible reason is that the initial material used in this study was an as-rolled sheet, which possessed finer microstructure and intense deformation texture, so it was more difficult to deform plastically during hot compression. Besides, the activation energy in α + β dual-phase region was greater than the activation energy in β single-phase region, which should be caused by lower deformation temperature and less slip system of α phase (hexagonal close-packed structure, HCP) than β phase (body-centered cubic structure, BCC).

3.3. Microstructure Evolution and Softening Mechanism

Figure 5 shows the microstructures of the as-rolled sheet and specimens deformed at different conditions. It is evident that the strip-like microstructure of as-rolled Ti-55 alloy was elongated in the rolled direction, as shown in Figure 5a. Due to the relatively lower deformation temperature, the initial large grains were elongated along the flow direction, exhibiting obvious deformation feature under the temperature of 700 °C and strain rate of 0.01 s^{-1} (Figure 5b). When the deformation temperature increased to 800 °C, the deformed microstructure was locally globalized, indicating the occurrence of dynamically recrystallization (Figure 5c). As the deformation temperature reached 900 °C, sufficient recrystallization took place, which contributed to the refinement of initial microstructure of the as-rolled

Ti-55 alloy (Figure 5d). Besides this, the lath-shaped β-transus microstructure (see Figure 6d) appeared because of the phase transformation from α to β during hot deformation and then the re-precipitation of secondary α phase in β phase region during the cooling process. After deformation at 950 °C/0.01 s^{-1}, the amount of initial α phase further decreased obviously and the amount of β-transus microstructure further increased dramatically due to the relatively higher deformation temperature close to β-transus temperature and relatively lower strain rate (see Figure 5e). When the deformation further increased to 1000 °C in β phase region, the initial α phase totally transformed to coarsened lath-shaped β-transus microstructure (Figure 5e).

Under the same deformation temperature (900 °C), as the strain rate increased to 0.1 s^{-1}, the elongated α grains could be clearly observed (Figure 5g), indicating less sufficient dynamic recrystallization. When the strain rate decreased to 0.001 s^{-1}, the dynamically recrystallized grains grew slightly and the volume fraction of β-transus microstructure increased since longer deformation time contributed to the coarsening of recrystallized grains as well as more sufficient phase transformation of α to β during hot deformation, as shown in Figure 5h. Hence, both strain rate and deformation temperature exhibited remarkable influence on the microstructure evolution, including dynamic recrystallization and phase transformation of the as-rolled Ti-55 alloy.

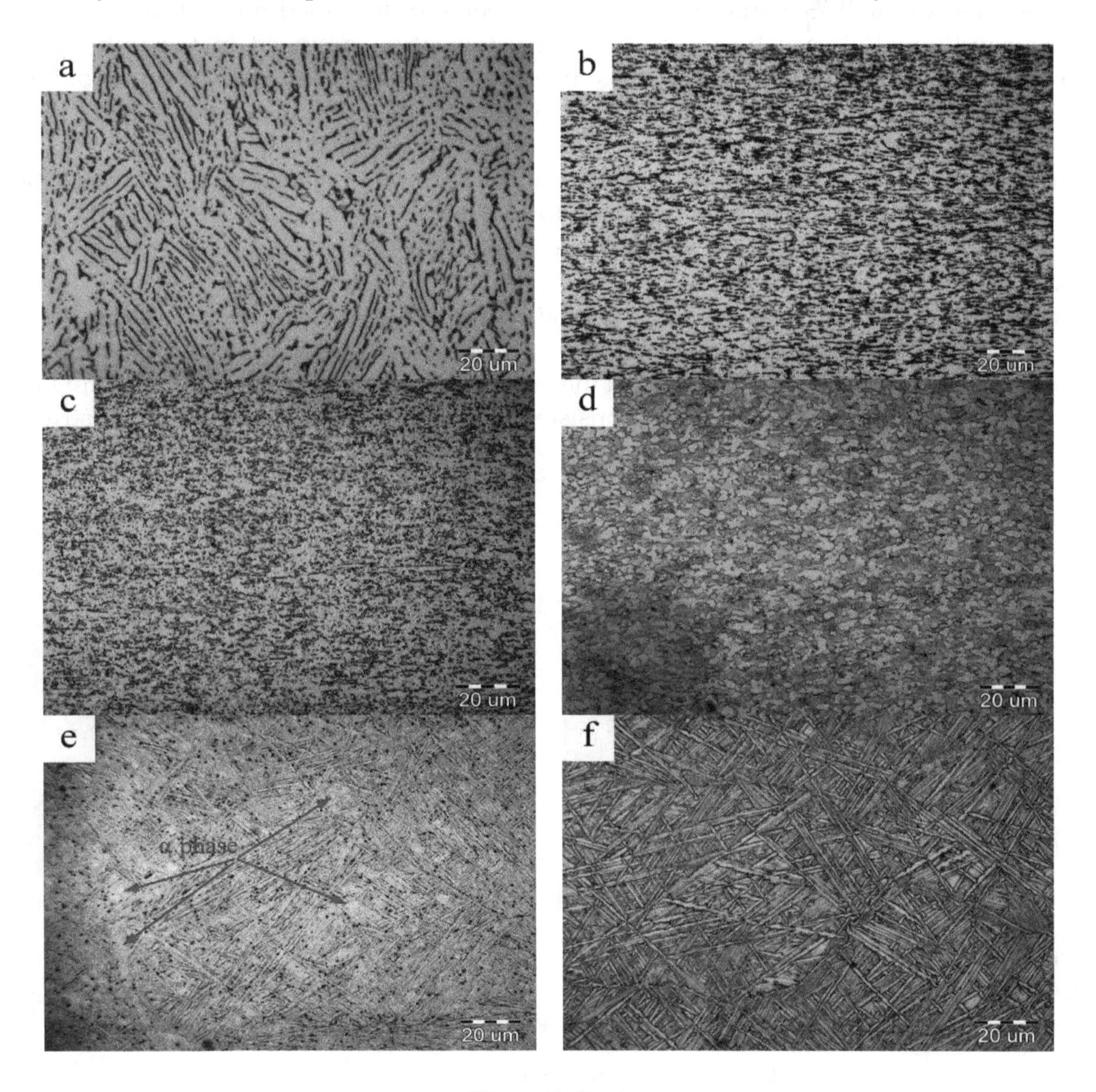

Figure 5. *Cont.*

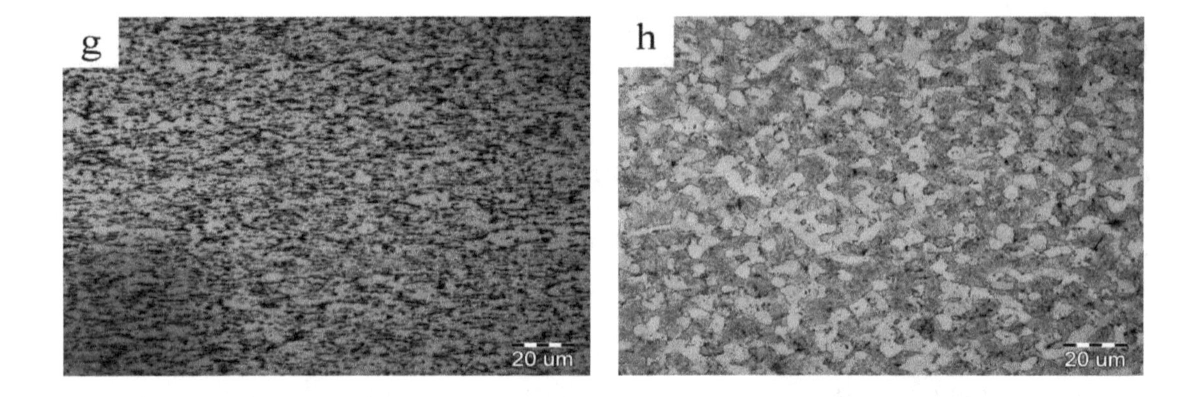

Figure 5. Microstructures of as-rolled sheet and specimens deformed at different temperatures and strain rates of the as-rolled Ti-55 alloy: (**a**) As-rolled sheet; (**b**) 700 °C/0.01 s^{-1}; (**c**) 800 °C/0.01 s^{-1}; (**d**) 900 °C/0.01 s^{-1}; (**e**) 950 °C/0.01 s^{-1}; (**f**) 1000 °C/0.01 s^{-1}; (**g**) 900 °C/0.1 s^{-1}; (**h**) 900 °C/0.001 s^{-1}.

Figure 6 shows the EBSD images of as-rolled sheet and deformed microstructure in different processing conditions. It can be found that recrystallization occurred in the initial elongated microstructure along the grain boundaries, as shown in Figure 6a. But there were a lot of low angle boundaries and small amounts of β phase retained in α phase matrix because of hot rolling deformation at relatively low temperature in the final pass. After hot compression at 800 °C, compared to the initial rolled microstructure, the fraction of low angle boundary reduced obviously and the area fraction of β phase changed slightly because of the relatively low deformation temperature below the β-transus (see Figure 6b,c). As the deformation temperature increased to 900 °C at the strain rate 0.01 s^{-1}, the fraction of low angle boundary continued to reduce, while the area fraction of β-transus microstructure increased significantly due to higher deformation temperature (see Figure 6d). Besides, with the increase of deformation temperature (Figure 6c,d) and the decrease of strain rate (Figure 6b,c), the fraction of low angle boundary decreased and the area fraction of β-transus microstructure increased, as shown in Figure 6e, which was basically consistent with the observation of OM microstructures.

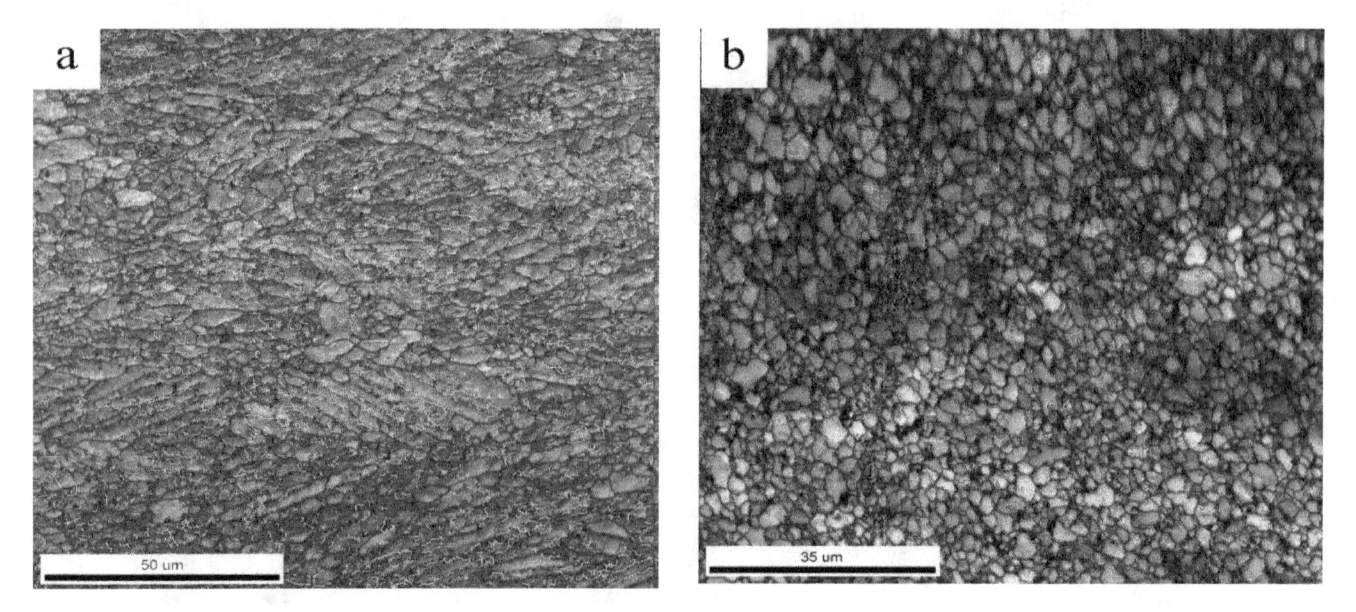

Figure 6. *Cont.*

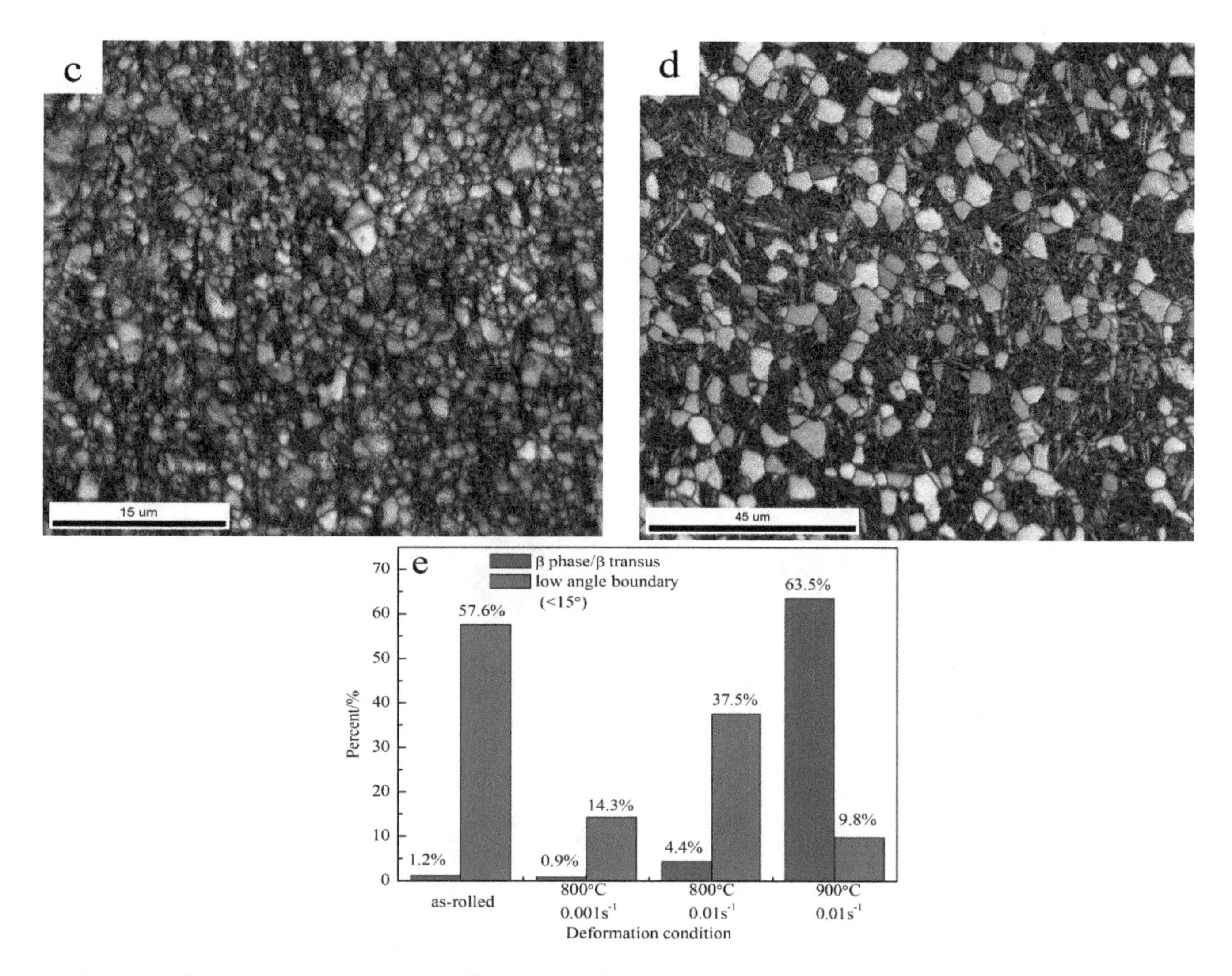

Figure 6. Electron back scattering diffraction (EBSD) images of the specimens deformed at different deformation conditions of the as-rolled Ti-55 alloy: (**a**) As-rolled; (**b**) 800 °C, 0.001 s^{-1}; (**c**) 800 °C, 0.01 s^{-1}; (**d**) 900 °C, 0.01 s^{-1}; (**e**) Percent of β phase/β transus and low angle boundary.

Figure 7 shows the TEM images of the as-rolled sheet and hot compressed microstructure at different deformation conditions. It can be seen that dynamic recrystallization occurred in the microstructure of as-rolled sheet and the dislocations density was relatively low (Figure 7a). Besides, some secondary phase particles rich in Mo, Sn, Nb and Ta elements existed at the grain boundaries, as shown in Figure 7a and Table 1. When deformed at the strain rate of 0.01 s^{-1} under the temperature lower than 900 °C, there was no β-transus microstructure occurring because of relatively low deformation temperature, as shown in Figure 7b,c. Especially at the temperature of 700 °C, the α phase was elongated perpendicular to the compression direction and amounts of dislocations existed in the titanium matrix (see Figure 7b). As the deformation temperature increased to 800 °C, some refined dislocation cells and dynamically recrystallized grains appeared, accompanied with the decrease of dislocations density, which indicated incomplete dynamic recrystallization appeared (see Figure 7c). Obviously, dynamic recrystallization proceeded more sufficiently and β-transus microstructure came into being at the temperature of 900 °C because of relatively high deformation temperature, as shown in Figure 7d. The phase transformation of α to β occurred during hot compression, and the secondary α phase re-precipitated as needle shape in the cooling process. Generally, α + β-transus microstructure was considered to be better for hot workability and mechanical property. But as the deformation temperature increased to β phase region, the needle-shaped β-transus microstructure would be coarsened to lath-typed shape and no initial α phase could be found, thus the hot workability was worsened, as shown in Figure 7e.

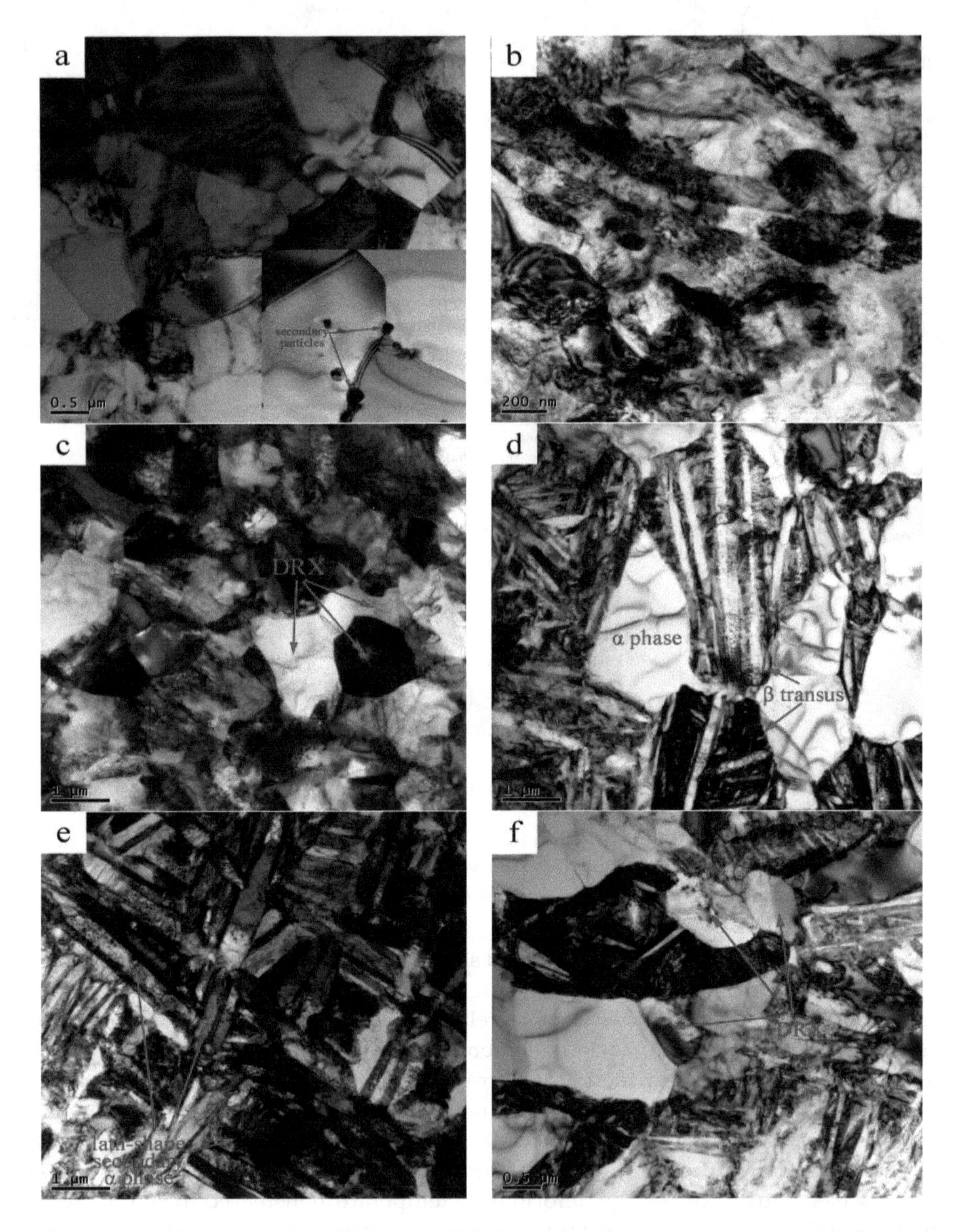

Figure 7. Transmission electron microscopy (TEM) microstructure of the specimens deformed at different deformation conditions of the as-rolled Ti-55 alloy: (**a**) As-rolled; (**b**) 700 °C/0.01 s^{-1}; (**c**) 800 °C/0.01 s^{-1}; (**d**) 900 °C/0.01 s^{-1}; (**e**) 1000 °C/0.01 s^{-1}; (**f**) 900 °C/1 s^{-1}. (DRX: Dynamic recrystallization.)

When the strain rate increased to 1 s^{-1} at the temperature of 900 °C, the volume fraction of β-transus microstructure obviously decreased, in which the re-precipitated needle-shaped α phase became finer (see Figure 7d,f). Albeit deformed at higher strain rate, dynamic recrystallization could be observed within the initial α phase due to relatively high deformation temperature, as shown in Figure 7f.

Table 1. The composition of secondary phase particles at the grain boundary.

Elements	Ti	Al	Sn	Zr	Mo	Ta	Nb	Si
Weight %	67.86	1.26	4.83	2.32	16.98	2.55	3.41	0.75
Atomic %	79.39	2.61	2.28	1.42	9.91	0.79	2.05	1.50

It should be noted that the maximum yield drop appeared at the deformation temperature of 950 °C, as shown in Figure 3, which should be ascribed to the transformation of α to β phase. Since the β-transus temperature was about 990 °C; there was a dramatic increase in β phase as the deformation temperature increased from 900 °C to 950 °C, which could be seen in Figure 5d,e. The β phase with body-centered cubic (BCC) structure possessed more operative slip systems than α phase with hexagonal close-packed (HCP) structure. Owing to high stress concentration at grain boundary of β and α phase, more mobile dislocations were easily generated to enhance the yield drop at the deformation temperature of 950 °C. When the deformation temperature increased over 1000 °C, the primary α phase totally transformed to β phase, so the vanishing of α/β interphase boundary would reduce the magnitude of yield drop.

3.4. Processing Map

3.4.1. Processing Map Theory

The processing map has been established recently by Prasad et al. [24–28] on the basis of the dynamic material model, aiming at studying the microstructure evolution and avoiding flow instability of many materials. In the dynamic materials model (DMM), the workpiece subjected to hot working is considered as a nonlinear dissipator of power. The instantaneous total power dissipation (P) at a given strain consists of two parts G and J, wherein the G represents the power dissipation for plastic deformation and J co-content is related to the power dissipation through metallurgical mechanisms, such as dynamic recovery, dynamic recrystallization and phase transformation, which can be described as a function of flow stress and strain rate:

$$P = \sigma\dot{\varepsilon} = G + J = \int_0^{\dot{\varepsilon}} \sigma d\dot{\varepsilon} + \int_0^{\sigma} \dot{\varepsilon} d\sigma. \tag{6}$$

The power dissipation characteristics of workpiece usually depend on the materials' flow behavior, which follows the power law equation:

$$\sigma = K\dot{\varepsilon}^m, \tag{7}$$

where K is the material constant; σ is the flow stress; $\dot{\varepsilon}$ is the strain rate; m is the strain rate sensitivity, by which the content G and J can be related in the phenomenological model, and can be described as follows:

$$m = \frac{dJ}{dG} = \frac{\partial(\ln\sigma)}{\partial(\ln\dot{\varepsilon})}. \tag{8}$$

The J co-content can be expressed as:

$$J = \sigma\dot{\varepsilon}\text{m}/(m+1). \tag{9}$$

For the ideal linear dissipation body, $m = 1$ and J co-content reaches to the maximum: $J_{\max} = \frac{1}{2}\sigma\dot{\varepsilon}$.

The power dissipation capacity of the material can be evaluated by the efficiency of power dissipation, η, which can be defined as:

$$\eta = \frac{J}{J_{\max}} = \frac{2m}{m+1}. \tag{10}$$

For the flow instability, Prasad developed a criterion from the extremum principle, which can be expressed as follows:

$$\xi(\dot{\varepsilon}) = \frac{\partial \ln(\frac{m}{m+1})}{\partial(\ln \dot{\varepsilon})} + m < 0. \tag{11}$$

The variation of the instability parameter $\xi(\dot{\varepsilon})$ with temperature and strain rate constitutes the instability map, from which the instability region can be obtained.

Predictably, for some materials, especially for metals with high content of alloying element and composites with high volume fraction of reinforcements, if the flow stress with respect to $\dot{\varepsilon}$ does not obey the power law in Equation (7), the computation of η and ξ in terms of m from Equations (10) and (11) becomes erroneous [29]. Hence, the DMM is further modified (MDMM) by Murty et al. [30], who suggests that the strain rate sensitivity parameter m is a variable and redefined the efficiency of power dissipation in terms of J co-content as:

$$\eta = J/J_{\max} = 2(1 - \frac{1}{\sigma\dot{\varepsilon}}\int_0^{\dot{\varepsilon}} \sigma d\dot{\varepsilon}), \tag{12}$$

$$\int_0^{\dot{\varepsilon}} \sigma d\dot{\varepsilon} = G = \int_0^{\dot{\varepsilon}_{\min}} \sigma d\dot{\varepsilon} + \int_{\dot{\varepsilon}_{\min}}^{\dot{\varepsilon}} \sigma d\dot{\varepsilon} = \left(\frac{\sigma\dot{\varepsilon}}{m+1}\right)_{\dot{\varepsilon}=\dot{\varepsilon}_{\min}} + \int_{\dot{\varepsilon}_{\min}}^{\dot{\varepsilon}} \sigma d\dot{\varepsilon}. \tag{13}$$

The condition for the metallurgical instability is given as:

$$2m < \eta. \tag{14}$$

The variation of η and ξ with deformation temperature and strain rate constitutes the power dissipation map and instability map. Hence, the processing map can be obtained through superimposing the instability map on the power dissipation map. Figure 8 shows the processing maps at the true strain of 0.8 of as-rolled Ti-55 alloy derived from different instability criteria.

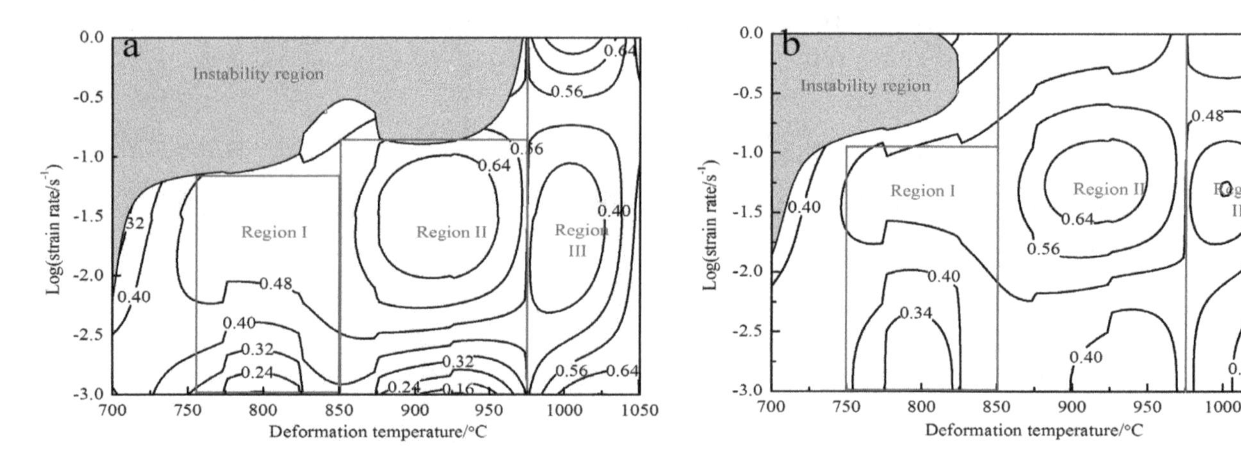

Figure 8. Processing maps at true strain of 0.8 of the as-rolled Ti-55 alloy derived from different instability criteria: (**a**) Dynamic materials model (DMM) and Prasad's instability criterion; (**b**) modified dynamic materials model (MDMM) and Murty's instability criterion.

3.4.2. Instability Region

Clearly, the flow instability region predicted by Prasad's instability criterion was located in the temperature region of 700–975 °C within the strain rate range of 0.1–1 s^{-1}, while the instability region predicted by Murty's instability criterion was significantly narrower within the temperature range of 700–825 °C and strain rate range of 0.1–1 s^{-1}, as shown in Figure 8. Generally, the mechanism of flow instability should be related to cracking or localized plastic flow [31]. Obviously, shear cracking exhibiting the orientation of ~45° with the compression direction appeared when the specimen was deformed at 700 °C/1 s^{-1}, as shown in Figure 9. In this case, dynamic softening was difficult to take place completely or even operate due to low temperature and short deformation time, which was prone to induce flow instability. However, the microstructure at 900 °C/1 s^{-1} and 900 °C/0.1 s^{-1} exhibited partial dynamic recrystallization, restraining flow instability effectively, as shown in Figures 5g and 7f. It indicated that Murty's criterion was more precise in predicting the flow instability of the as-rolled Ti-55 alloy compared to Parasad's criterion. Hence, the flow instability region of the as-rolled Ti-55 alloy was located in the temperature range of 700–825 °C and strain rate range of 0.1–1 s^{-1}, and thus the processing map of as-rolled Ti-55 alloy derived from Murty's criterion was only discussed in the following section, which was thought to have a wider application range for the type of flow stress versus strain rate curves [29,31].

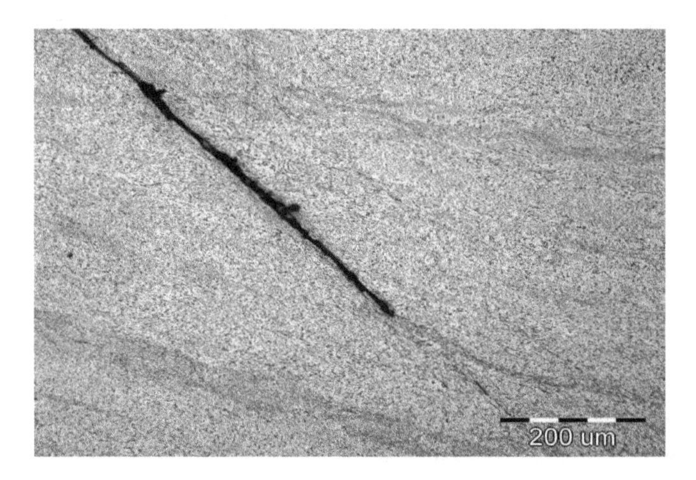

Figure 9. Microstructure of specimen deformed at 700 °C/1 s^{-1} of the as-rolled Ti-55 alloy.

3.4.3. Stability Region

Figure 8b shows the processing maps at the true strain of 0.8 of the as-rolled Ti-55 alloy derived from Murty's instability criterion. It can be seen that the processing map exhibited three higher power dissipation regions with peak dissipation efficiency of 48–64%: Region I: 750–850 °C/0.001–0.1 s^{-1}, Region II: 850–975 °C/0.001–1 s^{-1} and Region III: 975–1050 °C/0.001–1 s^{-1}. Although the stacking fault energy of the as-rolled Ti-55 alloy was relatively higher, the softening mechanism of the three steady deformation domains should be DRX because of the relatively higher dissipation efficiency of 48–64% [32]. Besides this, the occurrence of phase transformation also increased the efficiency of power dissipation [33]. Hence, the greater efficiency of power dissipation in Region II and III indicated the occurrence of phase transformation of α to β, which could be verified by the microstructure evolution during hot compression.

The microstructures of the as-rolled Ti-55 alloy at 800 °C/0.01 s^{-1} (Region I), 900 °C/0.01 s^{-1} and 950 °C/0.01 s^{-1} (Region II) and 1000 °C/0.01 s^{-1} (Region III) are shown in Figure 5c–f. Clearly, the dynamic recrystallization in Region I was the least insufficient due to relatively low deformation temperature. When deformed at 900 °C/0.01 s^{-1} (Region II), the microstructure of hot compressed specimens was consisted of α phase and β-transus microstructure, in which dynamic recrystallization took place sufficiently. Even when the strain rate increased to 1 s^{-1} at 900 °C, dynamic recrystallization still occurred relatively sufficiently (see Figure 7f). However, the coarse β grains formed and precipitated as coarsened lath-shaped secondary α microstructure after hot compression at 1000 °C/0.01 s^{-1} (Region III), which was harmful for the mechanical properties of Ti-55 alloy. Therefore, Region II (850–975 °C/0.001–1 s^{-1}) was considered to be the optimum deformation region.

4. Conclusions

The hot compression experiment of as-rolled Ti-55 alloy was conducted in the temperature range of 700–1050 °C and strain rate range of 0.001–1 s^{-1}. The hot deformation behavior and workability of the as-rolled Ti-55 alloy were studied and the optimized hot deformation parameters were obtained through analyzing microstructure evolution and establishing hot processing map, the following conclusions can be drawn:

(1) The flow stress decreased gradually with the increase of temperature and decrease of strain rate. The deformation softening mechanism was primarily controlled by DRX of α phase and DRV of β phase. The apparent activation energy Q was determined to be 453.00 KJ/mol and 279.88 KJ/mol in α + β dual-phase region and β single-phase region, respectively, which should be caused by lower deformation temperature and less slip system of α phase (HCP) than β phase (BCC). The constitutive equation for hot deformation in α + β dual-phase region and β single-phase region, respectively, was

$$\dot{\varepsilon} = 5.0129 \times 10^{18}[\sinh(5.73 \times 10^{-3}\sigma_p)]^{3.3851} \exp(-\frac{453000}{8.314T}), \tag{15}$$

and

$$\dot{\varepsilon} = 2.2101 \times 10^{9}[\sinh(2.69 \times 10^{-2}\sigma_p)]^{2.6125} \exp(-\frac{279880}{8.314T}). \tag{16}$$

(2) Discontinuous yielding behavior occurred mainly in the temperature range of 950–1050 °C and strain rate range of 0.01–1 s^{-1}. With the increase of deformation temperature, the yield drop decreased gradually due to the decrease of dislocation density and stress concentration, which restrained the generation of new mobile dislocation. Much higher strain rate could cause the reduction of yield drop since intense work hardening at higher strain rate may conceal the discontinuous yielding behavior. The increasing of β phase content could enhance the yield drop in α + β phase region and the vanishing of α/β interphase boundary may reduce the yield drop in β phase region.

(3) The processing map derived from Murty's instability criterion was more precise in predicting the hot workability of Ti-55 alloy compared to that based on Prasad's instability criterion. The processing map exhibited the optimized hot working region with sufficient dynamic recrystallization and α + β-transus microstructure: 850–975 °C/0.001–1 s^{-1}. A coarsened lath-shape β-transus microstructure was formed at higher temperature, while at lower temperature, dynamic recrystallization was not sufficient, which contributed to appearance of shear cracking at higher strain rate ($\geq$0.1 s^{-1}) and resulted in flow instability.

Acknowledgments: This work was supported by the National Natural Science Foundation of China (No.51275131).

Author Contributions: Wenchen Xu and Xueze Jin conceived and designed the experiments; Xunmao Zhong and Xingjie Wan performed the experiments; Fengyong Wu analyzed the data and wrote the paper; Bin Guo and Debin Shan provided guidance and all sorts of support during the work.

References

1. Guan, S.X.; Kang, Q.; Wang, Q.J.; Liu, Y.Y.; Li, D. Influence of long-term thermal exposure on the tensile properties of a high-temperature titanium alloy Ti-55. *Mater. Sci. Eng. A* **1998**, *243*, 182–185. [CrossRef]
2. Fang, B.; Chen, Z.Y.; Chen, Z.Q.; Liu, J.H.; Wang, Q.J.; Liu, Y.; Feng, Z.W.; Liu, J.R.; Song, X.Y.; Wei, M.X.; et al. Continuous cooling transformation diagram and microstructure evolution of Ti-55 alloy. *Chin. J. Nonferrous Met.* **2010**, *20*, s32–s35.
3. Li, X.F.; Jiang, J.; Wang, S.; Chen, J.; Wang, Y.Q. Effect of hydrogen on the microstructure and superplasticity of Ti-55 alloy. *Int. J. Hydrogen Energy* **2017**, *42*, 6338–6349. [CrossRef]
4. Wanjara, P.; Jahazi, M.; Monajati, H.; Yue, S.; Immarigeon, J.P. Hot working behavior of near-α alloy IMI834. *Mater. Sci. Eng. A* **2005**, *396*, 50–60. [CrossRef]

5. Wanjara, P.; Jahazi, M.; Monajati, H.; Yue, S. Influence of thermomechanical processing on microstructural evolution in near-α alloy IMI834. *Mater. Sci. Eng. A* **2006**, *416*, 300–311. [CrossRef]
6. Liu, Y.; Baker, T.N. Deformation characteristics of IMI685 titanium alloy under β isothermal forging conditions. *Mater. Sci. Eng. A* **1995**, *197*, 125–131. [CrossRef]
7. Niua, Y.; Houb, H.L.; Li, M.Q.; Li, Z.Q. High temperature deformation behavior of a near alpha Ti600 titanium alloy. *Mater. Sci. Eng. A* **2008**, *492*, 24–28. [CrossRef]
8. Liu, Z.G.; Li, P.J.; Xiong, L.T.; Liu, T.Y.; He, L.J. High-temperature tensile deformation behavior and microstructure evolution of Ti-55 titanium alloy. *Mater. Sci. Eng. A* **2017**, *680*, 259–269. [CrossRef]
9. Seshacharyulu, T.; Medeiros, S.C.; Frazier, W.G.; Prasad, Y.V.R.K. Hot working of commercial Ti-6 Al-4 V with an equiaxed α-β microstructure: Materials modeling considerations. *Mater. Sci. Eng. A* **2000**, *284*, 184–194. [CrossRef]
10. Jia, W.J.; Zeng, W.D.; Zhou, Y.G.; Liu, J.R.; Wang, Q.J. High-temperature deformation behavior of Ti60 titanium alloy. *Mater. Sci. Eng. A* **2011**, *528*, 4068–4074. [CrossRef]
11. Balasubrahmanyam, V.V.; Prasad, Y.V.R.K. Deformation behaviour of beta titanium alloy Ti-10 V-4.5 Fe-1.5 Al in hot upset forging. *Mater. Sci. Eng. A* **2002**, *336*, 150–158. [CrossRef]
12. Philippart, I.; Rack, H.J. High temperature dynamic yielding in metastable Ti-6.8 Mo-4.5 F-1.5 Al. *Mater. Sci. Eng. A* **1998**, *243*, 196–200. [CrossRef]
13. Zhu, Y.C.; Zeng, W.D.; Zhao, Y.Q.; Shu, Y.; Zhang, X.M. Effect of processing parameters on hot deformation behavior and microstructural evolution during hot compression of Ti40 titanium alloy. *Mater. Sci. Eng. A* **2012**, *552*, 384–391. [CrossRef]
14. Li, L.X.; Lou, Y.; Yang, L.B.; Peng, D.S.; Rao, K.P. Flow stress behavior and deformation characteristics of Ti-3 Al-5 V-5 Mo compressed at elevated temperatures. *Mater. Des.* **2002**, *23*, 451–457. [CrossRef]
15. Wang, K.; Li, M.Q. Characterization of discontinuous yielding phenomenon in isothermal compression of TC8 titanium alloy. *Trans. Nonferrous Met. Soc. China* **2016**, *26*, 1583–1588. [CrossRef]
16. Robertson, D.G.; McShane, H.B. Isothermal hot deformation behaviour of (α + β) titanium alloy Ti-4 Al-4 Mo-2 Sn-0.5 Si (IMI 550). *Mater. Sci. Technol.* **1997**, *13*, 459–468. [CrossRef]
17. Fan, J.K.; Kou, H.C.; Lai, M.J.; Tang, B.; Chang, H.; Li, J.S. High Temperature Discontinuous Yielding in a New Near β Titanium Alloy Ti-7333. *Rare Metal Mater. Eng.* **2014**, *43*, 0808–0812. [CrossRef]
18. He, G.A.; Liu, F.; Si, J.Y.; Yang, C.; Jiang, L. Characterization of hot compression behavior of a new HIPed nickel-based P/M superalloy using processing maps. *Mater. Des.* **2015**, *87*, 256–265. [CrossRef]
19. Zener, C.; Hollomon, J.H. Effect of strain rate upon plastic flow of steel. *J. Appl. Phys.* **1944**, *15*, 22–32. [CrossRef]
20. Wang, K.L.; Lu, S.Q.; Fu, M.W.; Li, X.; Dong, X.J. Identification of the optimal (α + β) forging process parameters of Ti-6.5 Al-3.5 Mo-1.5 Zr-0.3 Si based on processing-maps. *Mater. Sci. Eng. A* **2010**, *527*, 7279–7285. [CrossRef]
21. Zhao, H.Z.; Xiao, L.; Ge, P.; Sun, J.; Xi, Z.P. Hot deformation behavior and processing maps of Ti-1300 alloy. *Mater. Sci. Eng. A* **2014**, *604*, 111–116. [CrossRef]
22. Li, M.Q.; Pan, H.S.; Lin, Y.Y.; Luo, J. High temperature deformation behavior of near alpha Ti-5.6 Al-4.8 Sn-2.0 Zr alloy. *J. Mater. Process. Tech.* **2007**, *183*, 71–76. [CrossRef]
23. Seshacharyulu, T.; Medeiros, S.C.; Morgan, J.T.; Mala, J.C.; Frazier, W.G.; Prasad, Y.V.R.K. Hot deformation mechanisms in ELI grade Ti-6 A1–4 V. *Scripta Mater.* **1999**, *41*, 283–288. [CrossRef]
24. Dong, Y.Y.; Zhang, C.S.; Zhao, G.Q.; Guan, Y.J.; Gao, A.J.; Sun, W.C. Constitutive equation and processing maps of an Al-Mg-Si aluminum alloy: Determination and application in simulating extrusion process of complex profiles. *Mater. Des.* **2016**, *92*, 983–997. [CrossRef]
25. He, D.G.; Lin, Y.C.; Chen, M.S.; Chen, J.; Wen, D.X.; Chen, X.M. Effect of pre-treatment on hot deformation behavior and processing map of an aged nickel-based superalloy. *J. Alloy Compd.* **2015**, *649*, 1075–1084. [CrossRef]
26. Zhao, Z.L.; Li, H.; Fu, M.W.; Guo, H.Z.; Yao, Z.K. Effect of the initial microstructure on the deformation behavior of Ti60 titanium alloy at high temperature processing. *J. Alloy Compd.* **2014**, *617*, 525–533. [CrossRef]
27. Xia, X.S.; Chen, Q.; Li, J.P.; Shu, D.Y.; Hu, C.K.; Huang, S.H.; Zhao, Z.D. Characterization of hot deformation behavior of as-extruded Mg-Gd-Y-Zn-Zr alloy. *J. Alloy Compd.* **2014**, *610*, 203–211. [CrossRef]
28. Liu, J.; Cui, Z.S.; Li, C.X. Analysis of metal workability by integration of FEM and 3-D processing maps. *J. Mater. Process. Technol.* **2008**, *205*, 497–505. [CrossRef]

29. Gupta, R.K.; Narayana Murtya, S.V.S.; Panta, B.; Agarwalab, V.; Sinha, P.P. Hot workability of $\gamma+\alpha_2$ titanium aluminide: Development of processing map and constitutive equations. *Mater. Sci. Eng. A* **2012**, *551*, 169–186. [CrossRef]
30. Prasad, Y.V.R.K.; Gegel, H.L.; Doraivelu, S.M.; Malas, J.C.; Morgan, J.T.; Lark, K.A.; Barker, D.R. Modeling of Dynamic Material Behavior in Hot Deformation: Forging of Ti-6242. *Metall. Trans. A* **1984**, *15*, 1884–1891. [CrossRef]
31. Narayana Murty, S.V.S.; Nageswara Rao, B. Instability map for hot working of 6061 Al-10 vol% Al_2 O_3 metal matrix composite. *J. Phys. D* **1998**, *31*, 3306–3311. [CrossRef]
32. Xia, X.S.; Chen, Q.; Zhang, K.; Zhao, Z.D.; Ma, M.L.; Li, X.G.; Li, Y.J. Hot deformation behavior and processing map of coarse-grained Mg-Gd-Y-Nd-Zr alloy. *Mater. Sci. Eng. A* **2013**, *587*, 283–290. [CrossRef]
33. Kong, F.T.; Cui, N.; Chen, Y.Y.; Wang, X.P.; Xiong, N.N. Characterization of hot deformation behavior of as-forged TiAl alloy. *Intermetallics* **2014**, *55*, 66–72. [CrossRef]

Reaction Layer Analysis of In Situ Reinforced Titanium Composites: Influence of the Starting Material Composition on the Mechanical Properties

Isabel Montealegre-Meléndez [1], Cristina Arévalo [1,*], Ana M. Beltrán [1], Michael Kitzmantel [2], Erich Neubauer [2] and Eva María Pérez Soriano [1]

[1] Escuela Politécnica Superior, Universidad de Sevilla, Calle Virgen de África, 7, 41011 Sevilla, Spain; imontealegre@us.es (I.M.-M.); abeltran3@us.es (A.M.B.); evamps@us.es (E.M.P.S.)
[2] RHP Technology GmbH, 2444 Seibersdorf, Austria; m.ki@rhp.at (M.K.); e.ne@rhp.at (E.N.)
* Correspondence: carevalo@us.es

Abstract: This study aims at the analysis of the reaction layer between titanium matrices and reinforcements: B_4C particles and/or intermetallic Ti_xAl_y. Likewise, the importance of these reactions was observed; this was particularly noteworthy as regard coherence with the obtained results and the parameters tested. Accordingly, five starting material compositions were studied under identical processing parameters via inductive hot pressing at 1100 °C for 5 min in vacuum conditions. The results revealed how the intermetallics limited the formation of secondary phases (TiC and TiB) created from the B and C source. In this respect, the percentages of TiB and TiC slightly varied when the intermetallic was included in the matrix as prealloyed particles. On the contrary, if the intermetallics appeared in situ by the addition of Ti-Al powder in the starting blend, their content was lesser. The mechanical properties values and the tribology behaviour might deviate, depending on the percentage of the secondary phases formed and its distribution in the matrix.

Keywords: in situ composites; inductive hot pressing; tribology; titanium composites; reaction layer; intermetallics

1. Introduction

The interest of materials with high specific properties and good tribological behaviour raises the needs in terms of improving the properties of materials like titanium and its alloys. In the fields of aerospace, military industry and biomedicine, the use of these materials is very popular [1–4]; however, nowadays, there are limitations regarding their mechanical and poor tribological properties. Therefore, titanium matrix composites (TMCs) are valuated as materials that combine low density with mechanical properties [5–8]. Presently, there is a great variety of reinforcements employed in these composites. The most popular ones showed in diverse research works tend to increase the hardness, Young's modulus and the tribological behaviour without incrementing the density. As examples, it may include TiB_2, TiB, TiC, B, B_4C, carbon nanotubes, graphite and nanodiamonds [9–12]. Among these materials, B_4C ceramic particles have been presented in investigations as an optimal form to obtain B and C sources to origin in situ TiB and TiC [13–16]. In this regard, these secondary phases play a key role in the strengthening of the titanium matrices [17]. The main advantage of these in situ phases resides in the existence of a good interfacial bonding of the reinforcement-matrix. Moreover, the ceramic reinforcements formed during in -situ processing are finer, more thermodynamically stable and uniform in size distribution in the metal matrix [18,19]. Therefore, the study of the reaction layer between the matrix and the particles is important in order to achieve a better understanding of the cause that could promote or not these reactions and their products. While there is considerable literature on the grounds of strengthening and the reaction mechanisms [20,21], there are only a few studies where the reaction layer is analysed. It is also investigated how the presence of Ti_xAl_y as intermetallic in the titanium matrices could affect the final appearance of these secondary phases (TiB and TiC) [22–24]; however, the reaction layer in the presence of intermetallic, as well as its decomposition, has been little studied [25,26]. Hence, here lies the importance for understanding and determining the evolution of the reaction layer with B_4C particles and intermetallics Ti_xAl_y and how these layers and the final properties of the consolidated composites could be affected by the combination of the starting materials. Therefore, this research aims not only to study the reaction layer but also to investigate and characterise TMCs from five different blends. These blends were designed and processed considering interesting combinations of starting powders, in order to observe the regarding properties and the above-mentioned reaction layer phenomenon. Through the selection of the diverse intermetallic starting powders, variations in the behaviour of the TMCs may be expected.

In several investigations concerning the manufacture of titanium composites via powder metallurgy (PM) [27,28], inductive hot pressing (iHP) is considered as a suitable fabrication option due to its flexibility and short cycles. For that reason, and thanks to the experience of the authors on TMCs manufacturing, the fabrication route of the specimens was through iHP. In preliminary studies [29,30], an inflexion temperature, at which secondary phases were formed in situ, was observed. The analysis of the reaction layer has been carried out in specimens produced at this inflexion temperature (1100 °C) in vacuum conditions.

Hence, the five specimens were in detail characterised through scanning and scanning-transmission electron microscopies (SEM and (S)TEM) and by X-ray diffraction analysis (XRD). Furthermore, their tribological and physical properties were measured and evaluated. In this regard, the relation between the starting materials and the final properties was thoroughly investigated.

2. Materials and Methods

2.1. Materials

The starting blends were five different combinations of powders, using only one titanium matrix powder, as listed in Table 1.

Table 1. Summary of the composition of the tested specimens.

Blend	Materials	B_4C [Volume %]	Ti:Al [Volume %]
1	Ti + B_4C	30	-
2	Ti + B_4C + Ti-Al(1)	30	20
3	Ti + Ti-Al(1)	-	20
4	Ti + B_4C + Ti-Al(2)	30	20
5	Ti + Ti-Al(2)	-	20

The powders employed in this research were previously characterised through particle size distribution and morphological analysis, in order to verify their supplied information. The Mastersizer 2000 (Malvern Instruments, Malvern, UK) equipment was used to determinate the average particle size, and their morphology study was performed by SEM from FEI Teneo images (FEI, Eindhoven, Netherlands).

Titanium powder was produced by TLS GmbH (Bitterfeld, Germany); it showed a spherical morphology in agreement with the manufacturer's information, being its mean diameter of 109 μm. The B_4C particles were manufactured by ABCR GmbH & Co KG (Karlsruhe, Germany) with an average size diameter of 64 μm; the morphology of these ceramic particles was irregular and with slightly sharp edges.

Concerning the intermetallic powders, there were substantial differences between the two selected. The one named Ti-Al(1) was a prealloyed Ti_3Al and $TiAl_3$ intermetallic powder manufactured by TLS

GmbH (Bitterfeld, Germany). This Ti-Al(1) showed a spherical morphology, and its mean diameter was around 75 μm. The other intermetallic powder used was named Ti-Al(2). It was made from a blend of elementary Al (NMD GmbH, Heemsen, Germany; 9 μm) and Ti fine powder with a mean diameter of 29 μm in molar ratio 1:1 by TLS GmbH (Bitterfield, Germany) [22,30].

In addition to the starting materials' characterisations, XRD analysis was accomplished—Bruker D8 Advance A25 (Billerica, MA, USA) with Cu-K_α radiation. Figure 1 shows the patterns of the powders from which the five blends were prepared. Based on the resulted diffraction spectra, the composition of the powders agreed with the suppliers' information in the case of Ti powder, B_4C and Ti-Al(1) in Figure 1a–c, respectively. In the case of Ti-Al(2), the remarked peaks corresponded to the Ti and Al elements, matching the elementary blending of Ti and Al, whose molar ratio was 1:1 (Ti:Al), as it was above-mentioned.

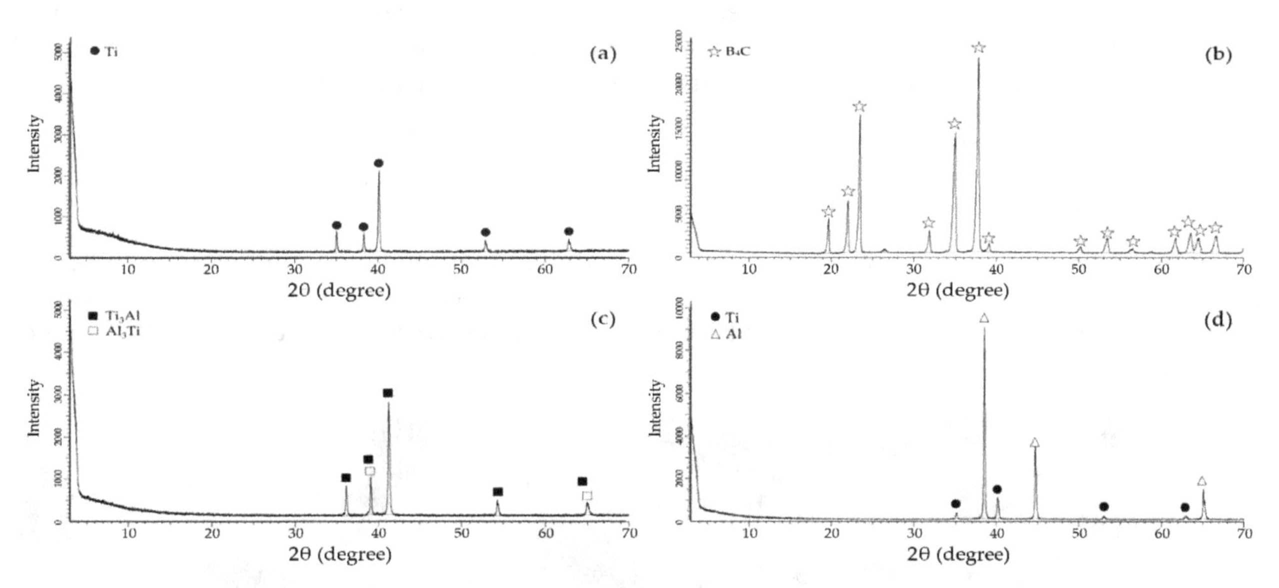

Figure 1. XRD (X-ray diffraction) patterns of the starting powders: (**a**) Ti powder, (**b**) B_4C particles, (**c**) Ti-Al(1) powder and (**d**) Ti-Al(2) powder.

2.2. Methods

After the blending preparation, similar as in previous authors' works [30], the consolidation of the specimens via iHP was performed, employing a self-made inductive hot-pressing machine; equipment from RHP-Technology GmbH (Seibersdorf, Austria).

This machine provided the time of the operational cycles to be reduced thanks to its advantageous high heating rate. The time set was 5 min in vacuum conditions at 1100 °C and 80 MPa. The same procedure to fill the die was carried out for the consolidation of each specimen [30]. The graphite die for all the iHP cycles had a diameter of 20 mm. A detailed description of the manufacturing process was reported in previous authors' work [22].

The consolidated composites were studied at length. After thorough metallographic preparation, their microstructures were examined by SEM at 15 kV. The (S)TEM characterisation was carried out using a FEI Talos F200S microscope (FEI, Eindhoven, Netherlands) operating at an accelerating voltage of 200 kV and equipped with a super-X energy-dispersive X-ray spectrometry (EDX) system, which includes two silicon drift detectors. The elemental mapping experiments were accomplished by combining high-angle annular dark-field imaging (HAADF) and EDX acquisition in (S)TEM mode. The mechanical grinding and ion milling procedures were performed for the (S)TEM studies following standard procedure for TEM lamella preparation. Moreover, an XRD analysis was conducted to identify the diverse crystalline phases in the TMCs. Ultrasonic method (Olympus 38 DL, Tokyo, Japan) was used to calculate Young's modulus by measuring longitudinal and transverse propagation velocities of acoustic waves [31]. The densification of the specimens was measured according to the Archimedes' method [32]. The relative density was computed as the ratio of the compacts' density to their theoretical values of the given material, determined by the rule of mixtures. A test model, Struers-Duramin A300 (Ballerup, Germany), was used to ascertain the Vickers hardness (HV2). The hardness measurements took place on the polished cross-section of the specimens. The reported values were the average of eight indentations.

After the characterisation of the specimens, the tribological behaviour was studied. Before running the tests, the specimens were prepared (grinded and polished), cleaned with acetone in an ultrasonic bath and, after that, dried. The wear behaviour of the TMCs was conducted in a ball-on-disc tribometer (Microtest MT/30/NI, Madrid, Spain) using alumina balls with a diameter of 6 mm. At room temperature, the normal load of 3 N on the ball with a sliding speed of 125 mm/s was employed to measure the wear properties. It was tested at a sliding distance of 500 m on the specimens' surface, with a circular path of 3 mm in radius. The results were analysed at similar conditions in order to compare and evaluate the influence of the starting materials. The morphology of the worn surfaces was characterised by optical microscopy with a Leica Zeiss DMV6 (Leica Microsystems, Heerbrugg, Switzerland).

3. Results and Discussion

3.1. *Microstructural Study and XRD Analysis*

The microstructural study revealed significant differences related to the starting powders employed. Figure 2 displayed a microstructure general overview of the specimens. The circular backscattered (CBS) SEM images in Figure 2a,b,d showed how the B_4C particles were homogeneously distributed in the titanium matrices.

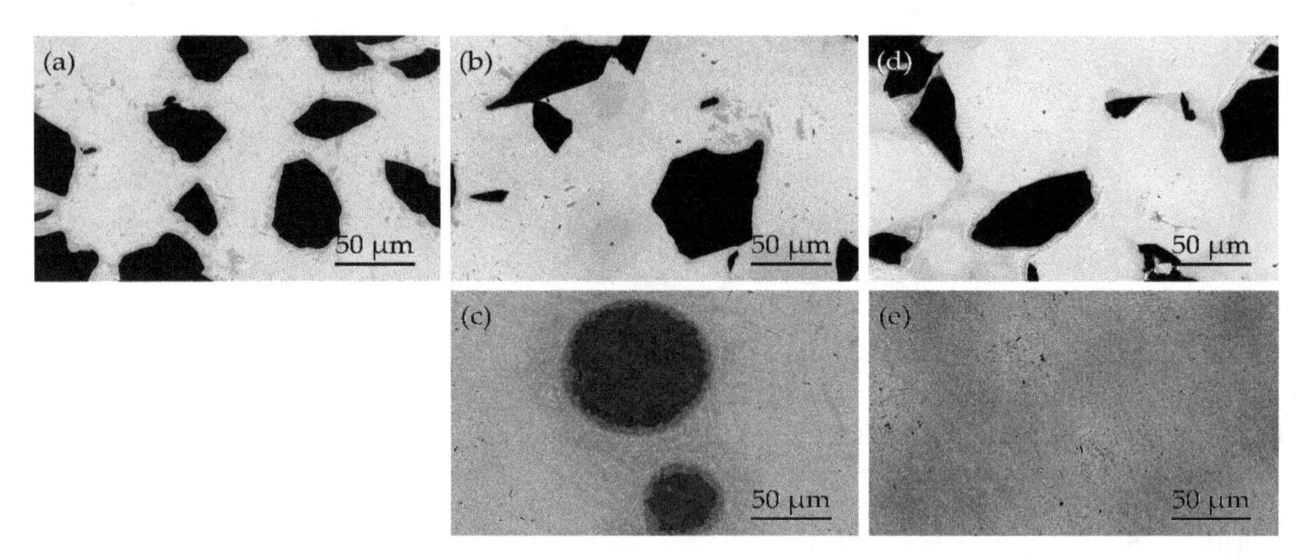

Figure 2. CBS-SEM (circular backscattered-scanning electron microscopies) images: (**a**) Ti + 30 vol. % B_4C, (**b**) Ti + 30 vol. % B_4C + 20 vol. % Ti-Al(1), (**c**) Ti + 20 vol. % Ti-Al(1), (**d**) Ti + 30 vol. % B_4C + 20 vol. % Ti-Al(2) and (**e**) Ti + 20 vol. % Ti-Al(2).

The different shades of grey close to these ceramic particles suggested the presence of in situ secondary phases. However, at this scale, that was not easily appreciated. Conversely, in the case of the specimen made from Ti only with Ti-Al(1) (Figure 2c), the dark grey areas corresponded to the intermetallic phases (Figure 2c). The intermetallics in the matrix were visible and recognised as spherical precipitates with degradation around. However, if there were Ti-Al powders employed in the starting mixing as Ti-Al(2), the location of the phases rich in Al was not appreciated (Figure 2e). Although, in the bibliography has been reported the formation of new phases due to the liquid metal dealloying reaction [33,34], in this specimen fabricated at this temperature, it seems that no intermetallics were in situ formed, having the aluminium in solid solution with the titanium matrix. These observations were verified by the XRD analysis.

The microstructural characterisation at higher magnification is presented in Figure 3, revealing several phases depending on the starting materials. Evaluating the area surrounding the B_4C particles, there were significant variations in the volume and thickness of the in situ secondary phases formed in the reaction layer. In the TMC without intermetallic addition, more in situ secondary phases were formed (Figure 3a). The incorporation of Ti_xAl_y in the starting powders of the TMCs not only modified the microstructure of the specimens but also affected the secondary phases formed, whereupon the final properties of the specimens could suffer variations. A detailed study of the TMCs' microstructure evolution contributed to a better understanding of in situ formed phases, as well as the distribution of the Ti_xAl_y in the titanium matrices. In Figure 3a,b,d, the reaction layer around the B_4C could be observed, being mainly composed of B and TiB. These TiB phases could also appear further from the B_4C particles when no intermetallic was used, as well as TiC (Figure 3a,b). It suggested that the employ of Ti-Al intermetallic in the initial blending involved a slight obstacle in the origin of in situ formed phases. The Al could act as a barrier, blocking the reactions between the Ti and the B and C sourced by the B_4C particles. This phenomenon was more visible in the sample with Al as an elementary powder (TiAl(2), Figure 3d). The morphology of in situ phases will be discussed later.

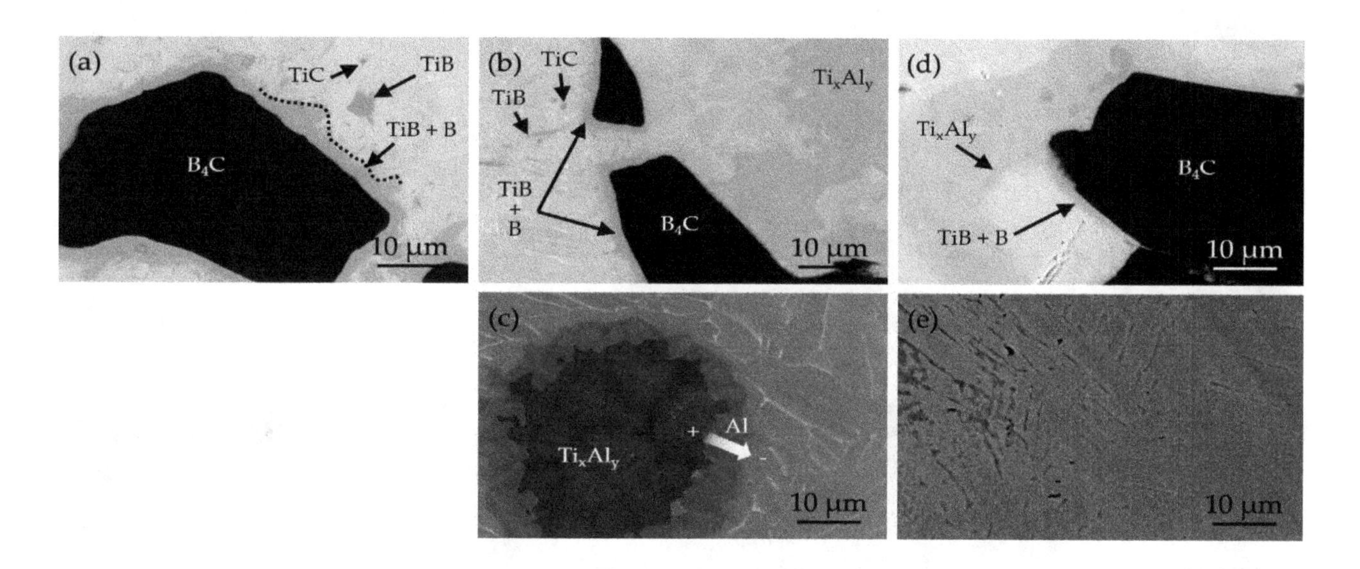

Figure 3. Phases identification via CBS-SEM images: (**a**) Ti + 30 vol. % B_4C, (**b**) Ti + 30 vol. % B_4C + 20 vol. % Ti-Al(1), (**c**) Ti + 20 vol. % Ti-Al(1), (**d**) Ti + 30 vol. % B_4C + 20 vol. % Ti-Al(2) and (**e**) Ti + 20 vol. % Ti-Al(2).

As previously mentioned, when no B_4C was added, how introducing Ti_xAl_y in the matrix caused differences. This can be clearly seen in Figure 3c,e. There was a reaction layer surrounding the intermetallic phases when prealloyed intermetallic was introduced as a starting powder (TiAl(1)). This reaction layer was due to the decomposition of the intermetallic being Al-incorporated into the matrix.

In order to understand TMCs with B_4C particles in detail, the reaction layer and secondary phases were analysed by TEM and (S)TEM (Figures 4–6). In the reaction layer, the in situ formed precipitates could be recognised. The more the presence of Al in the matrix, the thicker the reaction layer was observed. The reason, as previously discussed, could be the difficulted diffusion of B and C elements into the matrix. Next to the B_4C particles, the size of TiB and TiC in situ secondary phases created were smaller to grow as moving away from the ceramic particles. When Al was not present in the matrix, secondary phases were bigger in size (Figure 4a).

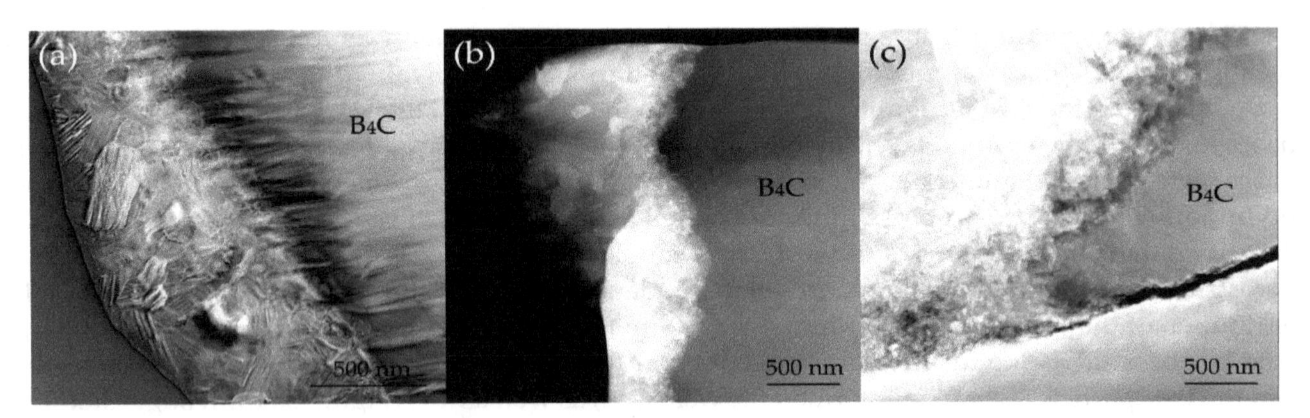

Figure 4. TEM (transmission electron microscopies) images: (**a**) Ti + 30 vol. % B_4C, (**b**) Ti + 30 vol. % B_4C + 20 vol. % Ti-Al(1) and (**c**) Ti + 30 vol. % B_4C + 20 vol. % Ti-Al(2).

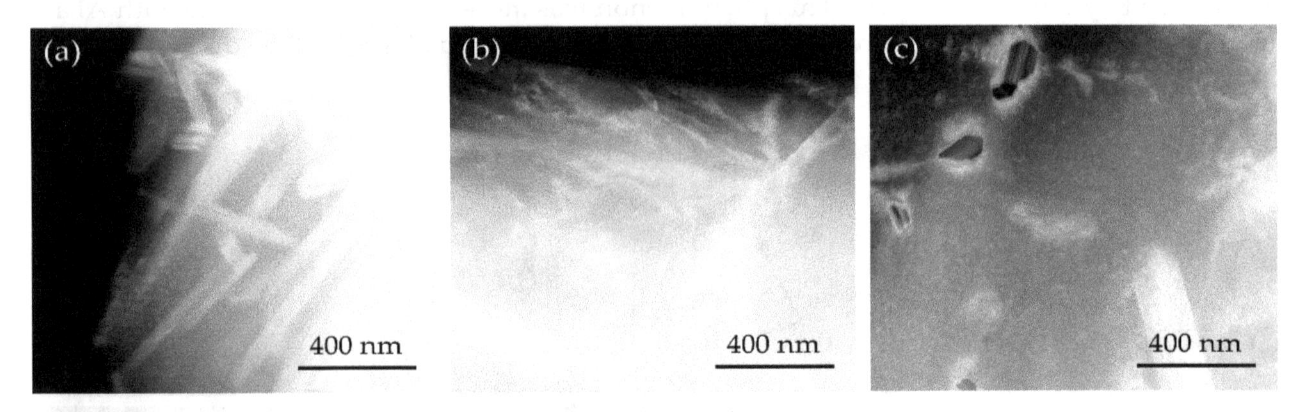

Figure 5. (S)TEM images 80,000x: (**a**) Ti + 30 vol. % B_4C, (**b**) Ti + 30 vol. % B_4C + 20 vol. % Ti-Al(1) and (**c**) Ti + 30 vol. % B_4C + 20 vol. % Ti-Al(2).

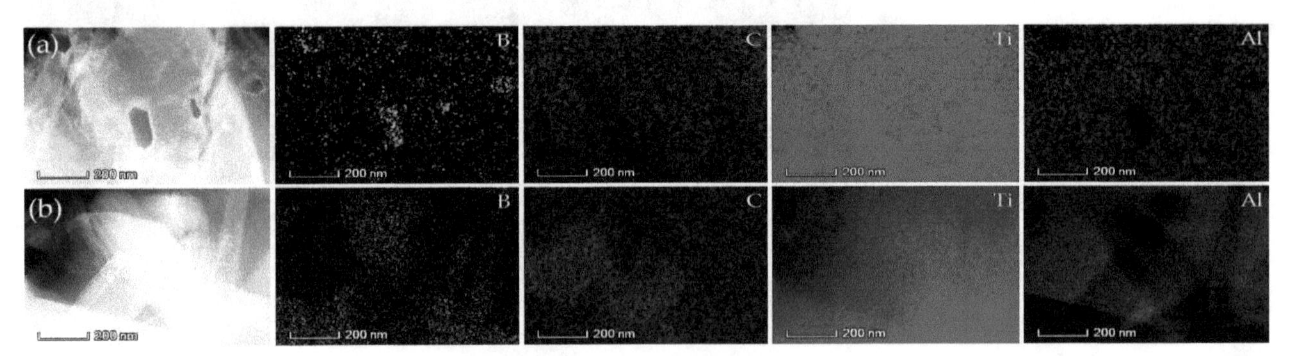

Figure 6. TEM images and compositional mappings: (**a**) Ti + 30 vol. % B_4C + 20 vol. % Ti-Al(1) and (**b**) Ti + 30 vol. % B_4C + 20 vol. % Ti-Al(2).

In Figures 4 and 5, the observed phases consisted in TiB and TiC were recognised due to their well-known morphology, as whiskers in the case of TiB and as isolated grey areas for TiC from its dendritic formation [30]. In TMCs reinforced only by B_4C, the morphologies of TiB and TiC phases could be more easily identified (Figure 4a), while for TMCs with B_4C and intermetallic, these precipitates were not clearly observed (Figure 4b,c). Figure 5 shows the formation of precipitates far from the B_4C particles. The lower the content of Al as a solid solution in the matrix, the higher the concentration of secondary phases in these areas was found. In this way, this was more significant with Ti-Al(2) (Figure 5c) than with Ti-Al(1) (Figure 5b), confirming SEM microscopy.

Concluding the microstructural characterisation, a compositional mapping of the main present elements in samples with B_4C and intermetallic was performed (Figure 6). The aim was to study the composition of the smallest precipitates to check if there were other formed secondary phases, apart from the expected TiB and TiC. The results confirmed that there was no reaction between the B and C particles with Al.

The XRD patterns of the specimens are illustrated in Figure 7. The one of TMCs without intermetallic (Figure 7a) was intended to serve as a reference on the comparison in terms of precipitates formation regarding the specimens with Ti-Al(1) (Figure 7b) and Ti-Al(2) (Figure 7d). The peaks of B_4C could clearly be identified in these three patterns. The variations in the peaks of the in situ formed TiB and TiC that were related to the intermetallic powder employed as the starting powder were meaningful in the results. It suggested that the reaction of the matrix and the ceramic particles was more decelerated if the starting powder was made from an elementary blending of Ti-Al. The peaks corresponding to the $TiAl_3$ phase were not visible in the XRD patterns. These results were in agreement with the microstructures observed by SEM and (S)TEM.

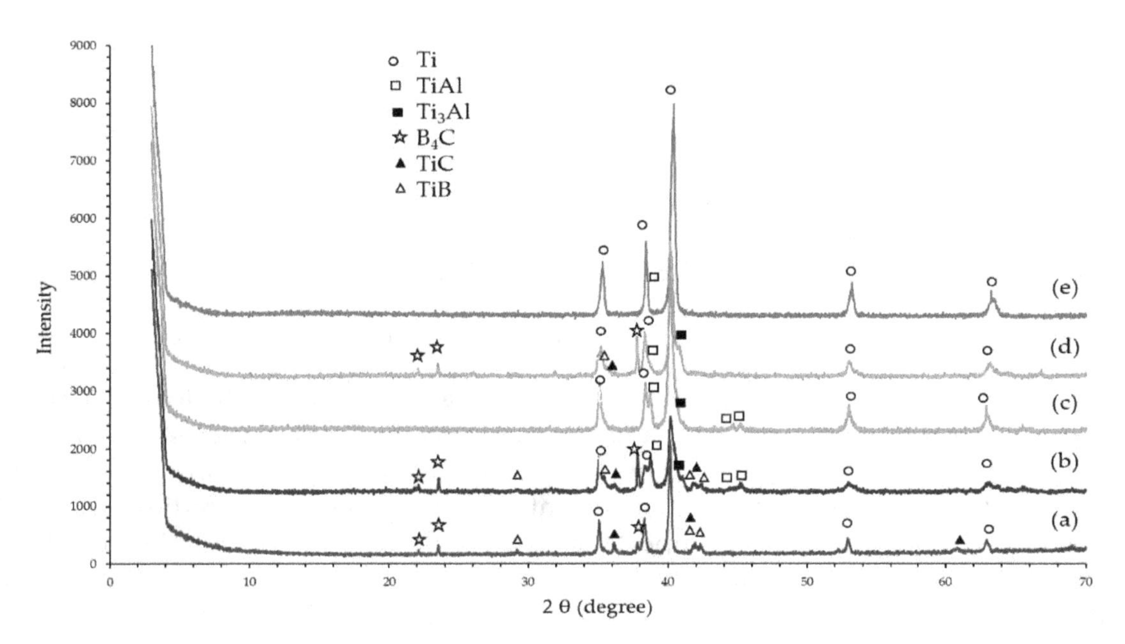

Figure 7. XRD patterns of the specimens: (**a**) Ti + 30 vol. % B_4C, (**b**) Ti + 30 vol. % B_4C + 20 vol. % Ti-Al(1), (**c**) Ti + 20 vol. % Ti-Al(1), (**d**) Ti + 30 vol. % B_4C + 20 vol. % Ti-Al(2) and (**e**) Ti + 20 vol. % Ti-Al(2).

When the starting blend was made without B_4C, while the pattern of the specimen produced from intermetallic powder Ti-Al(1) revealed the peaks of the Ti_3Al and TiAl phases (Figure 7c), the pattern of the specimens made from Ti-Al(2) powder showed mainly the displaced peaks of Ti and a small quantity of TiAl (Figure 7e). This could be due to the Al migrated into the titanium crystal lattice, avoiding the formation of Ti_xAl_y phases. The presence of B_4C and the secondary phases affected the Al diffusion; therefore, the peak of Ti_3Al was slightly detected in specimens made from B_4C and Ti-Al(2) (see Figure 7d).

3.2. Physical and Tribological Properties

The values of hardness in the matrix and Young's modulus are plotted in Figure 8. These results agreed with the previously commented microstructural study and XRD analysis. As it was expected, the B_4C addition led to the onset of TiB and TiC becoming in an increase of the hardness and Young's modulus. Whereas there was more evidence of the TiB and TiC precipitates in the composites made from Ti-Al(1), the hardness and Young's modulus of these specimens were similar to the composites made from elementary Ti-Al powder (Ti-Al(2)). The Al diffusion phenomenon into the titanium crystal lattice was helping to the matrix strengthening in specimens made from Ti-Al(2). Comparing samples without B_4C, the same trend was observed in the hardness and Young's modulus values (Figure 8). Therefore, this verified how, from different starting powders and diverse reactions phenomenon, the final properties could be, to some extent, similar; other authors have reported some positive or negative effects on the mechanical behaviour of the titanium alloys with other intermetallic additions [35,36].

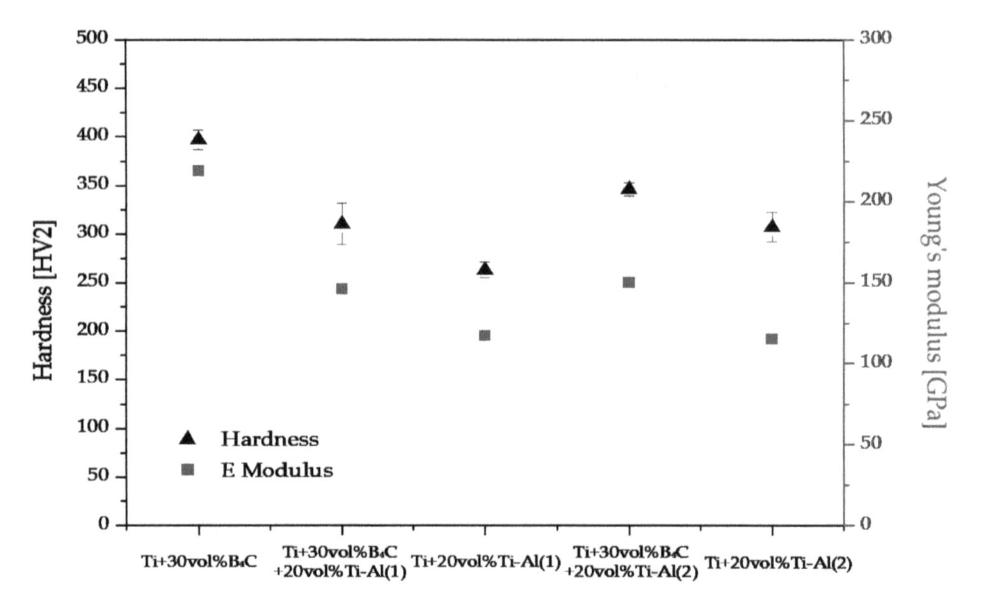

Figure 8. Hardness Vickers (HV2) and Young's modulus of the specimens.

As regards the tribological behaviour, the wear resistance and the coefficient of friction (COF) were measured and evaluated (Figures 9 and 10). The wear loss (mg) after sliding for 66.3 min (500 m, 400 rpm) is shown in Figure 9. This loss results differed significantly in their value due to the combination of B_4C and intermetallic content. The enhancement in fretting wear with only the addition of B_4C was relevant. It is important to note that the minor weight loss resulted in the TMC reinforced only with B_4C. The intermetallic incorporation in the starting powder involved that the weight loss increased by 50% and 53% in composites made from B_4C with Ti-Al(2) and Ti-Al(1), respectively. These phenomena were related to the amount of in situ formed TiB and TiC; if there were an obstacle to origin these secondary phases, the tribological properties could decline. Therefore, when the Al diffused into the matrix blocking these secondary reactions, there was less formation of TiB and TiC, which was reflected in the final tribological behaviour of the composites. These results were in agreement with the ones obtained by microscopy and XRD analysis.

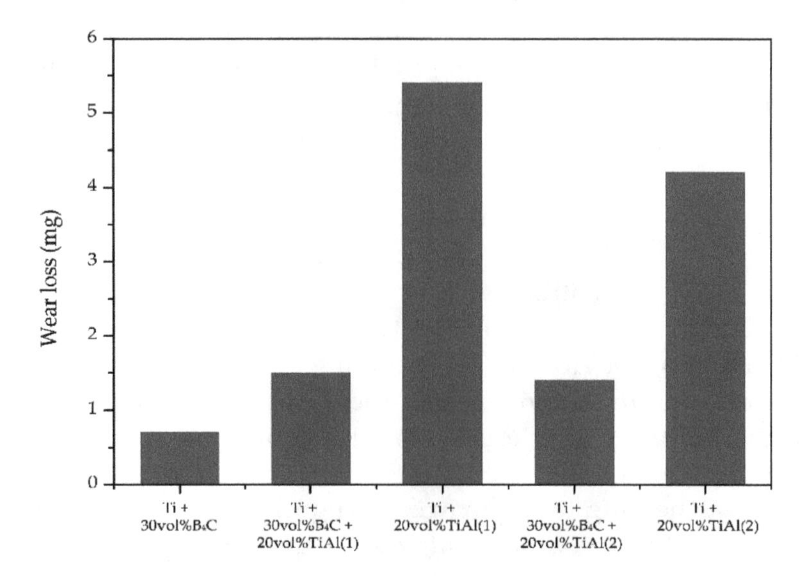

Figure 9. Wear loss vs. specimens.

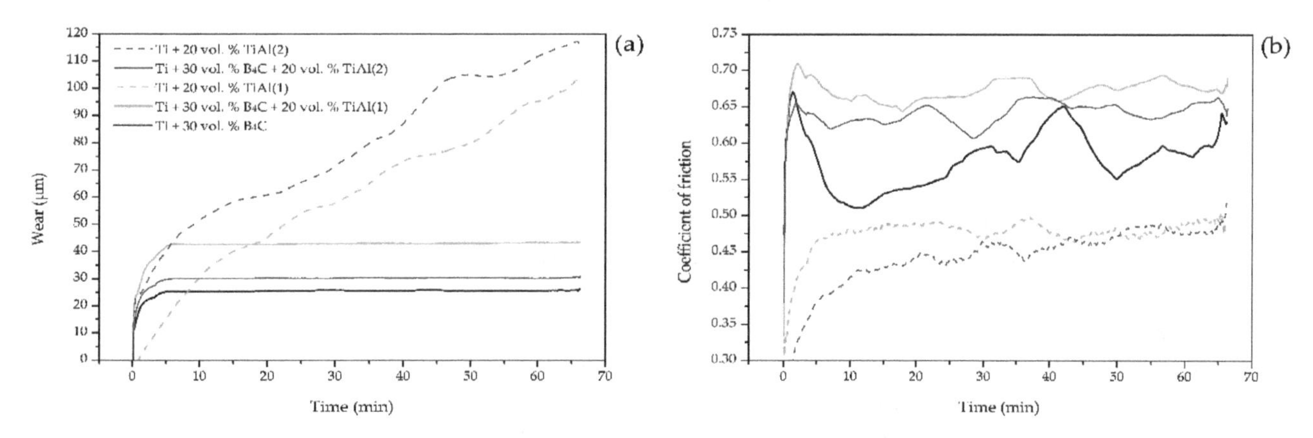

Figure 10. Tribology results: (**a**) wear vs. time and (**b**) coefficient of friction vs. time.

Figure 10 shows the wear and COF vs. time for all the specimens at the same wear conditions. In accordance with the commented above, the B_4C particles contributed to a lower penetration than in other specimens, as it can be appreciated in Figure 10a. In these specimens, the B_4C particles could participate in avoiding their pluck from the titanium matrix, maintaining the initial penetration corresponding to the applied load and speed. As expected, the absence of B_4C promoted the depth of the wear track, being the composite made from Ti-Al(2) the most affected. The COF vs. time is shown in Figure 10b. The coefficient values in the steady-state region were similar in specimens reinforced with B_4C. During the wear test, the particles and the hard precipitates in the titanium matrix could collide with the ball; therefore, COF was higher than in specimens without B_4C particles. Based upon the foregoing, when the intermetallic remained in the matrix, as in Ti-Al(1), there were also collisions between the ball and these precipitates, reflected in the COF values measured as seen in Figure 10b.

Optical micrographs in Figure 11 show representative worn track areas of the samples after the tribological characterisation. The composites with B_4C reinforcements seemed to be able to prevent the formation of severe worn surfaces, showing stronger behaviours. Moreover, in samples without ceramic reinforcements, Figure 11c,e, the material removal could be clearly appreciated with debris in the worn. The worn track surfaces were also significantly wider than in specimens with B_4C particles, being more noticeable when the intermetallic was added as an elementary blend.

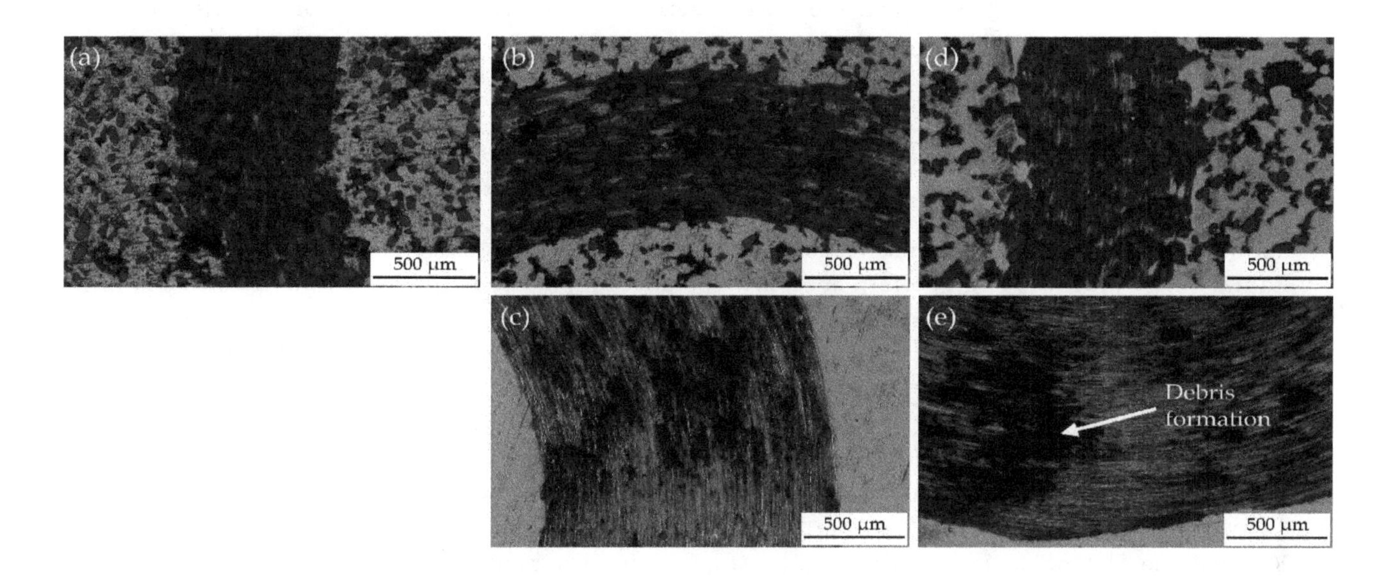

Figure 11. Worn surface: (**a**) Ti + 30 vol. % B_4C, (**b**) Ti + 30 vol. % B_4C + 20 vol. % Ti-Al(1), (**c**) Ti + 20 vol. % Ti-Al(1), (**d**) Ti + 30 vol. % B_4C + 20 vol. % Ti-Al(2) and (**e**) Ti + 20 vol. % Ti-Al(2).

4. Conclusions

The following conclusions were drawn from this research:

(1) The presence of intermetallic affected considerably the formation of the secondary phases TiC and TiB. If the aluminium was added as an elementary powder in the blend, it caused less formation of these secondary phases, in comparison to the effect of the prealloyed Ti-Al powder. The presence of Al in the matrix could block the C and B diffusion, leading to a weak reaction between these elements and the titanium.

(2) The reaction layers between B_4C and the matrix presented similar characteristics, regardless of the intermetallic powder in the blend. This reaction layer was narrower when only B_4C was used. When adding intermetallics, more TiB and TiC were accumulated around the ceramic particles due to the commented screen effect of Al in the matrix. The opposite occurred with the precipitates size of the secondary phases TiB and TiC being bigger when no intermetallic was added.

(3) In relation to the reaction layer of the prealloyed Ti_xAl_y with titanium, a decomposition occurred allowing the Al to be introduced into the matrix.

(4) Regarding the mechanical and tribological properties, composites with ceramic reinforcement showed excellent behaviour with high hardness values and good wear resistance. Without B_4C, the presence of prealloyed intermetallic in the matrix helped to a better performance than with the elementary Ti-Al blend.

Author Contributions: Authors have collaborated to obtain high-quality research work. Conceptualisation, E.N. and I.M.-M.; methodology, I.M.-M., E.M.P.S. and C.A.; formal analysis, I.M.-M.; investigation, C.A.; resources, M.K. and E.N.; data curation, E.M.P.S.; writing—original draft preparation, I.M.-M., E.M.P.S. and C.A.; writing—review and editing, C.A. and A.M.B.; visualisation, E.M.P.S. and funding acquisition, I.M.-M., E.M.P.S. and C.A. All authors have read and agreed to the published version of the manuscript.

References

1. Lütjering, G.; Williams, J.C. *Titanium*, 2nd ed.; Springer: Berlin, Germany, 2007.
2. Leyends, C.; Peters, M. *Titanium and Titanium Alloys: Fundamentals and Applications*, 1st ed.; Wiley-VCH Verlag GmbH & Co. KGaA: Weinheim, Germany, 2003.
3. Elias, C.N.; Lima, J.H.C.; Valiev, R.; Meyers, M.A. Biomedical applications of titanium and its alloys. *JOM* **2008**, *60*, 46–49. [CrossRef]
4. Leary, M. Design of titanium implants for additive manufacturing. *Titan. Med. Dent. Appl.* **2018**, 203–224. [CrossRef]
5. Tjong, S.C.; Mai, Y.-W. Processing-structure-property aspects of particulate- and whisker-reinforced titanium matrix composites. *Compos. Sci. Technol.* **2008**, *68*, 583–601. [CrossRef]
6. Ravi Chandran, K.S.; Panda, K.B.; Sahay, S.S. TiB_w-reinforced Ti composites: Processing, properties, application prospects, and research needs. *JOM* **2004**, *56*, 42–48. [CrossRef]
7. Zadra, M.; Girardini, L. High-performance, low-cost titanium metal matrix composites. *Mater. Sci. Eng. A* **2014**, *608*, 155–163. [CrossRef]
8. Neubauer, E.; Vály, L.; Kitzmantel, M.; Grech, D.; Rovira, A.; Montealegre-Meléndez, I.; Arévalo, C. Titanium Matrix Composites with High Specific Stiffness. *Key Eng. Mater.* **2016**, *704*, 38–43. [CrossRef]
9. Cao, Z.; Wang, X.; Li, J.; Wu, Y.; Zhang, H.; Guo, J.; Wang, S. Reinforcement with graphene nanoflakes in titanium matrix composites. *J. Alloys Compd.* **2017**, *696*, 498–502. [CrossRef]
10. Sabahi Namini, A.; Azadbeh, M.; Shahedi Asl, M. Effect of TiB_2 content on the characteristics of spark plasma sintered Ti–TiB_w composites. *Adv. Powder Technol.* **2017**, *28*, 1564–1572. [CrossRef]
11. Montealegre-Meléndez, I.; Neubauer, E.; Angerer, P.; Danninger, H.; Torralba, J.M. Influence of nano-reinforcements on the mechanical properties and microstructure of titanium matrix composites. *Compos. Sci. Technol.* **2011**, *71*, 1154–1162. [CrossRef]
12. Zhang, X.; He, M.; Yang, W.; Wu, K.; Zhan, Y.; Song, F. Structure and mechanical properties of in-situ titanium

matrix composites with homogeneous Ti_5Si_3 equiaxial particle-reinforcements. *Mater. Sci. Eng. A* **2017**, *698*, 73–79. [CrossRef]
13. Ni, D.R.; Geng, L.; Zhang, J.; Zheng, Z.Z. Effect of B_4C particle size on microstructure of in situ titanium matrix composites prepared by reactive processing of Ti–B_4C system. *Scr. Mater.* **2006**, *55*, 429–432. [CrossRef]
14. Zhang, Y.; Sun, J.; Vilar, R. Characterization of (TiB+TiC)/TC4 in situ titanium matrix composites prepared by laser direct deposition. *J. Mater. Process. Technol.* **2011**, *211*, 597–601. [CrossRef]
15. Arévalo, C.; Kitzmantel, M.; Neubauer, E.; Montealegre-Meléndez, I. Development of Ti-MMCs by the use of different reinforcements via conventional Hot-Pressing. *Key Eng. Mater.* **2016**, *704*, 400–405. [CrossRef]
16. Wang, X.; Wang, L.; Luo, L.; Yan, H.; Li, X.; Chen, R.; Su, Y.; Guo, J.; Fu, H. High temperature deformation behavior of melt hydrogenated (TiB+TiC)/Ti-6Al-4V composites. *Mater. Des.* **2017**, *121*, 335–344. [CrossRef]
17. Jia, L.; Li, S.; Imai, H.; Chen, B.; Kondoh, K. Size effect of B_4C powders on metallurgical reaction and resulting tensile properties of Ti matrix composites by in-situ reaction from Ti–B_4C system under a relatively low temperature. *Mater. Sci. Eng. A* **2014**, *614*, 129–135. [CrossRef]
18. Radhakrishna Bhat, B.V.; Subramanyam, J.; Bhanu Prasad, V.V. Preparation of Ti-TiB-TiC & Ti-TiB composites by in-situ reaction hot pressing. *Mater. Sci. Eng. A* **2002**, *325*, 126–130. [CrossRef]
19. AlMangour, B.; Grzesiak, D.; Yang, J.-M. In-situ formation of novel TiC-particle-reinforced 316L stainless steel bulk-form composites by selective laser melting. *J. Alloys Compd.* **2017**, *706*, 409–418. [CrossRef]
20. Jia, L.; Wang, X.; Chen, B.; Imai, H.; Li, S.; Lu, Z.; Kondoh, K. Microstructural evolution and competitive reaction behavior of Ti-B_4C system under solid-state sintering. *J. Alloys Compd.* **2016**, *687*, 1004–1011. [CrossRef]
21. Li, S.; Kondoh, K.; Imai, H.; Chen, B.; Jia, L.; Umeda, J. Microstructure and mechanical properties of P/M titanium matrix composites reinforced by in-situ synthesized TiC–TiB. *Mater. Sci. Eng. A* **2015**, *628*, 75–83. [CrossRef]
22. Arévalo, C.; Montealegre-Meléndez, I.; Ariza, E.; Kitzmantel, M.; Rubio-Escudero, C.; Neubauer, E. Influence of Sintering Temperature on the Microstructure and Mechanical Properties of In Situ Reinforced Titanium Composites by Inductive Hot Pressing. *Materials* **2016**, *9*, 919. [CrossRef]
23. Ma, F.; Shi, Z.; Liu, P.; Li, W.; Liu, X.; Chen, X.; He, D.; Zhang, K.; Pan, D.; Zhang, D. Strengthening effect of in situ TiC particles in Ti matrix composite at temperature range for hot working. *Mater. Charact.* **2016**, *120*, 304–310. [CrossRef]
24. Ma, F.; Wang, T.; Liu, P.; Li, W.; Liu, X.; Chen, X.; Pan, D.; Lu, W. Mechanical properties and strengthening effects of in situ (TiB+TiC)/Ti-1100 composite at elevated temperatures. *Mater. Sci. Eng. A* **2016**, *654*, 352–358. [CrossRef]
25. Zhao, Q.; Liang, Y.; Zhang, Z.; Li, X.; Ren, L. Effect of Al content on impact resistance behavior of Al-Ti-B_4C composite fabricated under air atmosphere. *Micron* **2016**, *91*, 11–21. [CrossRef] [PubMed]
26. Zhang, J.; Lee, J.-M.; Cho, Y.-H.; Kim, S.-H.; Yu, H. Effect of the Ti/B_4C mole ratio on the reaction products and reaction mechanism in an Al–Ti–B_4C powder mixture. *Mater. Chem. Phys.* **2014**, *147*, 925–933. [CrossRef]
27. Schmidt, J.; Boehling, M.; Burkhardt, U. Preparation of titanium diboride TiB_2 by spark plasma sintering at slow heating rate. *Sci. Technol. Adv. Mater.* **2007**, *8*, 376–382. [CrossRef]
28. Kondoh, K. 16—Titanium metal matrix composites by powder metallurgy (PM) routes. In *Titanium Powder Metallurgy*; Qian, M., Froes, F.H., Eds.; Butterworth-Heinemann: Oxford, UK, 2015; pp. 277–297.
29. Montealegre-Meléndez, I.; Neubauer, E.; Arévalo, C.; Rovira, A.; Kitzmantel, M. Study of Titanium Metal Matrix Composites Reinforced by Boron Carbides and Amorphous Boron Particles Produced via Direct Hot Pressing. *Key Eng. Mater.* **2016**, *704*, 85–93. [CrossRef]
30. Montealegre-Meléndez, I.; Arévalo, C.; Perez-Soriano, E.M.; Kitzmantel, M.; Neubauer, E. Microstructural and XRD Analysis and Study of the Properties of the System Ti-TiAl-B_4C Processed under Different Operational Conditions. *Metals* **2018**, *8*, 367. [CrossRef]
31. ASM-International. *Nondestructive Evaluation and Quality Control*, 9th ed.; ASM-International: Materials Park, OH, USA, 1989.
32. ASTM C373-14. *Standard Test Method for Water Absorption, Bulk Density, Apparent Porosity, and Apparent Specific Gravity of Fired Whiteware Products, Ceramic Tiles, and Glass Tiles*; ASTM International: West Conshohocken, PA, USA, 2014.
33. Okulov, A.V.; Volegov, A.S.; Weissmüller, J.; Markmann, J.; Okulov, I.V. Dealloying-Based Metal-Polymer

Composites for Biomedical Applications. *Scr. Mater.* **2018**, *146*, 290–294. [CrossRef]

34. Okulov, I.V.; Okulov, A.V.; Volegov, A.S.; Markmann, J. Tuning Microstructure and Mechanical Properties of Open Porous TiNb and TiFe Alloys by Optimization of Dealloying Parameters. *Scr. Mater.* **2018**, *154*, 68–72. [CrossRef]
35. Okulov, I.V.; Sarmanova, M.F.; Volegov, A.S.; Okulov, A.; Kühn, U.; Skrotzki, W.; Eckert, J. Effect of Boron on Microstructure and Mechanical Properties of Multicomponent Titanium Alloys. *Mater. Lett.* **2015**, *158*, 111–114. [CrossRef]
36. Okulov, I.V.; Bönisch, M.; Okulov, A.V.; Volegov, A.S.; Attar, H.; Ehtermam-Haghighi, S.; Calin, M.; Wang, Z.; Hohenwarter, A.; Kaban, I.; et al. Phase Formation, Microstructure and Deformation Behavior of Heavily Alloyed TiNb- and TiV-Based Titanium Alloys. *Mater. Sci. Eng. A* **2018**, *733*, 80–86. [CrossRef]

Experimental Characterization of the Primary Stability of Acetabular Press-Fit Cups with Open-Porous Load-Bearing Structures on the Surface Layer

Volker Weißmann [1,2,*], Christian Boss [3], Christian Schulze [2], Harald Hansmann [1] and Rainer Bader [2]

[1] Faculty of Engineering, University of Applied Science, Technology, Business and Design, Philipp-Müller-Str. 14, 23966 Wismar, Germany; h.hansmann@ipt-wismar.de
[2] Biomechanics and Implant Technology Research Laboratory, Department of Orthopedics, Rostock University Medicine, Doberaner Strasse 142, 18057 Rostock, Germany; christian_schulze@med.uni-rostock.de (C.S.); rainer.bader@med.uni-rostock.de (R.B.)
[3] Institute for Polymer Technologies e.V., Alter Holzhafen 19, 23966 Wismar, Germany; boss@ipt-wismar.de
* Correspondence: weissmann@ipt-wismar.de

Abstract: *Background:* Nowadays, hip cups are being used in a wide range of design versions and in an increasing number of units. Their development is progressing steadily. In contrast to conventional methods of manufacturing acetabular cups, additive methods play an increasingly central role in the development progress. *Method:* A series of eight modified cups were developed on the basis of a standard press-fit cup with a pole flattening and in a reduced version. The surface structures consist of repetitive open-pore load-bearing textural elements aligned right-angled to the cup surface. We used three different types of unit cells (twisted, combined and combined open structures) for constructing of the surface structure. All cups were manufactured using selective laser melting (SLM) of titanium powder (Ti6Al4V). To evaluate the primary stability of the press fit cups in the artificial bone cavity, pull-out and lever-out tests were conducted. All tests were carried out under exact fit conditions. The closed-cell polyurethane (PU) foam, which was used as an artificial bone cavity, was characterized mechanically in order to preempt any potential impact on the test results. *Results and conclusions:* The pull-out forces as well as the lever moments of the examined cups differ significantly depending on the elementary cells used. The best results in pull-out forces and lever-out moments are shown by the press-fit cups with a combined structure. The results for the assessment of primary stability are related to the geometry used (unit cell), the dimensions of the unit cell, and the volume and porosity responsible for the press fit. Corresponding functional relationships could be identified. The findings show that the implementation of reduced cups in a press-fit design makes sense as part of the development work.

Keywords: Ti6Al4V; selective laser melting; mechanical characterization; press-fit; primary stability

---

1. Introduction

Implants today are an important achievement of modern society and an indispensable part of daily life. To improve an implant design, it is important to build a knowledge base that allows insights gained to be integrated into new developments. Modern, generative manufacturing processes provide an excellent foundation for the support and acceleration of the knowledge required in the area of experimental development and for the transfer from result in application [1–4]. Developing implants beyond the current state of the art, for example in the field of orthopedics, is an interesting task for development engineers. Due to their outstanding mechanical and biocompatible properties, titanium and titanium alloys, in addition to other materials, are at the center of development work [5–7].

Of major interest is the implementation of open-porous structures in orthopedic implants. These structural elements provide excellent conditions to fulfil structural and functional requirements. Open-porous structures meet the mechanical requirements regarding surface quality as well as those regarding design conditions [8–10]. In addition, such structures offer a potential for solving the problems of different stiffnesses between human bone and full implants [11,12]. As a result of their geometry, open-pore structures offer the cells good conditions for nutrient supply, and consequently, the possibility to grow well into the pores. Characteristic features of open-pore structures like pore size and distribution as well as connectivity affect biological processes like cell migration and proliferation and as a result the regeneration process [3,13].

The applications of open-porous and load-bearing structures in orthopedic applications range from femoral stems, knee implants to artificial hip cups [3]. Harrison et al. developed a new surface architecture for orthopedic stem components to ensure a greater resistance against transverse motion. This allowed an enhanced primary fixation [14]. Jetté et al. designed a femoral stem with a diamond cubic lattice structure and assessed its potential as a biomimetic construct for load-bearing orthopedic implants [15]. Marin et al. evolved an acetabular cup with Trabecular Titanium^{TM} to increase osseointegration [16].

The design of the area between the implant and human bone or the transition boundary between the implant and human bone is crucial for the success of the substitution of bone with the implant. A large number of investigations are therefore concerned with the implementation of implant surfaces with biocompatible or bioactive properties [17–20]. The aim is to establish conditions that will optimally assist bone in growing in order to achieve maximum secondary stability [21–25].

Numerical simulations are also frequently used in the area of implant development as an indispensable link between constructive development ideas and experimental testing [26–30]. The success of an implantation is determined not only by secondary stiffness but also by primary anchoring strength [29,31,32]. Le Cann et al. investigated the influence of surface roughness on primary stability [33]. Goriainov et al. tested the interaction between the surface properties of the acetabular cup and its initial stability [34]. Gebert et al. studied the influence of press-fit parameters on the primary stability of uncemented femoral head resurfacing prostheses [35]. With this work, an influence of the surface roughness on the primary stability could be demonstrated. It is particularly remarkable that the primary stability can be improved up to a respective roughness value beyond which deterioration occurs is essentially influenced by the cup design. However, the influence of modifications to commercially available implants on primary stability must not be disregarded when considering the entire subject area [36,37]. Primary stability as a prerequisite for good osseointegration significantly influences the success of an implantation [29].

In the field of press-fit cups, experimental work evaluating the pull-out and lever-out behavior in preclinical as well as in post-clinical investigations is of particular interest for the assessment of anchoring strength [38–43]. Besides bones (cadavers) closed-cell foams are being used more and more often in their function as an artificial bone bed [37,44–46]. In addition to different PU (polyurethane) foams, EP-DUR polyurethane foams, polymethacrylamide (PMI) foams and a combination of a polyvinyl chloride (PVC) layer and a PMI foam have served as bone substitutes [47–49]. Although PU foam deviates from the properties of acetabular bone, it is well suited for experimental work due to its uniform cell structure and associated mechanical properties. This is mainly because of the reproducibility of the results, better availability and avoidance of ethical problems.

In the context of this work, standard acetabular cups in the press-fit version were constructively provided with a porous layer on the surface to experimentally determine the influence on primary stability. The porous structures were applied to a reduced-acetabular cup, the suitability of which for the characterization of primary stability has been evaluated in a previous study [50]. All acetabular cups were manufactured using additive manufacturing technology (Selective Laser Melting). The porous surface structures were varied constructively in order to generate different densities in the structural layer and to vary the structure-determining geometry. These constructively produced structures, though differing significantly, nevertheless aim to deliver bone-like properties as a load-bearing structural layer. Thus, forces occurring in the implant bed can be directly absorbed and transmitted by the implant. The porous structure, which has an osteoconductive effect and supports osteoinduction, can significantly improve primary stability [21,25].

The focus of the experimental work is the description of the impact of the applied structural geometry on the primary stability.

2. Materials and Methods

2.1. Cup Design

The modified cups (Figure 1) were designed on the basis of a conventional press-fit cup with a pole flattening. The suitability of a modified press-fit cup (reduced height) for the use in a development phase was verified in an earlier study [50]. All cups were designed in a reduced design with an equatorial cup diameter of 55.3 mm and a pole flattening of 1 mm. The height profiles of the cup were recorded (equatorial cup diameter 55.3 mm; pole flattening 1 mm) by means of a non-contact measuring microscope Mitutoyo—QVE-200 Pro (Mitutoyo Corporation, Kawasaki, Japan), transferred to a CAD model (PTC Creo, Version 3.0, Parametric Technology Corporation, Needham, MA, USA) and redesigned. The pattern used was an Allofit-IT 54/JJ (Zimmer GmbH; Winterthur; Switzerland). The surface structures consist of repetitive open-pore load-bearing textural elements aligned right-angled to the cup surface. The mechanical properties of the selected load-bearing open-pore structure were successfully ascertained in pretests [51–54]. The surface structure was adapted in its outer dimensions to the height profile of the Allofit IT-54/JJ. We have developed three different cup designs with three different types of unit cells (Table 1). Altogether, 8 different press-fit cups have been constructed.

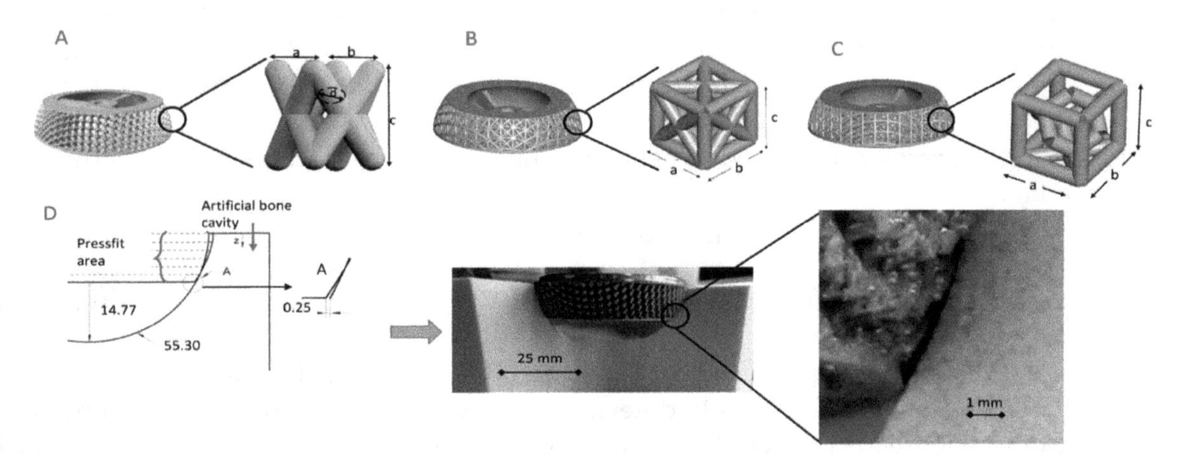

Figure 1. Designs of artificial acetabular cups with an open-porous load-bearing unit cell in a reduced variant; (**A**) Illustration—twisted unit cell, (**B**) Illustration—combined unit cell, (**C**) Illustration—combined open unit cell substitute, (**D**) Press-fit area and gap in case of reduced cup model (negative press-fit)—schematic figure and photograph, all units are in mm.

Cup variant *A* with a twisted unit cell geometry exists in five versions. The unit cells differ in depth *a* between 2.12 mm and 2.83 mm, in width *b* between 2.12 mm and 2.83 mm and in height *c* between 3 mm and 4 mm. The rod diameter *d* varied between 0.8 to 1.1 mm. Cup variant *B* with a combined unit cell geometry exists in two versions. The unit cells have a depth a of 4 mm, width *b* of 4 mm and height *c* of 4 mm. The rod diameter d varied between 0.8 and 0.9 mm. The combined unit cell geometry is designed with a cubic structure with transverse struts on the outer surfaces and a diamond-like structure. Regardless of the force acting on the unit cell, this structure offers very uniform strength. The structure is very suitable for use on the surface of a press-fit cup thanks to its direction-independent nature [54].

Table 1. Overview of the eight different cup-designs, the types of the unit cells (twisted, combined and combined open), the dimensions of the unit cells and porosities and volumes of the press-fit area. All values are derived from CAD data and are given in mm.

Unit Cell	Twisted (V)					Combined (D)		Combined Open (D_o)
Dimension	**V4_09**	**V4_10**	**V4_11**	**V3_09**	**V3_08**	**D4_09**	**D4_08**	**D_o_4_09**
Width-*a* (mm)	2.83	2.83	2.83	2.12	2.12	4.00	4.00	4.00
Depth-*b* (mm)	2.83	2.83	2.83	2.12	2.12	4.00	4.00	4.00
Height-*c* (mm)	4.00	4.00	4.00	3.00	3.00	4.00	4.00	4.00
Strut diameter-*d* (mm)	0.90	1.00	1.10	0.90	0.80	0.90	0.80	0.90
Porosity-Structure area (%)	72.50	67.40	60.60	58.80	65.50	61.10	66.90	74.80
Volume-Press-fit area (cm^3)	0.30	0.39	0.25	0.32	0.54	0.97	0.91	0.77

Cup variant C with a combined open unit cell geometry exists in one version. The unit cells have a depth *a* of 4 mm, width *b* of 4 mm and height c of 4 mm. The rod diameter *d* is 0.9 mm. The combined unit cell geometry is designed with a cubic and a diamond-like structure without transverse struts on the outer surfaces. Using the overall model of the cups as a basis, reduced designs were created. With the reduction of the acetabular cup, the pole near area was removed, but the press-fit was retained. Cup regions from the press-fit regions protrude so far that a gap of 0.25 mm is created between the artificial bone bed and the cup (negative press-fit-Figure 1-Area D).

The following expression was used to calculate the porosity of load-bearing structure volume from the CAD data:

$$\text{Porosity} - \text{structure area} = \left(1 - \frac{V_{\text{str}}}{V_{\text{full}}}\right) \cdot 100\% \tag{1}$$

where V_{str} is the volume of the area with the struts and V_{full} is the overall volume of this area in a closed manner.

The volume (Press-fit area) produced by the structured section of the cups was also calculated by CAD. For this intention, it was virtually determined how large the volume is that penetrates the artificial bone cavity (Figure 2). The acetabulum and artificial bone cavity were positioned in the CAD system in the same way as in the test situation. The results for every cup-design are given in Table 1.

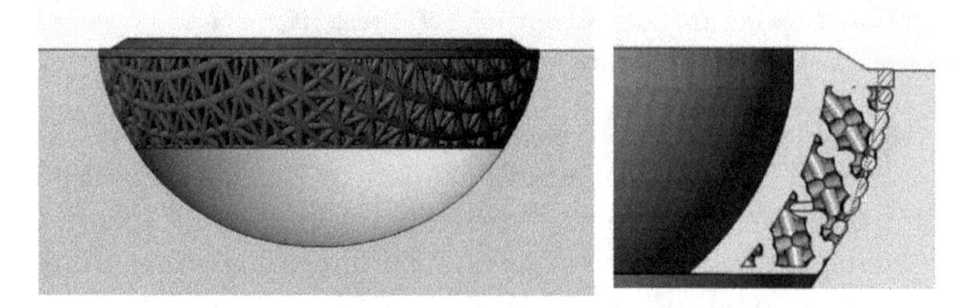

Figure 2. The cup is positioned in the artificial bone cavity (**left**) and the area virtually penetrates the artificial bone cavity-red hatched area (**right**). This area describes the Press-fit volume.

2.2. Fabrication

(1) The acetabular cups considered in this paper were manufactured by C. F. K. CNC-Fertigungstechnik Kriftel GmbH (C. F. K. CNC-Fertigungstechnik Kriftel GmbH, Kriftel, Germany) using selective laser melting with a SLM 280. Titanium powder (Ti6Al4V) with a mean particle size of 43.5 μm was used for their manufacture in a highly pure argon atmosphere. All parts were built using identical processing parameters (Table 2) in the same orientation and on a substrate plate with a support structure. The support structures were removed mechanically by hand.

Table 2. SLM process-energy-relevant process parameters.

Parameter	Description	Unit	Process Parameter
P	Laser power	W	275
v	Scan speed	mm/s	805
d	Hatch spacing	μm	120
t	Layer thickness	μm	50

(2) For the production of artificial bone cavities Sika Block M 330 (Sika GmbH, Stuttgart, Germany) was applied. This material, a thermosetting polyurethane with closed cells, is ideally suited for a comparative evaluation of the relevant acetabular cups. The properties comprise from a density of 0.24 g/cm^3 (according to test standard ISO 845) and a compressive strength of 4 MPa (according to test standard ISO 844) to an elastic modulus of 150 MPa (according to test standard ISO 850).

The material was provided in plate form in the dimensions 1000 × 500 mm. The artificial bone cavities were manufactured using a CNC milling machine i-mes-FLATCOM 50-VH (i-mes GmbH, Eiterfeld, Germany) using the plate.

The artificial bone cavities were manufactured as described in Weißmann et al. Since the mechanical properties of the plate vary across the width of the plate due to the manufacturing process, the cavities were used for each acetabulum from a corresponding material line (n = 5) [50].

2.3. Measurements

The measurements of the following points were carried out extensively as described in Weißmann et al. [50]. Here, the relevant points are briefly explained.

(1) The measurements of the acetabular cups as well as the artificial bone cavities, both being relevant for the press-fit, were performed with a non-contact measuring microscope (Mitutoyo-QVE-200 Pro; Mitutoyo Corporation, Kawasaki, Japan). Based on the measurement points, circles of best fit were determined using the method of least squares. The outlier identification and elimination from the measurement data due to light reflections and loose PUR particles was performed using a box plot (according to John W. Tukey) in a Matlab script. To verify the actual press-fits and for quality control, the resulting replacement diameters were used.

(2) In all cases, the assessment of the primary stability (anchoring strength) of the press-fit cups was realized by pull-out tests (Figure 3) with a universal testing machine (INSTRON E 10,000; Instron GmbH, Darmstadt, Germany). The cups were first press-fitted into the artificial bone cavities until they were flush with the edge of the cavity. Following this, the cups were pulled out of the cavity using a pull-out stamp. The speed for both the press-fit of the cup into the bone cavity and the pull-out of the cups was 5 mm/min. In the measurements, each performed 5 times per press-fit cup, the effective measurement data ($F_{\text{pull-out}}$) were recorded. As primary pull-out stability the first force maximum was used.

(3) The assessment of the initial tangential stability of the acetabular cups were realized by lever-out tests (Figure 4) with a universal testing machine (Zwick Z50; Zwick GmbH & Co. KG, Ulm, Germany). The cup was first pressed into the artificial bone cavity until the edge of the cup is flush with the bone bed. The cup was first pressed into the artificial bone cavity until the edge of the cup was flush with the bone bed. The cup was then vertically loaded with a force until it was released. The

first local maximum (F_L) load was evaluated as the primary lever-out stability, which at the same time indicates the beginning of the movement of the cup in the bone cavity. The speed for the press-fit of the cups into the bone cavity and the lever-out of the cups was 5 mm/min. A moment M_I of 0.62 Nm, resulting from the dead weight (0.87 kg) and length (178.3 mm) of the lever, was also integrated into the calculation.

Figure 3. Pull-out-test setup—(**A**) Complete experimental setup; (**B**) Cup ready for pressing in; (**C**) View from upside of the acetabular cup with artificial bone cavity and the pull-out stamp; (**D**) Cup completely press-fitted.

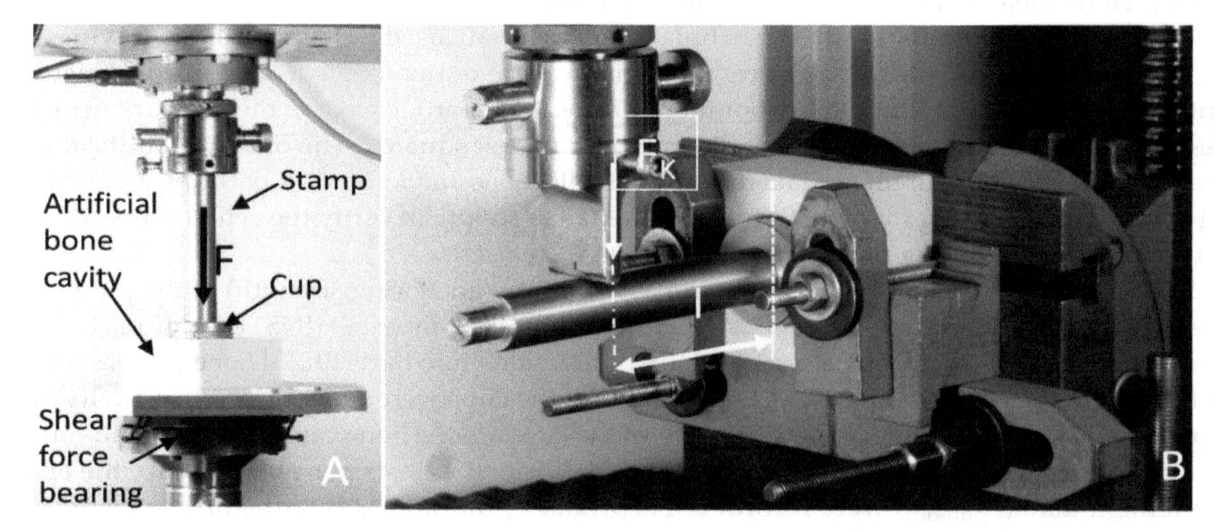

Figure 4. Pull-out-test setup—(**A**) Experimental setup-press-fitting; (**B**) Experimental setup-levering out.

The lever-out moment was calculated as follows:

$$M_L = F_L \cdot l + M_I \tag{2}$$

In the calculation is F_L the maximum lever-out tilting force, l the lever length and M_I the specific moment.

On the basis of the determined force F_L and the displacement of the cup in the bone cavity, it is possible to evaluate the work required to lever out the cup.

The lever-out work was calculated as

$$W = F_L \cdot s \tag{3}$$

from the lever-out tilting force F_L and the displacement s of the cup.

2.4. Statistical Analysis

All data listed in tables are expressed as mean values ± standard deviation (SD). A non-linear regression with Excel 2016 for Windows was used to display the relationships between the volume of the press-fit area and the lever-out moment as well as the pull-out force.

All statistical analyses were made using SPSS, software version 22 for Windows (SPSS® Inc. Chicago, IL, USA). For the pull-out force, the lever-out moment and the lever-out work, a one-way ANOVA followed by Dunn's T3 post-hoc test was made to statistically examine significant differences between the means. The results from this comparison were shown in a boxplot. A significance level of $p < 0.05$ was regarded as statistically significant.

3. Results and Discussion

3.1. Accuracy of Fabricated Samples

Table 3 lists the dimensions determined for the artificial bone cavity and the acetabular cups. The press-fit of the cups are calculated as the difference between the best fit circle of the press-fit cups and the best-fit circle of the artificial bone cavity

Table 3. Accuracy of fabricated bone cavities (diameter cavity) and acetabular cups (equatorial diameter) as well as the resulting press-fits of these combinations. The values from the bone cavities are given as the arithmetical average (n = 5).

Name	Press-Fit Cup		Artificial Bone Cavity		Press-Fit (mm)
	Best Fit Circle (mm)	Roundness (mm)	Best Fit Circle (mm)	Roundness (mm)	
V3_08	55.32	0.26	53.18 ± 0.02	0.14 ± 0.02	2.13 ± 0.02
V3_09	55.47	0.17	53.34 ± 0.02	0.13 ± 0.04	2.13 ± 0.01
V4_09	54.90	0.29	52.68 ± 0.01	0.15 ± 0.01	2.16 ± 0.01
V4_10	55.03	0.02	52.87 ± 0.01	0.14 ± 0.02	2.15 ± 0.01
V4_11	55.20	0.28	53.07 ± 0.01	0.12 ± 0.01	2.13 ± 0.01
D4_08	54.98	0.30	52.87 ± 0.01	0.14 ± 0.01	2.11 ± 0.01
D4_09	55.04	0.11	52.87 ± 0.01	0.14 ± 0.01	2.17 ± 0.01
D_o_4_09	55.03	0.25	52.87 ± 0.01	0.14 ± 0.01	2.16 ± 0.01

The processing values for the artificial bone cavities were determined based on the values for press-fit cups. The aim was to provide a constructive press-fit of 2 mm for all cup-bone cavity pairs.

For all pairings a press-fit was achieved between a minimum of 2.11 mm and a maximum of 2.17 mm. The deviations among each other amount to a maximum of 0.06 mm. With respect to the minimum possible press-fit, this is less than 3% (2.84%). The roundness values of the bone cavity of 0.12 to 0.15 demonstrate the high repeatability of the manufacturing method for artificial bone cavities. The roundness values of the press-fit cups from 0.02 to 0.30 vary slightly more. With respect to the additive manufacturing process, these are excellent results [55–58].

Dimensional deviations or differences in the produced press-fit can lead to different insertion forces. These differences would be the cause of stress differences in the bone cavity and unequal conditions for the contact of the press-fit cup with the surface of the bone cavity. The resulting deviations produce differences in tension in the bone cavity and create different conditions for the movements of the press-fit cup in the bone cavity [44,49]. Only if the conditions for the generation of a good primary stability are given, can corresponding good long-term results be expected [27].

Overall, it can be assumed that the differences between each other are so small that this will have no effect on the assessment of the primary stability of the artificial acetabular cups. The press-fit results are only so slightly different that the results in the pull-out test and the lever-out test are not affected.

3.2. Pull-Out Force

To determine the pull-out forces, the manufactured cups were stripped from the cavities after being press-fitted into the artificial bone cavity. The results are shown in Figure 5 and Table 4.

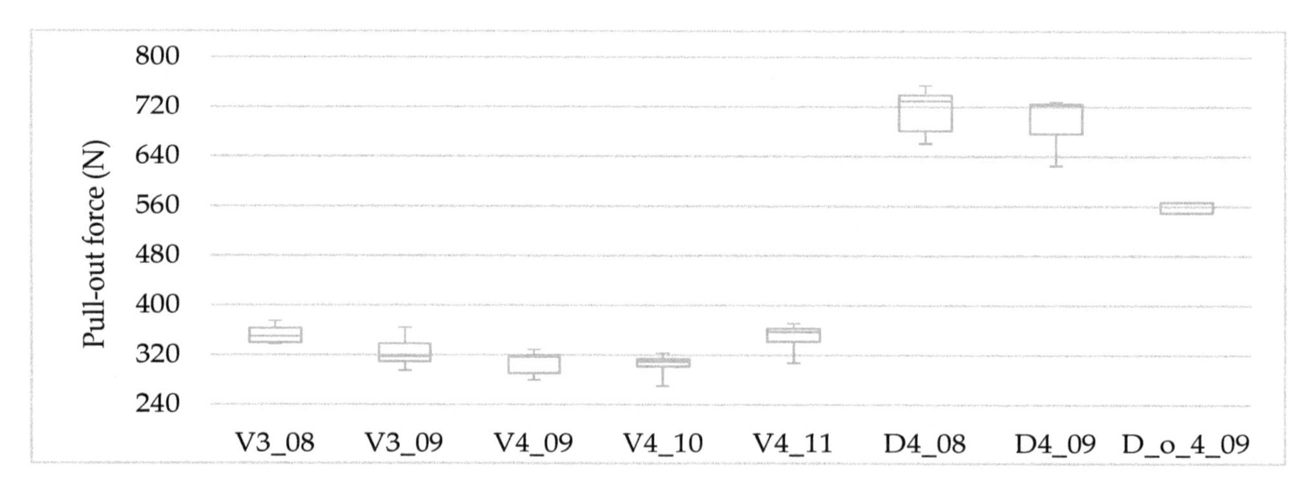

Figure 5. Boxplots of the measured pull-out force (N). Boxplots indicate the median value, the interquartile range (IQR: interval between the 25th and 75th percentile, blue rectangle) and the extremum values ($n = 5$).

Table 4. Significances of the determined pull-out-forces from the different press-fit cups. For statistical analysis one-way ANOVA with Dunn's T3 post-hoc test was conducted. Values of $p < 0.05$ were set to be significant (N.S.—not significant).

Cupversion	D4_09	D_o_4_09	V3_08	V3_09	V4_09	V4_10	V4_11
D4_08	N.S.	0.00438	<0.001	<0.001	<0.001	<0.001	<0.001
D4_09	-	0.00193	<0.001	<0.001	<0.001	<0.001	<0.001
D_o_4_09	-	-	<0.001	0.0006	<0.001	<0.001	<0.001
V3_08	-	-	-	N.S.	N.S.	0.0242	N.S.
V3_09	-	-	-	-	N.S.	N.S.	N.S.
V4_09	-	-	-	-	-	N.S.	N.S.
V4_10	-	-	-	-	-	-	N.S.

The results of the experiments carried out according to the measuring methodology reveal differences that are related to the structural elements used. Whereas the combined structures achieve the highest results (D4_08 = 708 N; D4_09 = 704 N), the pull-out forces for the twisted structures (Max: V3_08 = 351 N; Min: V4_10 = 308 N) are significantly lower. The combined open structure (550 N) lies between the two combined variants and the cups with the twisted structures.

After carrying out a statistical significance test using one-way Anova with Dunnett's T3 post-hoc test (multiple comparisons), the following relationships become clear. The two combined structures do not differ significantly from each other. However, the combined open structure is significantly below the combined structure (D4_08 to D_o_4_09/$p = 0.00438$; D4_09 to D_o_4_09/$p = 0.00193$). The differences in the twisted structures are consistently significant (values see Table 4). In the twisted structures only version V3_08 deviates significantly from version V4_10 ($p = 0.0242$). The differences between the combined open and twisted structures can mainly be explained by the existing differences in press-fit volume. The press-fit volumes of the combined (D4_08 = 0.91 cm^3; D4_09 = 0.97 cm^3) and the combined open structure with 0.77 cm^3 clearly differ from the twisting structures (<0.54 cm^3). However, this relationship is not identifiable in the twisting structures, since despite clear differences in the press-fit volume between the twisting structures, a significant difference could only be determined between the variants V3_08 and V4_10. It seems that in addition to the press-fit volume, other influencing factors such as the surface quality (roughness and manufacturing accuracy) of the struts of the structure and their dimensions (length, diameter, surface area) could play a role [55,59].

The pull-out behavior of the different cup models is shown in Figure 6. The representation of the force profiles over cup displacement in the artificial bone cavity additionally offers the possibility to evaluate the measured maximum force in relation to the reached cup displacement at that time. The curves show characteristic differences.

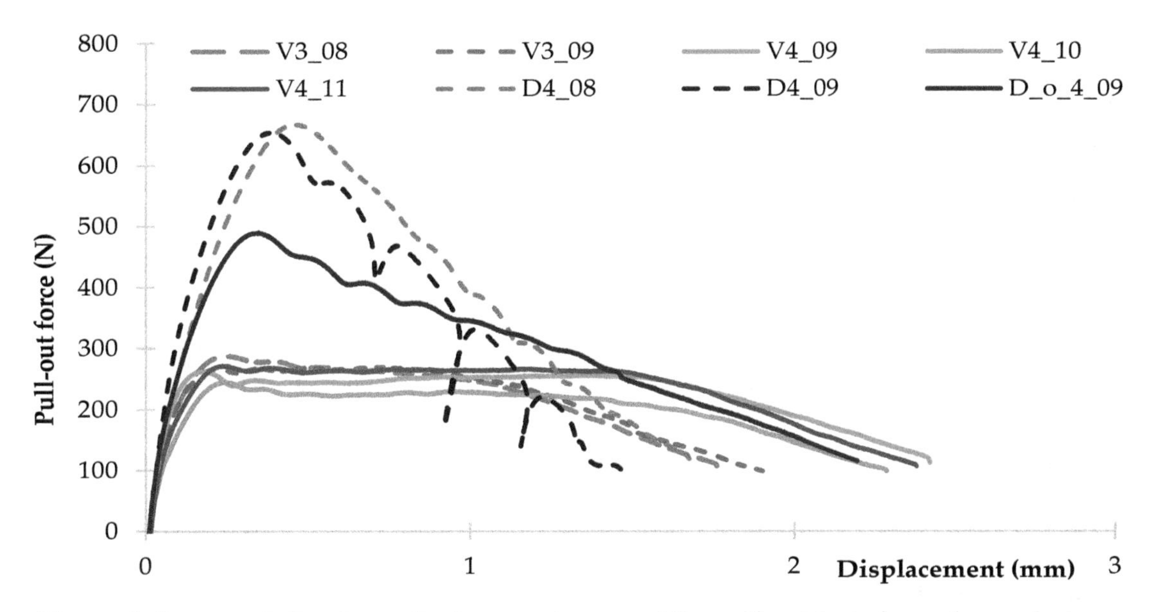

Figure 6. Representative force-displacement curve of the pull-out tests for each cup design.

The curve for the cups with a combined structure differs clearly from the curves for the cups with a combined open or twisting structure. The most striking feature here is the cascading force decrease after a maximum force has been exceeded. This cascade is characterized in that a renewed force increase is determined after a drop in force. This course reflects the loosening and re-jamming of the cup in the artificial bone cavity. These cascades are most pronounced in version D4_09. This cascade development is also evident in the combined open structure version D4_08, though weaker. Apparently, this cascade is due to the larger space between the individual struts or the greater porosity. Here, the material of the artificial bone cavity has the possibility to fill more space. The necessary release from this room requires force again.

This cascade is characterized in that a renewed force increase is determined after a drop in force. This course reflects the loosening and re-jamming of the cup in the artificial bone cavity. These cascades are most pronounced in version D4_09. This cascade development is also evident in the combined open structure version D4_08, though weaker. Apparently, this cascade is due to the larger space between the individual struts or the greater porosity. Here, the material of the artificial bone cavity has the possibility to fill more space. The necessary release from this room requires force again.

The number of cascades obviously results from the number of superficial, continuous struts (Figure 7—red lines). The maximum peak (and thus the first peak of force) results from overcoming the edge of the hip cup. The second to fifth peak results from the strut contours. Starting at the highest point of the continuous strut lines. The differences in cascade intensity of the cup variants are caused by the differences in the strut diameter. The strut with a rod diameter of 0.9 mm has a larger contact surface to the artificial bone bed. This requires more force to loosen from the artificial bone cavity. The differences between the open and closed variants (D_o_4_09 and D4_09) are due to the varying degrees of free space in the surface of the hip cups. More free space (D_o_4_09) requires less force than with the closed variant (D4_09).

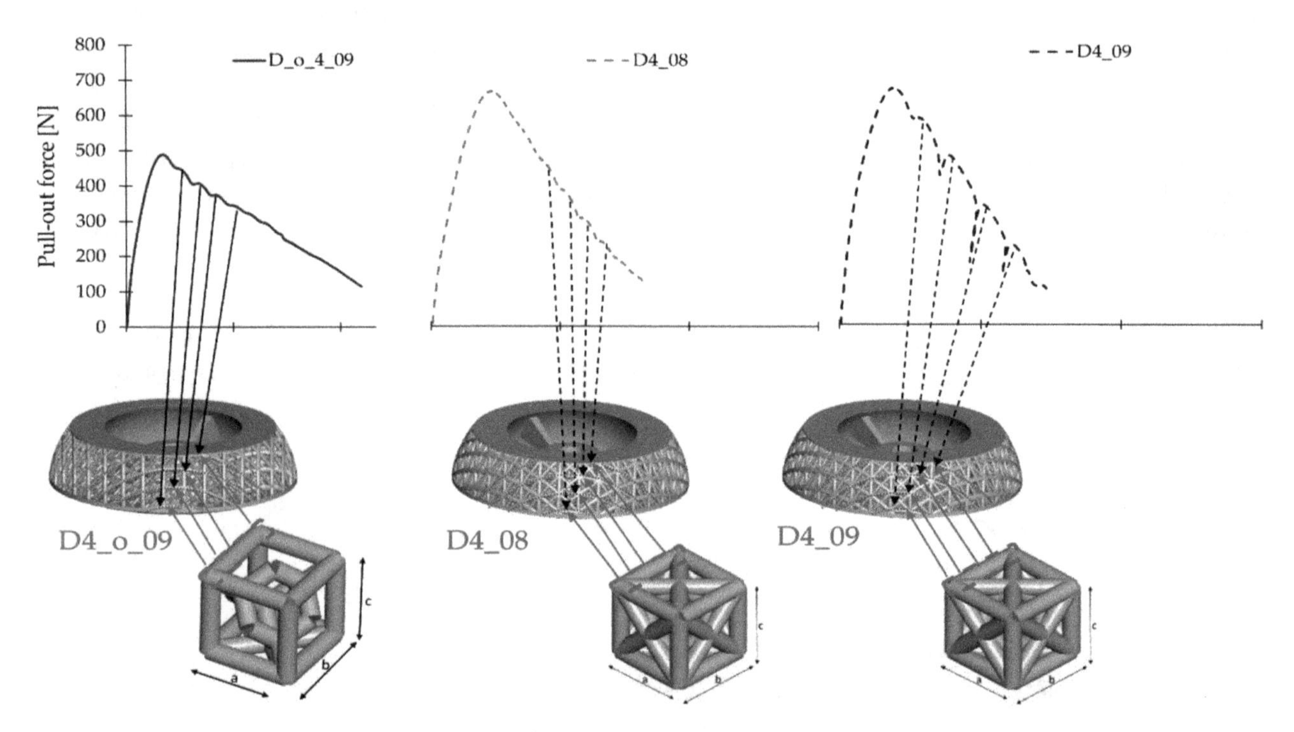

Figure 7. Representation of the cascades with reference to the structure on the cup surface.

The press-fit cups with the twisted structure show a completely different behavior. After reaching the force maximum, the corresponding force path continues at a uniform level of force. This applies to the twisted structure with a height of 3 mm as well as to the structure with a height of 4 mm. It is clearly shown here; however, that the versions in the 4 mm height maintain this level of force significantly longer. A weakening of the cup anchoring takes place here only after about 1.5 mm compared to about 1 mm in the variants with a height of 3 mm. Here, the cups with the structural elements whose individual elements have a height of 4 mm and an associated spacing of the bars of 2.83 mm, provide the artificial bone cavity material more space for anchoring than the variant of 3 mm height and a spacing of 2.12 mm. As a result, the force is maintained longer at one level.

In view of later desired ingrowth of the bone into the structural area as well as the formation of blood vessels, larger open areas have advantages over the smaller areas [22,25,60]. Here it is important to carefully observe the interaction of the geometric conditions (unit cell and macro-porosity) and the component properties influenced by the additive manufacturing process (e.g., roughness or micro-porosity, surface finish at intersections) [61–63].

While the diamond structures reach the maximum force required to pull out at approx. 0.6 to 0.7 mm, these values for the twisted structures are approx. 0.2 to 0.3 mm. The open combined structure shows a maximum at approx. 0.35 mm. In addition, it can be seen that the twisted version with a height of 3 mm as well as the combined structure D4_08 still require approximately 100 N after about 1.6 to 1.8 mm displacement for a further release.

In the case of the twisted versions with a height of 4 mm and the open combined structure, the cups have already experienced a displacement of approximately 2.5 mm at a force of 100 N. The progression curves of the press-fit cups are very similar. This value probably reflects the interaction between the artificial bone cavity and the surface of the additively manufactured cup.

As can be seen from Figure 8, all cups leave clear traces of an impression on the entire circumference of the artificial bone bed. The evaluation of these traces using this visual assessment of the contact surface has been described, for example, by Le Cann et al. to characterize how the roughness of a cup affects primary stability [33].

Figure 8. Representative pictures of the bone cavities after the pull-out test for each cup design.

All cups left distinct positioning traces in the press-fit region. The artificial bone cavity remained intact. The artificial bone cavities shown in Figure 8 exhibit clear marks of an anchorage. The damage patterns of the artificial bone cavity differ optically from each other.

All twisted versions show dot-like impressions in the cavity area. The cavity edges remain sharply intact. Differences caused by the different bar diameters (3 and 4 mm) and bar distances (2.83 and 2.12 mm) are optically present. With increasing bar diameter, the damage in the bone bed also increases. Variant V4_11 shows clearer and stronger traces than versions V4_10, V4_09, V3_08 and V3_09.

The combined structures (D4_08, D4_09) show rather flat impressions on the artificial bone cavity areas. The cavity edges tend to blur slightly, as a representation of slight material detachments. These detachments are much less pronounced in the diamond open structure.

The forces determined in the pull-out test and the traces in the bone bearing also allow the following conclusion to be drawn. The twisting structure already destroys the corresponding area in the bone bearing during the press fitting. Because of that, less force is required when pulling out of the bearing because the resistances against loosening are lower than with intact material. The combined structure, on the other hand, only damages the bone bearing when it is pulled out. Here, the resistance of predominantly intact material must be overcome. This leads to a higher power requirement.

In addition, the contacting of the structures with the bone bed takes place differently. The contact of the twisting structure is made punctually. The combined and combined open structure creates a two-dimensional contact to the surface of the bone bed. To overcome the press fit, more force is required for the two-dimensional contacts than for the punctual contacts.

3.3. Lever-Out Moment

After being press-fitted into the artificial bone cavities, all cup models were levered out from the cavities to determine the lever-out moments as described in 0. The results are shown in Figure 9 and Table 5. The course of the forces required to lever out the cups over the displacement is shown in Figure 10.

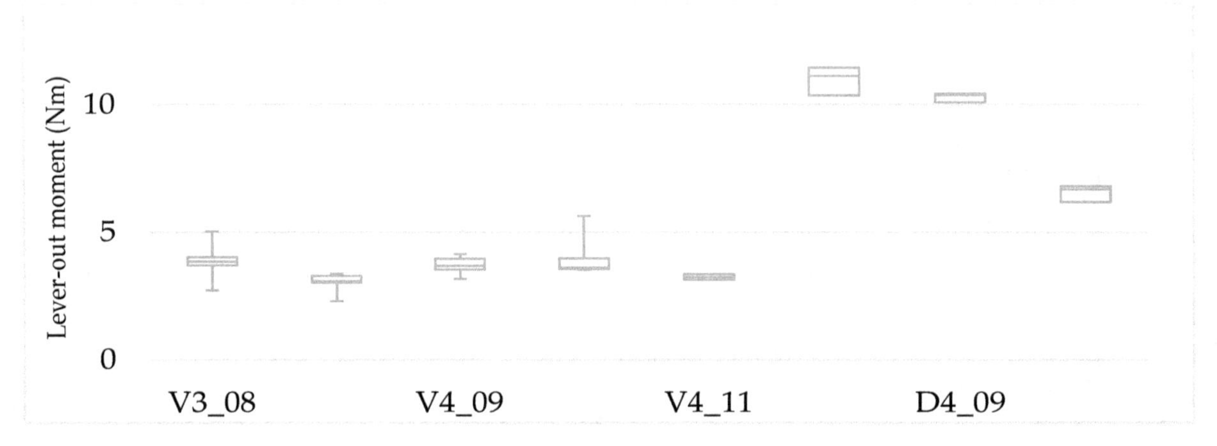

Figure 9. Boxplots of the measured lever-out moments (Nm). Boxplots indicate the median value, the interquartile range (IQR: interval between the 25th and 75th percentile, blue rectangle) and the extremum values ($n = 5$).

Table 5. Significances of the determined lever-out-moments for the different press-fit cups. For statistical analysis one-way ANOVA with Dunn's T3 post-hoc test was conducted. Values of $p < 0.05$ were set to be significant (N.S.—not significant).

Cupversion	D4_09	D_o_4_09	V3_08	V3_09	V4_09	V4_10	V4_11
D4_08	N.S.	<0.001	<0.001	<0.001	<0.001	<0.001	<0.001
D4_09	-	<0.001	<0.001	<0.001	<0.001	<0.001	<0.001
D_o_4_09	-	-	<0.001	<0.001	<0.001	<0.001	<0.001
V3_08	-	-	-	0.04619	N.S.	N.S.	0.04649
V3_09	-	-	-	-	N.S.	N.S.	N.S.
V4_09	-	-	-	-	-	N.S.	N.S.
V4_10	-	-	-	-	-	-	N.S.

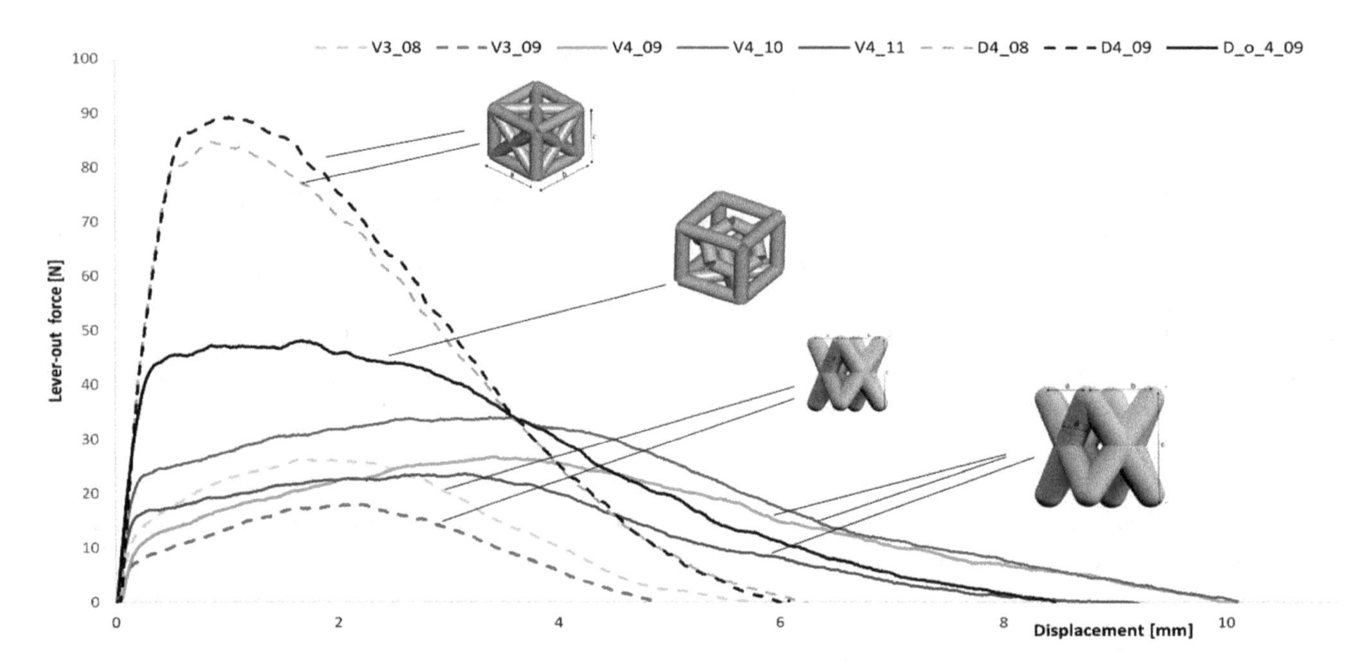

Figure 10. Representative lever-out force vs. displacement curve of the lever-out test for each cup design.

The influence of the applied structural elements on the behavior of the press-fit cups in the lever-out test can be clearly established on the basis of the experimentally determined lever-out moments. The best results were achieved by the combined structure (D4_08 = 10.9 Nm,

D4_09 = 10.3 Nm), followed by the combined open structure (6.5 Nm) and the twisted structure (Max: V3_08 = 3.9 Nm; Min: V3_09 = 3.1 Nm).

By carrying out a statistical significance test using one-way Anova with Dunnett's T3 post-hoc test (multiple comparisons) it is possible to describe the following relationships. The two combined structures do not differ significantly from each other. However, the combined open structure is significantly below the combined structure (D4_08 and D4_09 to D_o_4_09/$p < 0.001$).

The differences of the experimentally determined lever-out moments shown between the combined structures, the combined open structures and the twisted structures are significant in all cases ($p < 0.001$). For the twisted structures, only the version V3_08 deviates significantly from both version V3_09 ($p = 0.04619$) and version V4_11 ($p = 0.04649$). Similar to the pull-out tests, the differences between the combined and the combined open structures to the twisted structures can be explained by the existing differences in press-fit volume. The differences between the structure V3_08 and V3_09 and V4_11 also result from the differences in the press-fit volumes (V3_08 = 0.54 cm^3; V3_09 = 0.32 cm^3; V4_11 = 0.25 cm^3). The fact that variant V4_09 does not deviate significantly from variant V3_8 despite a lower press-fit volume (0.3 cm^3) is additional evidence that other factors are notoriously influencing the anchoring strength.

The lever-out behavior of the tested cup models is shown in Figure 10. All additively manufactured cups show curves which are characteristic for the structural elements used.

All models were preloaded with an initial moment of 0.62 Nm by the self-weight of the test setup. The representation of lever-out forces over displacement displays for the combined structure a maximum lever-out force (mean values: D4_08 = 90.3 N; D4_09 = 85 N) at a displacement of approx. 1 mm and then a decrease of the moment up to a displacement of 6 mm. The combined open structure reaches a lever-out force maximum (mean value: 51.6 N) after approx. 1.8 mm. This cup variant reduces the force to zero after a displacement of about 8.3 mm. The twisted structures show differences depending on the size of the structure. The twisted structures with dimensions of 3 mm height reach a lever-out force maximum (mean values: V3_08 = 29 N; V3_09 = 22 N) after about 1.8 to 2.2 mm. The twisted structures with dimensions of 4 mm height reach force maximums (mean values: V4_09 = 27.1 N; V4_10 = 27.5 N; V4_11 = 23.2 N) after about 3.5 to 3.7 mm. The force reduction continues in the V3-versions up to a displacement of approx. 4.8 to 5.8 mm. The V4 versions run to zero at about 9 to 10.5 mm.

Similar to the pull-out tests, it can be seen that, following a steep rise, the cups with the combined structure show a continuous force drop after reaching a lever-out force maximum. The combined open structure and the twisted structures behave differently. Here the maximum force is only reached after passing through a plateau phase. This plateau phase is much longer for the V4-variants than for the V3-variants.

This functional difference is related to the geometric design of the individual structures. As shown in Table 1, the combined structures are structures that produce a relatively uniformly shaped surface whose interstices engage only weakly in the bone bed. Here the press-fit is in the foreground.

In the combined open structure and the twisted structures, the shaped surface of the cups is much more open. These structures engage more clearly in the artificial bone stock. The differences between the V3 and V4 variants are due to the geometric dimensions of the individual rods. The larger-sized rods of the V4 variant have larger gaps than the V3-variants (V4-2.83 mm and V3-2.12 mm). Thus, a hooking of the structural elements in the bone cavity in the V4-variant is possible across a longer distance than in the V3 variant.

This leads to differences in the height of the moments determined due to the structure design. In addition, it becomes clear that the twisted structures in the artificial bone bed produce deeper punctual impressions. During the lever-out test, the struts move along these impressions. This behavior is recognizable for all twisted structure variants by traces between the punctual impressions. The illustrations of the bone beds after the pull-out test (Figure 8) do not show these traces. Therefore, due to the already damaged surface, less force is required to lever-out. The twisted structures thereby show overall lower moments than the combined and combined open structure due to the different nature of the unit cell.

A larger structural design is helpful in terms of the positive effects for bone ingrowth [25]. In addition to good primary stability, the bone-like properties of the load-bearing structural layer are an essential prerequisite for good secondary stability of the implant [64]. Secondary stability is essentially characterized by the ability of bone to grow onto the implant surface and thereby firmly anchor the implant. The use of open-pore structures enlarges the implant surface and thus improves the prerequisite for the formation of sufficiently high secondary stability. In addition, a high primary anchoring strength is the prerequisite for creating a sufficiently high secondary stiffness, since only then is sufficient growth of the bone on the surface possible. Only if a load transfer via the implant into the surrounding bone is possible without stress-shielding can a successful use of the implants be ensured. With regard to the geometric selection of structural elements, this circumstance must be taken into account [65]. The combined structures, which are more direction-independent in their properties, show slight advantages here [52,66].

The artificial bone cavities show distinct traces left by the lever-out of the cups. In the following Figure 11 the cup models are shown with representative examples of the artificial bone cavity. The artificial bone cavity is intact despite clear traces of anchoring. The damage patterns of the bone cavities differ optically from each other, as in the case of the pull-out experiments. The twisted versions show, as expected, punctually impressions in the bone cavities. The edge of the cavity remains sharp. The different strut diameters and spaces of the struts in the structure produce visually recognizable representative patterns (dot-like impressions). The combined structures leave flat traces on the bone cavities. The edge of the bone cavities tends to blur slightly, as a representation of slight material detachments. These detachments are significantly less pronounced in the combined open structure.

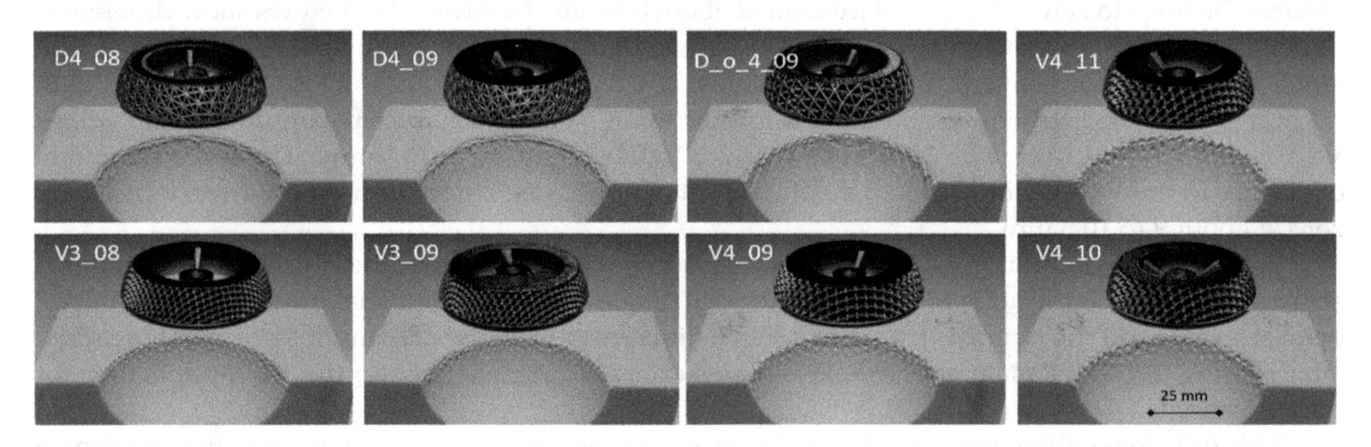

Figure 11. Representative imaging of mechanical deformations in the artificial bone cavity for cup design usage in lever-out test.

The use of an artificial bone cavity has a positive effect on the characterization of primary stability. This speaks in favor of the experimental results determined here since possible property variations, as they occur in the use of cadaveric models, have been omitted. Goldman et al. compared the effect of component surface roughness at the bone implant interface and the quality of the bone on initial press-fit stability [67]. They found no significant differences between the bending moment at 150 m for two kind of press-fit cups with different coefficients of friction. They made clear in the discussion that the results from the use of the cadaveric models represent a realistic representation of surgical interventions, but are also associated with corresponding scatter of the results. For the purpose of this study, which is to evaluate structurally differently designed press-fit cups, the artificial bone bed is the better choice. The uniform mechanical properties of the artificial bone bed provide a much better basis for a comparative consideration of the different cup designs.

3.4. Lever-Out Momentmechanical Work

The lever-out work shown in Figure 12 illustrates the individual force differences required to loosen the cups from the artificial bone cavities. The moment of relaxation thus represents the beginning of the failure.

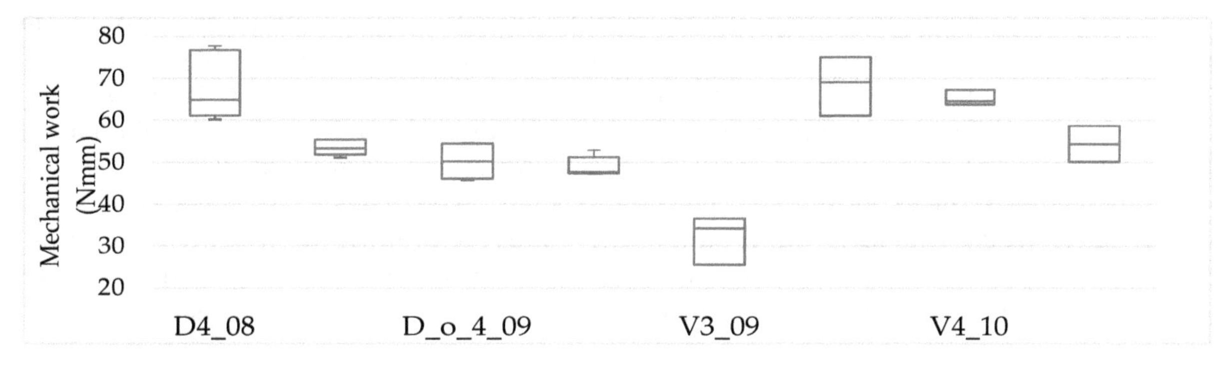

Figure 12. Boxplots of the measured mechanical work (Nmm) during the lever-out test. Boxplots indicate the median value, the interquartile range (IQR: interval between the 25th and 75th percentile, blue rectangle) and the extremum values ($n = 5$).

The best results were achieved with the cup versions V4_10 (69.9 Nmm), D4_08 (68.8 Nmm) and V4_09 (66.9 Nmm), followed by versions V4_11 (54.8 Nmm), D4_09 (53.1 Nmm) and D_o_4_09 (52.5 Nmm). Much less work was afforded for the loosening of versions V3_08 (48 Nmm) and V3_09 (32 Nmm).

After carrying out a statistical significance test (results Table 6) using one-way Anova with Dunnett's T3 post-hoc test (multiple comparisons), the following coherences become clear. The combined structures D4_08 ($p = 0.04296$) and D4_09 ($p = 0.01733$) deviate significantly from version V3_09. The twisted structure V3_09 deviates significantly from versions V4_09 ($p = 0.01595$), V4_10 ($p = 0.03089$) and V4_11 ($p = 0.01335$).

Table 6. Significances of the determined lever-out work for the different press-fit cups. For statistical analysis one-way ANOVA with Dunn's T3 post-hoc test was conducted. Values of $p < 0.05$ were set to be significant (N.S.—not significant).

Cupversion	D4_09	D_o_4_09	V3_08	V3_09	V4_09	V4_10	V4_11
D4_08	N.S.	N.S.	N.S.	0.04296	N.S.	N.S.	N.S.
D4_09	-	N.S.	N.S.	0.01733	N.S.	N.S.	N.S.
D_o_4_09	-	-	N.S.	N.S.	N.S.	N.S.	N.S.
V3_08	-	-	-	N.S.	N.S.	N.S.	N.S.
V3_09	-	-	-	-	0.01595	0.03089	0.01335
V4_09	-	-	-	-	-	N.S.	N.S.
V4_10	-	-	-	-	-	-	N.S.

In the pull-out test (determined force) and lever-out test (determined moment), the twisted structures perform worse in the evaluation than the combined and combined open structure. However, in the mechanical work determined, the twisted structures with a strut diameter of 4 mm achieve equivalent results here. One reason seems to be that the struts pressed into the artificial bone bed material move along the entire lever-out process in the bone bed material. This means that permanent work has to be done to move the cup further out of the bone bed. This is clearly demonstrated by the curves in Figure 10. This is also supported by the fact that the variants V4_10 and V4_09 achieve the highest values in the work determined. These variants also have the largest gaps between the struts, followed by version V4_11. A larger gap also has a higher proportion of material in the gap than smaller gaps. More material at the same time means more work to overcome the resistance. In total, this means that the work performed for the cup variants D4_08 and V4_09 and V4_10 is comparable. This fact is supported by the results of the cup variant V3_09, which has the lowest porosity (58.8 %) compared to all the other variants tested.

3.5. Correlations—Lever-Out Moment and Pull-Out Force Versus Volume of the Press-Fit Area

Anchoring strength is significantly influenced by the structure used, with its open-porous design characterizing the area that represents the press-fit. When looking at the volume characteristic of each cup variant in relation to the pull-out force or the lever-out moment (Figure 13), it can be seen that the pull-out force and the lever-out moment could be determined by a direct functional relationship, which can be described using a non-linear regression. An exponential function was found which describes the results of the experimental investigations very well. The curve clearly shows that the pull-out forces as well as the lever-out moments are relatively uniform up to a press-fit volume of 0.39 cm^3, followed by a strong increase towards higher press-fit volumes. At high volumes (>0.9 cm^3), the results are very similar for the pull-out forces as well as for the lever-out moments.

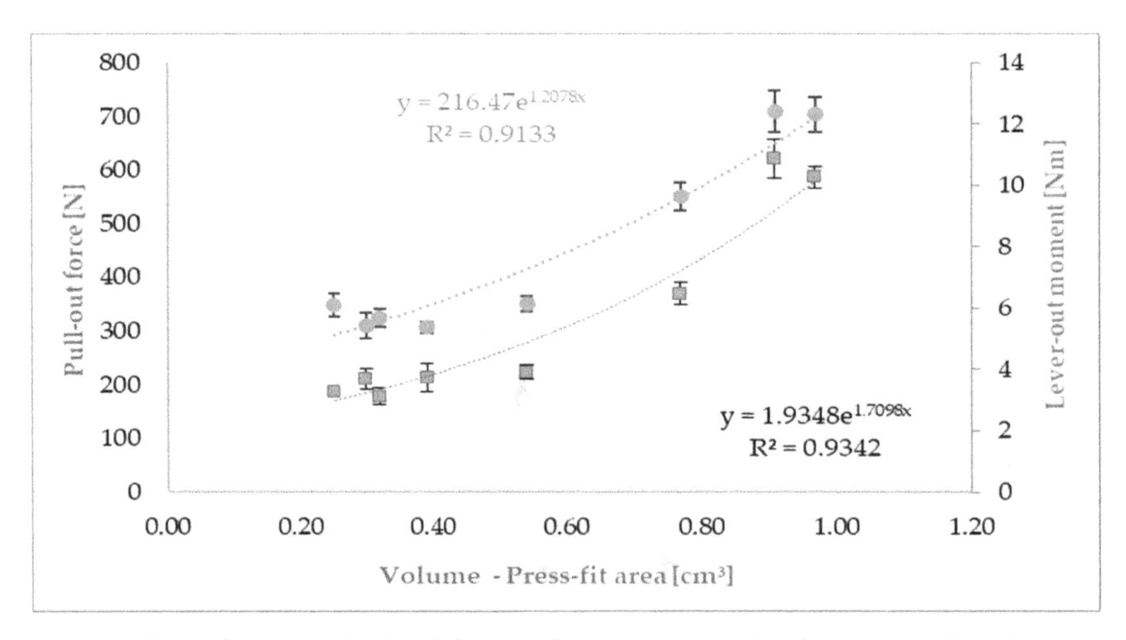

Figure 13. Pull-out force as calculated from pull-out testing and volume press-fit area as well as lever-out moment as calculated from lever-out testing for the eight cup-designs. Results are shown as mean values with the corresponding standard deviation ($n = 5$ for each design).

Both dimensions show an exponential functional relation to the press-fit volume, which is reflected by strong regression coefficients ($R^2 = 0.9342$ for pull-out force, $R^2 = 0.9133$ for lever-out moment). This makes it clear that an increase in anchoring strength can be achieved with increasing press-fit volume.

Although the press-fit volume used for this reference does not represent the full volume that actually penetrates the area of the artificial bone, it does directly represent the volume that creates the press-fit.

The determined functional relationships as well as the experimentally obtained measurement results provide a good basis for the selection of appropriate structural elements for the final development of press-fit acetabular cups, which ensure an increase in primary anchoring strength. In particular, the geometric design of the structural elements can thus be used in a targeted manner in conjunction with the mechanical properties and porosity [52,66,68–70]. Also, the determined functional relationships prove that the influence of the volume responsible for the actual press-fit is significantly greater than the porosity. However, since porosity is a measure relevant to secondary anchoring strength, it must not be disregarded.

Structured press-fit cups present an interesting solution, especially with regard to strong pelvic defects (D'Antonio type II). Due to the geometric freedom in structure design and possible size variations, these types of press-fit cups could offer advantages over non-structured cups in anchoring strength [71].

The characterization of the structurally differently designed press-fit cups with two test methods as well as the evaluation of the results in relation to different influencing factors makes a distinctive estimation of the types of cups possible. While the evaluation of anchoring strength with only one procedure or from one aspect is being discussed controversially, a good summary can be made in this study [67]. Several factors, such as material and surface structure (e.g., bead or wire) have been shown to be responsible for bone ingrowth [72]. The press-fit cups used here in this study have almost identical properties so that these can be neglected in the consideration.

The characterization of the cup variants based on the experimentally determined results offers the possibility to capture significant influences and thus show differences. The functional relationships also offer the opportunity to actively intervene in the constructive process and influence the structure design based on the results.

4. Conclusions

In this study, acetabular press-fit cups with a porous, load-bearing structural layer were examined for primary stability. The press-fit cup used was a design developed and evaluated in a previous study.

The porous, load-bearing structural layer was formed from geometrically differently designed unit cells. The preparation was carried out by means of selective laser melting of TiAl6V4. As an artificial bone cavity a PU foam was used, which was characterized experimentally in terms of mechanical properties.

The results show significant differences in the experimentally determined pull-out force, lever-out moment and lever-out-work results. The best results in pull-out and lever-out moments are achieved by the press-fit cups made in the combined structure (denoted D4_08 and D4_09). When looking at the work required to lever out the press-fit cups, it is noticeable that the press-fit cups designated as D4_08, V4_09 and V4_10 achieved the best results.

Overall, it becomes clear that the results for the evaluation of primary stability are related to the geometry used (unit cell), the dimensions of the unit cell, and the volume and porosity which are responsible for the press fit. Corresponding functional relationships could be determined.

The results of the work provide an excellent starting point for the development of press-fit acetabular cups with increased primary stability as a basis for high secondary stability.

Author Contributions: We point out that all authors were fully involved in the study and in preparing the manuscript. V.W. and C.S. designed the study. V.W. generated the CAD samples with support of C.B. and was involved in the manufacturing process of the scaffolds. V.W. and C.S. performed the experiments, analyzed the data with support of C.B. and wrote the initial manuscript. H.H. and R.B. organized the research funding. All authors ensured the accuracy of the data and the analyses and reviewed the manuscript in its current state.

References

1. Harrison, N.; Field, J.R.; Quondamatteo, F.; Curtin, W.; McHugh, P.E.; Mc Donnell, P. Preclinical trial of a novel surface architecture for improved primary fixation of cementless orthopaedic implants. *Clin. Biomech.* **2014**, *29*, 861–868. [CrossRef] [PubMed]
2. Levine, B. A new era in porous metals: Applications in orthopaedics. *Adv. Eng. Mater.* **2008**, *10*, 788–792. [CrossRef]
3. Murr, L.E. Open-cellular metal implant design and fabrication for biomechanical compatibility with bone using electron beam melting. *J. Mech. Behav. Biomed. Mater.* **2017**, *76*, 164–177. [CrossRef] [PubMed]

4. Sing, S.L.; An, J.; Yeong, W.Y.; Wiria, F.E. Laser and electron-beam powder-bed additive manufacturing of metallic implants: A review on processes, materials and designs. *J. Orthop. Res.* **2016**, *34*, 369–385. [CrossRef] [PubMed]
5. Geetha, M.; Singh, A.K.; Asokamani, R.; Gogia, A.K. Ti based biomaterials, the ultimate choice for orthopaedic implants—A review. *Prog. Mater. Sci.* **2009**, *54*, 397–425. [CrossRef]
6. Tan, X.P.; Tan, Y.J.; Chow, C.S.L.; Tor, S.B.; Yeong, W.Y. Metallic powder-bed based 3D printing of cellular scaffolds for orthopaedic implants: A state-of-the-art review on manufacturing, topological design, mechanical properties and biocompatibility. *Mater. Sci. Eng. C* **2017**, *76*, 1328–1343. [CrossRef] [PubMed]
7. Schulze, C.; Weinmann, M.; Schweigel, C.; Keßler, O.; Bader, R. Mechanical Properties of a Newly Additive Manufactured Implant Material Based on Ti-42Nb. *Materials* **2018**, *11*, 124. [CrossRef] [PubMed]
8. Murr, L.E.; Amato, K.N.; Li, S.J.; Tian, Y.X.; Cheng, X.Y.; Gaytan, S.M.; Martinez, E.; Shindo, P.W.; Medina, F.; Wicker, R.B. Microstructure and mechanical properties of open-cellular biomaterials prototypes for total knee replacement implants fabricated by electron beam melting. *J. Mech. Behav. Biomed. Mater.* **2011**, *4*, 1396–1411. [CrossRef] [PubMed]
9. Do Prado, R.F.; De Oliveira, F.S.; Nascimento, R.D.; De Vasconcellos, L.M.R.; Carvalho, Y.R.; Cairo, C.A.A. Osteoblast response to porous titanium and biomimetic surface: In vitro analysis. *Mater. Sci. Eng. C* **2015**, *52*, 194–203. [CrossRef] [PubMed]
10. Wang, X.; Zhou, S.; Xu, W.; Leary, M.; Choong, P.; Qian, M.; Brandt, M.; Xie, Y.M.; Xu, S. Topological design and additive manufacturing of porous metals for bone scaffolds and orthopaedic implants: A review. *Biomaterials* **2016**, *83*, 14. [CrossRef] [PubMed]
11. Limmahakhun, S.; Oloyede, A.; Sitthiseripratip, K.; Xiao, Y.; Yan, C. Stiffness and strength tailoring of cobalt chromium graded cellular structures for stress-shielding reduction. *Mater. Des.* **2017**, *114*, 633–641. [CrossRef]
12. Simoneau, C.; Terriault, P.; Jetté, B.; Dumas, M.; Brailovski, V. Development of a porous metallic femoral stem: Design, manufacturing, simulation and mechanical testing. *Mater. Des.* **2017**, *114*, 546–556. [CrossRef]
13. Kumar, A.; Nune, K.C.; Murr, L.E.; Misra, R.D.K. Biocompatibility and mechanical behaviour of three-dimensional scaffolds for biomedical devices: Process-structure-property paradigm. *Int. Mater. Rev.* **2016**, *61*, 20–45. [CrossRef]
14. Harrison, N.; McHugh, P.E.; Curtin, W.; Mc Donnell, P. Micromotion and friction evaluation of a novel surface architecture for improved primary fixation of cementless orthopaedic implants. *J. Mech. Behav. Biomed. Mater.* **2013**, *21*, 37–46. [CrossRef] [PubMed]
15. Jetté, B.; Brailovski, V.; Dumas, M.; Simoneau, C.; Terriault, P. Femoral stem incorporating a diamond cubic lattice structure: Design, manufacture and testing. *J. Mech. Behav. Biomed. Mater.* **2018**, *77*, 58–72. [CrossRef] [PubMed]
16. Marin, E.; Fusi, S.; Pressacco, M.; Paussa, L.; Fedrizzi, L. Characterization of cellular solids in Ti6Al4V for orthopaedic implant applications: Trabecular titanium. *J. Mech. Behav. Biomed. Mater.* **2010**, *3*, 373–381. [CrossRef] [PubMed]
17. Le Guéhennec, L.; Soueidan, A.; Layrolle, P.; Amouriq, Y. Surface treatments of titanium dental implants for rapid osseointegration. *Dent. Mater.* **2007**, *23*, 844–854. [CrossRef] [PubMed]
18. Khanna, R.; Kokubo, T.; Matsushita, T.; Nomura, Y.; Nose, N.; Oomori, Y.; Yoshida, T.; Wakita, K.; Takadama, H. Novel artificial hip joint: A layer of alumina on Ti-6Al-4V alloy formed by micro-arc oxidation. *Mater. Sci. Eng. C* **2015**, *55*, 393–400. [CrossRef] [PubMed]
19. Ramsden, J.J.; Allen, D.M.; Stephenson, D.J.; Alcock, J.R.; Peggs, G.N.; Fuller, G.; Goch, G. The design and manufacture of biomedical surface. *CIRP Ann. Manuf. Technol.* **2007**, *56*, 687–711. [CrossRef]
20. Emmelmann, C.; Scheinemann, P.; Munsch, M.; Seyda, V. Laser additive manufacturing of modified implant surfaces with osseointegrative characteristics. *Phys. Procedia* **2011**, *12*, 375–384. [CrossRef]
21. Paris, M.; Götz, A.; Hettrich, I.; Bidan, C.M.; Dunlop, J.W.C.; Razi, H.; Zizak, I.; Hutmacher, D.W.; Fratzl, P.; Duda, G.N.; et al. Scaffold curvature-mediated novel biomineralization process originates a continuous soft tissue-to-bone interface. *Acta Biomater.* **2017**, *60*, 64–80. [CrossRef] [PubMed]
22. Wang, Z.; Wang, C.; Li, C.; Qin, Y.; Zhong, L.; Chen, B.; Li, Z.; Liu, H.; Chang, F.; Wang, J. Analysis of factors influencing bone ingrowth into three-dimensional printed porous metal scaffolds: A review. *J. Alloys Compd.* **2017**, *717*, 271–285. [CrossRef]

23. Schouman, T.; Schmitt, M.; Adam, C.; Dubois, G.; Rouch, P. Influence of the overall stiffness of a load-bearing porous titanium implant on bone ingrowth in critical-size mandibular bone defects in sheep. *J. Mech. Behav. Biomed. Mater.* **2016**, *59*, 484–496. [CrossRef] [PubMed]
24. de Wild, M.; Zimmermann, S.; Rüegg, J.; Schumacher, R.; Fleischmann, T.; Ghayor, C.; Weber, F.E. Influence of microarchitecture on osteoconduction and mechanics of porous titanium Scaffolds generated by selective laser melting. 3D Print. *Addit. Manuf.* **2016**, *3*, 142–151. [CrossRef]
25. Taniguchi, N.; Fujibayashi, S.; Takemoto, M.; Sasaki, K.; Otsuki, B. Effect of pore size on bone ingrowth into porous titanium implants. *Mater. Sci. Eng. C* **2016**, *59*, 690–701. [CrossRef] [PubMed]
26. Jetté, B.; Brailovski, V.; Simoneau, C.; Dumas, M.; Terriault, P. Development and in vitro validation of a simplified numerical model for the design of a biomimetic femoral stem. *J. Mech. Behav. Biomed. Mater.* **2017**, *77*, 539–550. [CrossRef] [PubMed]
27. Bellini, C.M.; Galbusera, F.; Ceroni, R.G.; Raimondi, M.T. Loss in mechanical contact of cementless acetabular prostheses due to post-operative weight bearing: A biomechanical model. *Med. Eng. Phys.* **2007**, *29*, 175–181. [CrossRef] [PubMed]
28. Souffrant, R.; Zietz, C.; Fritsche, A.; Kluess, D.; Mittelmeier, W.; Bader, R. Advanced material modelling in numerical simulation of primary acetabular press-fit cup stability. Comput. *Methods Biomech. Biomed. Engin.* **2012**, *15*, 787–793. [CrossRef] [PubMed]
29. Small, S.R.; Berend, M.E.; Howard, L.A.; Rogge, R.D.; Buckley, C.A.; Ritter, M.A. High initial stability in porous titanium acetabular cups: A biomechanical study. *J. Arthroplast.* **2013**, *28*, 510–516. [CrossRef] [PubMed]
30. Udofia, I.; Liu, F.; Jin, Z.; Roberts, P.; Grigoris, P. The initial stability and contact mechanics of a press-fit resurfacing arthroplasty of the hip. *J. Bone Jt. Surg. Br.* **2007**, *89*, 549–556. [CrossRef] [PubMed]
31. Chang, J.-D.; Kim, T.-Y.; Rao, M.B.; Lee, S.-S.; Kim, I.-S. Revision total hip arthroplasty using a tapered, press-fit cementless revision stem in elderly patients. *J. Arthroplast.* **2011**, *26*, 1045–1049. [CrossRef] [PubMed]
32. Chanlalit, C.; Fitzsimmons, J.S.; Shukla, D.R.; An, K.-N.; O'Driscoll, S.W. Micromotion of plasma spray versus grit-blasted radial head prosthetic stem surfaces. *J. Shoulder Elb. Surg.* **2011**, *20*, 717–722. [CrossRef] [PubMed]
33. Le Cann, S.; Galland, A.; Rosa, B.; Le Corroller, T.; Pithioux, M.; Argenson, J.N.; Chabrand, P.; Parratte, S. Does surface roughness influence the primary stability of acetabular cups? A numerical and experimental biomechanical evaluation. *Med. Eng. Phys.* **2014**, *36*, 1185–1190. [CrossRef] [PubMed]
34. Goriainov, V.; Jones, A.; Briscoe, A.; New, A.; Dunlop, D. Do the cup surface properties influence the initial stability? *J. Arthroplast.* **2014**, *29*, 757–762. [CrossRef] [PubMed]
35. Gebert, A.; Peters, J.; Bishop, N.E.; Westphal, F.; Morlock, M.M. Influence of press-fit parameters on the primary stability of uncemented femoral resurfacing implants. *Med. Eng. Phys.* **2009**, *31*, 160–164. [CrossRef] [PubMed]
36. Ries, M.D.; Harbaugh, M.; Shea, J.; Lambert, R. Effect of cementless acetabular cup geometry on strain distribution and press-fit stability. *J. Arthroplast.* **1997**, *12*, 207–212. [CrossRef]
37. Adler, E.; Stuchin, S.A.; Kummer, F.J. Stability of press-fit acetabular cups. *J. Arthroplast.* **1992**, *7*, 295–301. [CrossRef]
38. Macdonald, W.; Carlsson, L.V.; Charnley, G.J.; Jacobsson, C.M. Press-fit acetabular cup fixation: Principles and testing. *Proc. Inst. Mech. Eng. Part H J. Eng. Med.* **1999**, *213*, 33–39. [CrossRef] [PubMed]
39. Morlock, M.; Götzen, N.; Sellenschloh, K. Bestimmung der Primärstabilität von künstlichen Hüftpfannen. In *DVM Bericht 314—Eigenschaften und Prüftechniken mechanisch Beanspruchter Implantate*; DVM: Berlin, Germany, 2002; pp. 221–229.
40. Toossi, N.; Adeli, B.; Timperley, A.J.; Haddad, F.S.; Maltenfort, M.; Parvizi, J. Acetabular components in total hip arthroplasty: Is there evidence that cementless fixation is better? *J. Bone Jt. Surg.* **2013**, *95*, 168–174. [CrossRef] [PubMed]
41. Roth, A.; Winzer, T.; Sander, K.; Anders, J.O.; Venbrocks, R.-A. Press fit fixation of cementless cups: How much stability do we need indeed? *Arch. Orthop. Trauma Surg.* **2006**, *126*, 77–81. [CrossRef] [PubMed]
42. Tabata, T.; Kaku, N.; Hara, K.; Tsumura, H. Initial stability of cementless acetabular cups: Press-fit and screw fixation interaction—An in vitro biomechanical study. *Eur. J. Orthop. Surg. Traumatol.* **2015**, *25*, 497–502. [CrossRef] [PubMed]

43. Takao, M.; Nakamura, N.; Ohzono, K.; Sakai, T.; Nishii, T.; Sugano, N. The results of a press-fit-only technique for acetabular fixation in hip dysplasia. *J. Arthroplast.* **2011**, *26*, 562–568. [CrossRef] [PubMed]
44. Amirouche, F.; Solitro, G.; Broviak, S.; Gonzalez, M.; Goldstein, W.; Barmada, R. Factors influencing initial cup stability in total hip arthroplasty. *Clin. Biomech.* **2014**, *29*, 1177–1185. [CrossRef] [PubMed]
45. Clarke, H.J.; Jinnah, R.H.; Warden, K.E.; Cox, Q.G.; Curtis, M.J. Evaluation of acetabular stability in uncemented prostheses. *J. Arthroplast.* **1991**, *6*, 335–340. [CrossRef]
46. Klanke, J.; Partenheimer, A.; Westermann, K. Biomechanical qualities of threaded acetabular cups. *Int. Orthop.* **2002**, *26*, 278–282. [CrossRef] [PubMed]
47. Baleani, M.; Fognani, R.; Toni, A. Initial stability of a cementless acetabular cup design: Experimental investigation on the effect of adding fins to the rim of the cup. *Artif. Organs.* **2001**, *25*, 664–669. [CrossRef] [PubMed]
48. Olory, B.; Havet, E.; Gabrion, A.; Vernois, J.; Mertl, P. Comparative in vitro assessment of the primay stbility of cementless press-fit acetabular cups. *Acta Orthop. Belg.* **2004**, *70*, 31–37. [PubMed]
49. Fritsche, A.; Zietz, C.; Teufel, S.; Kolp, W.; Tokar, I.; Mauch, C.; Mittelmeier, W.; Bader, R. In-vitro and in-vivo investigations of the impaction and pull-out behavior of metal-backed acetabular cups. *Br. Ed. Soc. Bone Jt. Surg.* **2011**, *93*, 406.
50. Weißmann, V.; Boss, C.; Bader, R.; Hansmann, H. A novel approach to determine primary stability of acetabular press-fit cups. *J. Mech. Behav. Biomed. Mater.* **2018**, *80*, 1–10. [CrossRef] [PubMed]
51. Markhoff, J.; Wieding, J.; Weissmann, V.; Pasold, J.A.; Jonitz-Heincke, R. Bader, Influence of different three-dimensional open porous titanium scaffold designs on human osteoblasts behavior in static and dynamic cell investigations. *Materials* **2015**, *8*, 5490–5507. [CrossRef] [PubMed]
52. Weißmann, V.; Bader, R.; Hansmann, H.; Laufer, N. Influence of the structural orientation on the mechanical properties of selective laser melted Ti6Al4V open-porous scaffolds. *Mater. Des.* **2016**, *95*, 188–197. [CrossRef]
53. Weißmann, V.; Wieding, J.; Hansmann, H.; Laufer, N.; Wolf, A.; Bader, R. Specific yielding of selective laser-melted Ti6Al4V open-porous scaffolds as a function of unit cell design and dimensions. *Metals* **2016**, *6*, 166. [CrossRef]
54. Weißmann, V.; Hansmann, H.; Bader, R.; Laufer, N. Influence of the Structural Orientation on the Mechanical Properties of Selective Laser Melted TiAL6V4 Open-Porous Scaffold. In Proceedings of the 13th Rapid Tech Conference Erfurt, Erfurt, Germany, 14–16 June 2016.
55. Fox, J.C.; Moylan, S.P.; Lane, B.M. Effect of process parameters on the surface roughness of overhanging structures in laser powder bed fusion additive manufacturing. *Procedia CIRP* **2016**, *45*, 131–134. [CrossRef]
56. Rashed, M.G.; Ashraf, M.; Mines, R.A.W.; Hazell, P.J. Metallic microlattice materials: A current state of the art on manufacturing, mechanical properties and applications. *Mater. Des.* **2016**, *95*, 518–533. [CrossRef]
57. Suard, M.; Martin, G.; Lhuissier, P.; Dendievel, R.; Vignat, F.; Blandin, J.J.; Villeneuve, F. Mechanical equivalent diameter of single struts for the stiffness prediction of lattice structures produced by Electron Beam Melting. *Addit. Manuf.* **2015**, *8*, 124–131. [CrossRef]
58. Weißmann, V.; Drescher, P.; Bader, R.; Seitz, H.; Hansmann, H.; Laufer, N. Comparison of single Ti6Al4V struts made using selective laser melting and electron beam melting subject to part orientation. *Metals* **2017**, *7*, 91. [CrossRef]
59. Triantaphyllou, A.; Giusca, C.L.; Macaulay, G.D.; Roerig, F.; Hoebel, M.; Leach, R.K.; Tomita, B.; Milne, K.A. Surface texture measurement for additive manufacturing. *Surf. Topogr. Metrol. Prop.* **2015**, *3*, 024002. [CrossRef]
60. Frosch, K.; Barvencik, F.; Viereck, V.; Lohmann, C.H.; Dresing, K.; Breme, J.; Brunner, E.; Stürmer, K.M. Growth behavior, matrix production, and gene expression of human osteoblasts in defined cylindrical titanium channels. *J. Biomed. Mater. Res. Part A* **2004**, *68*, 325–334. [CrossRef] [PubMed]
61. Knychala, J.; Bouropoulos, N.; Catt, C.J.; Katsamenis, O.L.; Please, C.P.; Sengers, B.G. Pore geometry regulates early stage human bone marrow cell tissue formation and organization. *Ann. Biomed. Eng.* **2013**, *41*, 917–930. [CrossRef] [PubMed]
62. Kienapfel, H.; Sprey, C.; Wilke, A.; Griss, P. Implant fixation by bone ingrowth. *J. Arthroplast.* **1999**, *14*, 355–368. [CrossRef]
63. Kawai, T.; Takemoto, M.; Fujibayashi, S.; Tanaka, M.; Akiyama, H.; Nakamura, T.; Matsuda, S. Comparison between alkali heat treatment and sprayed hydroxyapatite coating on thermally-sprayed rough Ti surface

in rabbit model: Effects on bone-bonding ability and osteoconductivity. *J. Biomed. Mater. Res. Part B Appl. Biomater.* **2015**, *103*, 1069–1081. [CrossRef] [PubMed]

64. Grimal, Q.; Haupert, S.; Mitton, D.; Vastel, L.; Laugier, P. Assessment of cortical bone elasticity and strength: Mechanical testing and ultrasound provide complementary data. *Med. Eng. Phys.* **2009**, *31*, 1140–1147. [CrossRef] [PubMed]
65. Niinomi, M.; Nakai, M. Titanium-based biomaterials for preventing stress shielding between implant devices and bone. *Int. J. Biomater.* **2011**, *2011*. [CrossRef] [PubMed]
66. Wauthle, R.; Vrancken, B.; Beynaerts, B.; Jorissen, K.; Schrooten, J.; Kruth, J.-P.; Humbeeck, J. Effects of build orientation and heat treatment on the microstructure and mechanical properties of selective laser melted Ti6Al4 V lattice structures. *Addit. Manuf.* **2014**, *5*, 6–13. [CrossRef]
67. Goldman, A.H.; Armstrong, L.C.; Owen, J.R.; Wayne, J.S.; Jiranek, W.A. Does increased coefficient of friction of highly porous metal increase initial stability at the acetabular interface? *J. Arthroplast.* **2016**, *31*, 721–726. [CrossRef] [PubMed]
68. Ahmadi, S.M.; Campoli, G.; Amin Yavari, S.; Sajadi, B.; Wauthle, R.; Schrooten, J.; Weinans, H.; Zadpoor, A.A. Mechanical behavior of regular open-cell porous biomaterials made of diamond lattice unit cells. *J. Mech. Behav. Biomed. Mater.* **2014**, *34*, 106–115. [CrossRef] [PubMed]
69. Lopez-Heredia, M.A.; Goyenvalle, E.; Aguado, E.; Pilet, P.; Leroux, C.; Dorget, M.; Weiss, P.; Layrolle, P. Bone growth in rapid prototyped porous titanium implants. *J. Biomed. Mater. Res. Part A* **2008**, *85*, 664–673. [CrossRef] [PubMed]
70. Hedayati, R.; Sadighi, M.; Mohammadi-Aghdam, M.; Zadpoor, A.A. Mechanics of additively manufactured porous biomaterials based on the rhombicuboctahedron unit cell. *J. Mech. Behav. Biomed. Mater.* **2016**, *53*, 272–294. [CrossRef] [PubMed]
71. Gollwitzer, R.; Gradinger, H. *Ossäre Integration*; Springer Medizin Verlag: Heidelberg, Germany, 2006.
72. Swarts, E.; Bucher, T.A.; Phillips, M.; Yap, F.H.X. Does the ingrowth surface make a difference? A retrieval study of 423 cementless acetabular components. *J. Arthroplast.* **2015**, *30*, 706–712. [CrossRef] [PubMed]

Permissions

All chapters in this book were first published by MDPI; hereby published with permission under the Creative Commons Attribution License or equivalent. Every chapter published in this book has been scrutinized by our experts. Their significance has been extensively debated. The topics covered herein carry significant findings which will fuel the growth of the discipline. They may even be implemented as practical applications or may be referred to as a beginning point for another development.

The contributors of this book come from diverse backgrounds, making this book a truly international effort. This book will bring forth new frontiers with its revolutionizing research information and detailed analysis of the nascent developments around the world.

We would like to thank all the contributing authors for lending their expertise to make the book truly unique. They have played a crucial role in the development of this book. Without their invaluable contributions this book wouldn't have been possible. They have made vital efforts to compile up to date information on the varied aspects of this subject to make this book a valuable addition to the collection of many professionals and students.

This book was conceptualized with the vision of imparting up-to-date information and advanced data in this field. To ensure the same, a matchless editorial board was set up. Every individual on the board went through rigorous rounds of assessment to prove their worth. After which they invested a large part of their time researching and compiling the most relevant data for our readers.

The editorial board has been involved in producing this book since its inception. They have spent rigorous hours researching and exploring the diverse topics which have resulted in the successful publishing of this book. They have passed on their knowledge of decades through this book. To expedite this challenging task, the publisher supported the team at every step. A small team of assistant editors was also appointed to further simplify the editing procedure and attain best results for the readers.

Apart from the editorial board, the designing team has also invested a significant amount of their time in understanding the subject and creating the most relevant covers. They scrutinized every image to scout for the most suitable representation of the subject and create an appropriate cover for the book.

The publishing team has been an ardent support to the editorial, designing and production team. Their endless efforts to recruit the best for this project, has resulted in the accomplishment of this book. They are a veteran in the field of academics and their pool of knowledge is as vast as their experience in printing. Their expertise and guidance has proved useful at every step. Their uncompromising quality standards have made this book an exceptional effort. Their encouragement from time to time has been an inspiration for everyone.

The publisher and the editorial board hope that this book will prove to be a valuable piece of knowledge for researchers, students, practitioners and scholars across the globe.

List of Contributors

Xiaoguang Fan, Qi Li, Anming Zhao, Yuguo Shi and Wenjia Mei
State Key Laboratory of Solidification Processing, School of Materials Science and Engineering, Northwestern Polytechnical University, Xi'an 710072, China

Vasile Danut Cojocaru, Alexandru Dan, Raluca Irimescu and Doina Raducanu
Materials Science and Engineering Faculty, University Politehnica of Bucharest, 060042 Bucharest, Romania

Anna Nocivin
Mechanical, Industrial and Maritime Faculty, Ovidius University of Constanta, 900527 Constanta, Romania

Corneliu Trisca-Rusu
National Institute for Research and Development in Micro-technologies, 077190 Bucharest, Romania

Bogdan Mihai Galbinasu
Dental Medicine Faculty, University of Medicine and Pharmacy "Carol Davila" Bucharest, 020021 Bucharest, Romania

Khaja Moiduddin, Syed Hammad Mian, Usama Umer, Hisham Alkhalefah and Wadea Ameen
Advanced Manufacturing Institute, King Saud University, Riyadh 11421, Saudi Arabia

Naveed Ahmed
Advanced Manufacturing Institute, King Saud University, Riyadh 11421, Saudi Arabia
Department of Industrial and Manufacturing Engineering, University of Engineering and Technology, Lahore 54000, Pakistan

Yanxin Qiao, Jian Chen, Yuxin Wang and Huiling Zhou
School of Materials Science and Engineering, Jiangsu University of Science and Technology, Zhenjiang 212003, China

Daokui Xu and Yingjie Ma
Institute of Metal Research, Chinese Academy of Sciences, Shenyang 110016, China

Shuo Wang
School of Materials Science and Engineering, Jiangsu University of Science and Technology, Zhenjiang 212003, China
Institute of Metal Research, Chinese Academy of Sciences, Shenyang 110016, China
School of Materials Science and Engineering, Northeastern University, Shenyang 110004, China

Changzhou Yu, Peng Cao and Mark Ian Jones
Department of Chemical and Materials Engineering, The University of Auckland, Private Bag 92019, Auckland 1142, New Zealand

Guangbao Mi
Key Laboratory of Science and Technology on Advanced Titanium Alloys, AECC Beijing Institute of Aeronautical Materials, Beijing 100095, China

Kai Yao, Pengfei Bai, Congqian Cheng and Xiaohua Min
School of Materials Science and Engineering, Dalian University of Technology, Dalian 116024, China

Rosa Rojo and Juan Carlos Prados-Frutos
Department of Medicine and Surgery, Faculty of Health Sciences, Rey Juan Carlos University, 28922 Alcorcon, Spain

María Prados-Privado
Department Continuum Mechanics and Structural Analysis Higher Polytechnic School, Carlos III University, 28911 Leganes, Spain
Asisa Dental (Engineering Researcher), José Abascal 32, 28003 Madrid, Spain

Antonio José Reinoso
Department of ICT Engineering, Alfonso X El Sabio University, 28691 Madrid, Spain

Suzan Meijs and Nico J.M. Rijkhoff
Department of Health, Science and Technology, Center for Sensory-Motor Interaction (SMI), Aalborg University, 9220 Aalborg, Denmark

Kristian Rechendorff and Søren Sørensen
Materials Department, Danish Technological Institute, 8000 Århus, Denmark

Qi Liu, Zhaotian Wang, Hao Yang and Yongquan Ning
School of Materials Science and Engineering, Northwestern Polytechnical University, Xi'an 710072, China

Damon Kent
School of Science and Engineering, University of the Sunshine Coast, Maroochydore DC 4558, Australia
Queensland Centre for Advanced Materials Processing and Manufacturing (AMPAM), The University of Queensland, St. Lucia 4072, Australia
ARC Research Hub for Advanced Manufacturing of Medical Devices, St. Lucia 4072, Australia

Rizwan Rahman Rashid
School of Engineering, Faculty of Science, Engineering and Technology, Swinburne University of Technology, Victoria 3122, Australia
Defence Materials Technology Centre, Victoria 3122, Australia

Michael Bermingham and Matthew Dargusch
Queensland Centre for Advanced Materials Processing and Manufacturing (AMPAM), The University of Queensland, St. Lucia 4072, Australia
ARC Research Hub for Advanced Manufacturing of Medical Devices, St. Lucia 4072, Australia

Hooyar Attar
Queensland Centre for Advanced Materials Processing and Manufacturing (AMPAM), The University of Queensland, St. Lucia 4072, Australia

Shoujin Sun
School of Engineering, Faculty of Science, Engineering and Technology, Swinburne University of Technology, Victoria 3122, Australia

Ryan Cottam, Suresh Palanisamy and Rizwan Abdul Rahman Rashid
School of Engineering, Faculty of Science, Engineering and Technology, Swinburne University of Technology, Hawthorn, VIC 3122, Australia
Defence Materials Technology Centre, Hawthorn, VIC 3122, Australia

Maxim Avdeev
The Bragg Institute, Australian Nuclear Science and Technology Organisation (ANSTO), Lucas Heights, NSW 2234, Australia

Tom Jarvis
Monash Centre for Additive Manufacturing, Monash University, Notting Hill, VIC 3168, Australia

Chad Henry
Commonwealth Scientific and Industrial Research Organization (CSIRO), Clayton, VIC 3168, Australia

Dominic Cuiuri
School of Mechanical, Materials, and Mechatronic Engineering, Faculty of Engineering and Information Sciences, University of Wollongong, Wollongong, NSW 2522, Australia

Levente Balogh
Department of Mechanical and Materials Engineering, Queen's University, Kingston, ON K7L 3N6, Canada

Nina Radishevskaya, Olga Lepakova, Anastasiya Nazarova and Nikolai Afanasiev
Tomsk Scientific Centre SB RAS, Tomsk 634055, Russia

Natalia Karakchieva
Physical-Technical Institute, Tomsk State University, Tomsk 634050, Russia

Anna Godymchuk
Department of Nanomaterials and Nanotechnologies, National Research Tomsk Polytechnic University, Tomsk 634050, Russia
Department of Functional Nanosystems and High-Temperature Materials, National University of Science and Technology MISIS, Moscow 119991, Russia

Alexander Gusev
Department of Functional Nanosystems and High-Temperature Materials, National University of Science and Technology MISIS, Moscow 119991, Russia
Research Institute of Environmental Science and Biotechnology, G.R. Derzhavin Tambov State University, Tambov 392000, Russia

Le Chang, Chang-Yu Zhou and Xiao-Hua He
School of Mechanical and Power Engineering, Nanjing Tech University, Nanjing 211816, China

Fengyong Wu, Wenchen Xu, Xueze Jin, Debin Shan and Bin Guo
School of Materials Science and Engineering & National Key Laboratory for Precision Hot Processing of Metals, Harbin Institute of Technology, Harbin 150001, China
National Key Laboratory for Precision Hot Processing of Metals, Harbin Institute of Technology, Harbin 150001, China

Xunmao Zhong and Xingjie Wan
School of Materials Science and Engineering & National Key Laboratory for Precision Hot Processing of Metals, Harbin Institute of Technology, Harbin 150001, China

Isabel Montealegre-Meléndez, Cristina Arévalo, Ana M. Beltrán and Eva María Pérez Soriano
Escuela Politécnica Superior, Universidad de Sevilla, Calle Virgen de África, 7, 41011 Sevilla, Spain

Michael Kitzmantel and Erich Neubauer
RHP Technology GmbH, 2444 Seibersdorf, Austria

Volker Weißmann
Faculty of Engineering, University of Applied Science, Technology, Business and Design, Philipp-Müller-Str. 14, 23966Wismar, Germany
Biomechanics and Implant Technology Research Laboratory, Department of Orthopedics, Rostock University Medicine, Doberaner Strasse 142, 18057 Rostock, Germany

Christian Boss
Institute for Polymer Technologies e.V., Alter Holzhafen 19, 23966 Wismar, Germany

Christian Schulze and Rainer Bader
Biomechanics and Implant Technology Research Laboratory, Department of Orthopedics, Rostock University Medicine, Doberaner Strasse 142, 18057 Rostock, Germany

Harald Hansmann
Faculty of Engineering, University of Applied Science, Technology, Business and Design, Philipp-Müller-Str. 14, 23966 Wismar, Germany

Index